COMBATING DESERTIFICATION WITH PLANTS

COMBATING DESERTIFICATION WITH PLANTS

Edited by

Dov Pasternak

and

Arnold Schlissel

International Program for Arid Land Crops
Ben-Gurion University of the Negev
Beer Sheva, Israel

SPRINGER SCIENCE+BUSINESS MEDIA, LLC

Library of Congress Cataloging-in-Publication Data

Combating desertification with plants/edited by Dov Pasternak and Arnold Schlissel.
p. cm.
Papers from a conference held in Beer Sheva, Israel from November 2–5, 1999.
Includes bibliographical references (p.).
ISBN 978-1-4613-5499-4 ISBN 978-1-4615-1327-8 (eBook)
DOI 10.1007/978-1-4615-1327-8
1. Plants for soil conservation—Congresses. 2. Soil conservation—Congresses. 3. Plants for soil conservation—Africa—Congresses. 4. Soil conservation—Africa—Congresses. I. Pasternak, D. (Dov), 1940- II. Schlissel, Arnold, 1948-

S627.P55 C66 2001
631.4′52′096—dc21

2001041334

Proceedings of the conference Combating Desertification with Plants, held November 2–5, 1999, in Beer Sheva, Israel

Originally published by Kluwer Academic / Plenum Publishers, New York in 2001
Softcover reprint of the hardcover 1st edition 2001

10 9 8 7 6 5 4 3 2 1

A C.I.P. record for this book is available from the Library of Congress

I will put in the desert the cedar and the acacia, the myrtle and the olive. I will set pines in the wasteland, the fir and the cypress together, so the people may see and know, may consider and understand, that the hand of the Lord has done this, and the Holy One of Israel has created it.

Isaiah 41:19–20

FOREWORD

The conference "Combating Desertification with Plants" was held in Beer Sheva, Israel, from November 2-5, 1999, and was attended by 70 participants from 30 countries and/or international organisations.

Desertification – the degradation of soils in drylands – is a phenomenon occurring in scores of countries around the globe. The number of people (in semiarid regions) affected by the steady decline in the productivity of their lands is in the hundred millions. The measures required to halt and reverse the process of desertification fall into many categories – policy, institutional, sociological–anthropological, and technical.

Although technical "solutions" are not currently in vogue, the conference organizers felt that perhaps the pendulum had swung too far in the direction of "participatory approaches." Hence IPALAC – The International Program for Arid Land Crops – whose function is to serve as a catalyst for optimizing the contribution of plant germplasm to sustainable development in desertification-prone regions – felt the time was opportune for providing a platform for projects where the "plant-driven" approach to development finds expression.

Some 45 papers were delivered at the conference, falling into the categories of this volume: Overview, Potential Germplasm for Arid Lands, Introduction, Domestication and Dissemination of Arid Land Plants, Land Rehabilitation, and Mechanisms of Plant Transfer.

The conference was funded by UNESCO (Division of Ecological Sciences), the Ministry of Foreign Affairs of Finland, and MASHAV, Israel's Center for International Development Cooperation.

A number of collaborative projects that are already being implemented between participants is the best indication that the conference was more than a transient success.

Finally, preparing the manuscript of this volume was a challenging task. While the electronic era frequently enhances capacities substantially, its downside can be unmerciful, so a very special thanks goes to Marie-Laure Cohen, who displayed a range of talents of which dealing with downloaded formats is but one. No less important were the wonderful relations she enjoyed with the authors and the staff at Kluwer Academic/Plenum Publishers.

Arnold Schlissel
Secretary
Conference Organizing Committee

PREFACE

Plants are both the creators and the guardians of the soil. Plant cover prevents the formation of soil crusts under the impact of raindrops, thus ensuring smooth infiltration of rainwater. Plant roots bind the soil to minimize erosion, while plant litter provides the "glue" that keeps soil aggregates together. Removal of plant cover results in soil erosion, destruction of soil structure, and, in warm places, loss of organic matter and nitrogenous compounds as well. Thus, in most cases, "desertification" – the loss of soil fertility in semiarid and dry subhumid regions – is a direct result of the disruption of the natural plant cover, be it through the action of climate or man. It therefore stands to reason that plants are the most effective and efficient "weapon" for combating desertification. Plants can not only prevent further loss of soil fertility, they can also restore degraded lands to productivity.

Nowadays, destruction of plant cover and depletion of the plants' gene reserve is in most cases a result of the actions carried out by impoverished and ever-expanding rural populations. This section of society is continually mining the environment in order to survive. The rural society lives from the land, from plants. Provide this society with plants and cropping systems that significantly raise their income and improve their nutritional status, and they will stop mining the environment. Desertification will then cease.

This book deals with the role of plants in combating desertification. It provides the reader with the experience of scientists from 30 countries and international institutions around the globe, working in cold and warm dry regions to bring plant-based solutions to one of the gravest problems of our times. Africa is the place which is most severely plagued by desertification, and therefore emphasis was given in the conference to work being carried on in that continent.

The book starts with an introduction composed of two articles. The first analyzes the world's main agricultural production systems in terms of the farmers' nutritional levels and correlates the farmer's daily calorie consumption with his receptivity to technological changes. The second gives an overview of the socioeconomic problems causing present day desertification and offers proposals for plant-based solutions to them.

The book is further subdivided into four sections. In the first, "Potential Germplasm for Arid Lands," the contributions describe various uses of arid land germplasm, from desert mushrooms (terfezias), through nectar and pollen-producing trees, to medicinal plants and indigenous species.

The second section, called "Introduction, Domestication and Dissemination of Arid Land Plants," is the largest, presenting current work carried out in the area of domestication of arid land germplasm. The first contribution presents the results of 40 years of research during which thousands of plant species were systematically screened for adaptability to the soil and climate of the Middle East deserts. The list of drought- and salt-tolerant plants useful for afforestation and landscaping in arid lands provides a wealth of information with potential for widespread application. The section closes by describing the work on domestication of Jatropha in Mali, an outstanding example of a well planned project which led to the development of a new multipurpose income-generating arid land crop with a significant positive impact on the environment.

The third section, "Land Rehabilitation," consists of nine articles dealing with rehabilitation of salinized and eroded soils, the arrest of sand dune movement, and the rehabilitation of biodiversity. The story of seabuckthorn and its role in the rehabilitation of eroded loess soils in semiarid China perhaps best exemplifies the title of this book. When the ecological balance is radically disturbed and soil erosion proceeds unabated, a very aggressive colonizing plant may be the best weapon to fight desertification. The case study describing the successful use of seabuckthorn brings a ray of hope to the community of "desertification fighters." The last contribution of this section introduces a highly original approach to restoration of biodiversity. It could be called "Combating Desertification with Animals." The author argues that in ancient times the world's savannas were crowded with grazing animals which, through heavy trampling, grazing, dunging, and urinating, helped maintain a rich and diversified plant cover. These processes can nowadays be mimicked through the use of domestic animals such as large, herding ungulates.

The fourth section, "Mechanisms of Plant Transfer," presents the efforts of international and national organizations which use plants for desertification control. An outstanding example is the work of the Uganda Women Tree Planting Movement, a volunteer grass-roots organization which after only a few years began to have a significant impact on desertification control in Uganda. The article on obstacles to adopting Mucuna highlights the centrality of economics in each of the programs dealing with introduction of desertification-fighting crops.

We hope that the diversity of projects and experiences presented in this volume will lead to a growth in the legions of those who have opted to "combat desertification with plants" and will serve them well in their battles. May it prove a sustainable alternative to the failures of the past.

Dov Pasternak
Director
IPALAC

CONTENTS

INTRODUCTION

SOME OBSERVATIONS RELATING TO THE GLOBAL AGRICULTURAL SITUATION

Yitzhak Abt
Peres Center for Peace, 2 Ha-Shalom Rd., Tel Aviv 67892, Israel, Tel: 972 3 568 0657, Fax: 972 3 562 7265, email: info@peres-center.org

Over the last century world agriculture has undergone a major transition from pre-dominantly precarious subsistence agriculture to a highly capitalised farm industry. Nevertheless, the overall contemporary situation of world agriculture, especially in developing countries, is still often very precarious with alarming socio-economic implications.

There are four main agricultural situations identified with a combined presence virtually in every developing country. The first situation of **subsistence agriculture** has and is the most problematic one, due to a vicious circle of deficient nutrition, shift agriculture and a low level of inputs used with very negative effects on land use and the advance of desertification processes. There seems to be a clear correlation between the level of nutrition and the ability of peasant communities to break out of a subsistence situation to a more advanced one. It is my belief that nutrition deficiency in the intake of calories, proteins, fats and vitamins is one of the most limiting factors to break this viscous circle. Some five decades ago, a renowned development statistician, Dr. Colin Clark (1956) related the economic and caloric values of crops grown and expressed these values in economic grain equivalents related to the nutritional value of wheat. This tool was found to be very useful to identify the actual situation of peasant communities and their ability to progress into a more sustainable agricultural situation. Thus, for instance, the minimum grain equivalent consumption per capita for basic survival was found to be approximately 350-kg grain-equivalent per capita per year. Such low-level consumption is only enough to maintain an adult on

Combating Desertification with Plants, edited by Dov Pasternak and Arnold Schlissel.
Kluwer Academic/Plenum Publishers, New York, 2001.

a 4-day work schedule. In other words, at this level the peasant is lacking the required energy to seek a better future. The following details some major characteristics of subsistence agriculture and represents over 20% of the world's active rural population, even to this day:

Subsistence Agriculture

- Deficient nutrition
  (350 kg grain equiv./cap./year)
- Shift cultivation
- Hand implements only
- Limited cultivated area
- Predominance of rain-fed agriculture
- Use of degenerated seeds
- Low input level used
- Very low yield output
- Overgrazing
- Land degradation
- Traditional agroforestry systems
- Desertification

The challenge therefore is to break the vicious circle of subsistence agriculture by devising ways of improving, first and foremost, nutritional levels in those populations. In this respect I strongly **favour well-managed food-for-work programs** that can be leveraged in order to implement through sweat-capital, land and water conservation projects (including terracing and mulching) as well as the development of agroforestry systems, which IPALAC is effectively promoting.

This International Program for Arid Land (IPALAC) is a practical and professional way to upgrade natural resources owned by subsistence communities based on developing water-sheds, micro catchments, protective buns (for water harvesting) and the planting of tested crops, bushes, shrubs and trees of economic value. The activity can be organised either as a public works program or as community based activities financed from food donations. The element of remuneration should not be food exclusively and part of the payment should be on a cash basis. Thus for instance, if a project is to be developed through the payment of food (say, 1.5 kg. wheat/adult/day). The remuneration should be supplemented by a minimum of, say, $1/cap./day. Such a policy achieves three objectives:

1. The use of sweat labour to improve the natural resources base of the community.
2. To meet the immediate caloric needs of people (ranging from 1,700 cal./day - 2,200 cal./day depending on climate and food availability from other sources).

3. The modest cash remuneration becomes the first stage to introduce population into a market economy.

Shift cultivation was once the most **economic** way to produce food, when cropping took place every three to four years on the same land, allowing for fallow lands to build up soil nutrients to produce a crop without adding fertilisers. As population pressures on land increase the shift cultivation is carried out on the same land area every year or season causing a depletion of essential soil nutrients. The other major damaging effect of population pressure is the deforestation of virgin lands to expand the cultivated area under shift cultivation. This has become a major global problem related to global warming and the diminishment of "green lungs".

Hand implements are the only accessories used to produce a crop, which limits the aeration of soils and land preparation. The use of hand implements only restricts the ability of the peasant to "cultivate" his lands. In global scale terms, subsistence agriculture is based on rain-fed agriculture and in the absence of irrigation production is limited to one crop a year grown in the rainy season.

The use of seeds kept from the previous season has over time degenerated in quality and germination. The introduction of open pollinated improved seed is probably the most effective way to alleviate the agricultural subsistence condition.

The absence of secure land tenure and/or any other guarantees to procure critical inputs is a major constraint with the result that yield output is very low, often being only one tenth of what can be attained in commercial agriculture. For example, in Latin America subsistence yields in maize only reach 800 kg per hectare compared to 8 tons/ha in commercial enterprises.

The problem of overgrazing in dry savannah regions of the world is a problem of disproportionate numbers of livestock to grazing resources. It is also caused by erroneous grazing practices. Whereby soils are "hammered" in specific areas particularly around water holes where herds and flocks are concentrated for the sake of management convenience. This has, over the centuries, become a major cause of land degradation and specifically in Middle East and African countries where the black goat is herded.

Traditional agroforestry systems are a "light at the end of the tunnel" and much can be learnt from the indigenous traditions and the maintenance of bio-diversity. It can be a link in the chain to break the vicious circle of subsistence.

Subsistence agriculture in a world of population explosion is not sustainable as it was until a century ago and remains a situation of food insecurity and impoverishment. A small illustration of the meaning of subsistence - one hectare of rain-fed maize production under hand implement cultivation only, requires about 150 labour days with a yield of about 1 ton - a family of six persons will require approximately 1.8 tons per year of grain equivalent to survive. Therefore, the subsistence agriculture will cope with difficulty to produce enough grain equivalent of 1.8 Tons on 2 hectares representing an intake of about 2,200 calories per capita per day and requiring approximately 300 labour days over a six-month period. About 10-15% of the yield will remain as surplus reserve in case of drought, floods, plant pests or diseases. Conflicts, wars, epidemics of chronic illness can affect the delicate balance between family labour availability and cultivable area. This illustrates the vicious circle.

In my travels and encounters with communities in subsistence conditions I can identify the "real" situation by quantifying the food reserve stored in the farm of community yard. A typical aspiration of a peasant practising subsistence agriculture is to have on farm storage of 5-6 months of stored grain to cater for food security.

The second situation in world agriculture, which represents over 50% of the world's farm population, can be defined as "**improved traditional agriculture**":

Improved Traditional Agriculture

- Improved Nutrition
  (450-550 kg grain equiv./cap/year)
- Sedentary farming
- Partial irrigation (including rain-harvest systems)
- Terraced and/or mulch farming
- Double cropping (when applicable)
- Improved seed and plant material
- Use of traction animals and/or mini-tractors
- Relative low level other inputs
- Low cost habitat / settlement

This situation is characterised by practising traditional crop production under a sedentary farm system that provides some land tenure security. Often it is based on rain-fed conditions but by applying partial irrigation facilities the land holding can produce two crops per year utilising the dry season.

Aside from providing partial irrigation facilities from an existing water resource, rain-harvest systems to channelise and use storm waters can be effectively used for the same objective of double cropping. The improvement is certainly attained by the use of improved seed- and plant-material as one component of a low but critical level of purchased inputs. The use of traction animals for land preparation, cultivation and transport is characteristic of the situation due to an improved level of human nutrition, which at 550 kg grain equivalent per capita does not pose competition with the feed inputs required by traction animals. In societies that are not accustomed to working with traction animals the mini tractor becomes the means for cultivating family farm plots, which for instance today is very prominent in Asian agriculture. My own experience in Asia is that water buffaloes are used for contract cultivation and are owned by operators that enjoy a nutrition level of over 550-kg equivalent and therefore can afford to feed the animals (often with broken rice) during the heavy working season. In other words, the traction animal does not compete with the peasant and his family for the same rice bowl.

The most critical aspect of improved traditional crop farming is to provide a financial and logistic system to supply **critical inputs** and an **effective farm extension-delivery system**. The T&V (training and visits) system in extension was effectively introduced in Asia, particularly to promote the introduction of high-yielding varieties as part of the green revolution of the 70s. The bulk of active farm population in this category is characterised by a significant reduction of hunger but nevertheless many suffer from poverty and lack of basic amenities. The intervention of the public sector for the provision of basic service centres to avail paramedical care, primary education and other socio-economic element for developing a healthy civic society is mandatory. This also includes community-driven agroforestry and conservation programs and rudimentary support for the improvement of rural housing.

The third situation of world agriculture is the diversification of agriculture that **is market- and cash-oriented**. Farm populations practising a diversified basket of farm production usually are above basic food consumption needs (550-800 kg grain equivalent), partially produced on the farm holding and partially purchased:

Diversified Agriculture

- Balanced food basket
  (550-800 kg grain equiv./cap/year)
- Cash and market-oriented
- Credit delivery systems
- Irrigation (where relevant)
- Double and multi-cropping
- Semi-intensive livestock husbandry
- Use of by-products
- Pre-processing facilities
- Employment generation
- Higher capitalised habitat and / or settlement

The main thrust of this situation is to produce commodities for the **national food basket**. The diversity of supply is governed by market demand factors in urban centres as fresh or processed products. The diversification of production is influenced both by the market and the availability of credit and extension delivery systems. The role of irrigation becomes vital to ensure both seasonal as well as out-of-season supply. Livestock husbandry, particularly relating to milk, poultry, pigs and small ruminant production is semi-intensive, using by-products on the farm of purchased from the processing industries. Supply to urban markets inevitably requires pre-processing facilities, such as dairy, cooling and pasteurisation plants, slaughter houses, grading and chilling, grain cleaning, storage, etc. The role of government is to promote through providing incentives to the private sector the establishment of such facilities, which are often critical to advance policies of diversification.

Diversified agriculture implies the intensification of agricultural production and land-use, which a-priory can generate employment opportunities as well as employment in secondary post-harvest activities, agricultural processing and market services.

Israel entered this stage close to its independence in 1948, when the challenge was to feed a growing population, including the absorption of hundreds of thousands of people, the remnants from the concentration camps in Europe. The government hastily sought to provide incentives and capital to cater for the necessary components of an indicative-planned national food basket. At first, the basic components of the food basket were provided under a strict food rationing system, corresponding to an intake 2,500 calories of per adult. This policy was phased out after about six years due to the ability of Israel's developing agriculture to supply vegetables, fruits, milk derivatives, and poultry products to replace imports. The policy of diversification also included the production of industrial crops, such as cotton to supply a growing textile industry, sugarbeet for local sugar consumption, groundnuts for exports, sorghum for the feed industry and selected vegetables for processing. It also included the upgrading and diversification of citrus production both for processing and accelerated sales of different citrus products abroad. This diversification was attained by a combination of rural extension, farm investments, seasonal credit and water resource and irrigation systems combined with a policy of minimum price support acting as a social security net for the inexperienced farm population. There was an early realisation that "know-how" is as important as physical resources (land and water and capital). Hence, the establishment of extension at the village regional and national levels.

The role of applied and/or adaptive research, which included the extension workers and farmers themselves, made it possible to reduce the time period for introducing agro-technical innovations from the accepted period of 7 years down to 3-4 years. Another aspect of the effort to promote agricultural diversification was the introduction of innovative technologies that made it possible to produce farm products under plastic cover outside the normal season.

Agriculture co-operatives and second tier associations played an important role to ensure the timely supply of inputs and the establishment of processing and marketing systems. This would not have happened if Israel would not have had a comprehensive but pragmatic agricultural settlement policy entailing the establishment of regional integrated water resource programs, planned villages, rural service centres and agri-townships within functionally planned development regions.

The fourth situation, defined as **market-oriented specialised agriculture**, representing about 15% of the world's active farm population, is the basis today for providing farm commodities to global demand. It is the dominant characteristic of agriculture in the developed industrialised countries but can also be identified in the form of commercial enclaves in most developing countries. Commodities produced under progressive technologies with specialised post harvest and marketing systems in developing countries include tea, coffee, cocoa, bananas, pineapples, sugar, rubber, etc. This category implies that farm holding engaged only in a limited number of farm branches using highly mechanised production methods and/or highly specialised labour. Some branches in specialised agriculture are labour intensive, such as floriculture and vegetables, whereas others, such as field crops (e.g. cotton, maize, wheat, sugarbeet, sugarcane, etc.) are highly capitalised with farm machinery.

Market-Oriented Specialised Agriculture

- Food basket
  (550-800 kg grain equiv./cap/year)
- Specialised agriculture in limited number of farm branches
- Improved pressure irrigation systems (where relevant)
- Protected agriculture: tunnels, greenhouses, tissue culture
- Labour intensive or highly mechanised
- Advanced post-harvest infrastructures and facilities
- Export oriented agribusiness
- Highly capitalised

A key for specialised agriculture is the improvement of irrigation and fertilisation by using improved pressure irrigation systems combining water and fertiliser applications through the same system, known as fertigation. Another aspect that has contributed to specialised agriculture is the application of protected agriculture technologies including the use of plastic cover in the form of greenhouses and tunnels and the eradication of soil pathogens through solarization of soils.

A major key today is to base production systems on disease-free genetic material, partially produced by tissue-culture laboratories according to specific plant protocol market demands acclimatised to varying ecosystems. Therefore, specialised agriculture must go hand in hand with advanced seed production enterprises.

Israel's agriculture today is highly specialised and totally market-oriented. Over 30% of its production in economic terms is to export markets in West Europe which requires very advanced post-harvest facilities and market infrastructure. Individual producers of high-value crops associate with each other to attain the required quality of production and marketing economies of scale. The focus is oriented towards agribusiness ventures requiring carefully produced feasibility studies and business plans. These studies and plans are then presented to the banking system for financing with some government intervention in financing the infrastructure required. The critical shortage of water resources in Israel compounds the constraints on specialised agriculture necessitating careful profit and loss analysis and the expected economic returns per unit water used. The Israel experience of developing its specialised market-oriented agriculture illustrates the outmost importance of constantly seeking innovations led by basic and applied research by the Agricultural Research Organisation, the Faculty of Agriculture, Ben Gurion University and the Technion.

The application of Area integrated Pest Management (APM) is another critical aspect of the equation. Its vigorous application (as a partnership between research, extension and the farmer) is the key to opening up markets in the industrialised countries. An example of this challenge is to develop production regions free of the Mediterranean fruit fly scourge.

Market-oriented specialised agriculture requires levels of investments per worker similar to those required in the medium-level industrial sector. The risks on investments often are higher than those in footloose industries bearing in mind climatic and biological uncertainties, but by developing such agribusiness initiatives, the multiplier effect for employment and economy may be substantially higher than footloose industries. A challenge of contemporary specialised agriculture is to widen the participation of the farm community in this "leap forward". This can be done by combining the efforts of the private sector that has a capability of supplying inputs with particular relevance to planting material and the post harvest care and marketing of the product. It is based on the principle of contract farming whereby commercial agricultural enterprises enter into contract with smallholders to produce the same commodity with the necessary technical assistance and market services. This satellite farm system, linking agricultural commercial enterprises with smaller holders in the same region is a way to reduce the intervention of government in consolidating the livelihood of smallholders, who statistically represent the overwhelming percentage of active farm population.

Finally, in this situation of specialised agriculture, there are clear indications that peri-urban agricultural development will become high on the national agenda, bearing in mind the tremendous growth of cities on all the continents. We can expect that perishable farm commodities will be grown as near as possible to the urban centres even if such areas will eventually become fully urbanised. It can be expected that the heaviest farm investments in protected agriculture will therefore be located on the periphery of cities and this applies in particular to milk, poultry, vegetable and floriculture production. The stages of peri-urban agriculture from a real-estate viewpoint are shown in the diagram below and are relevant to the on-going peri-urban processes in many countries including Africa. For instance, stage 1 relates to the production of vegetables even in the heart of a city (such as Kinshasa or Soweto in South Africa), but as real-estate development takes place farm activities move to a second stage in geographic corridors to the city and later to the wedges shown in the diagram and finally within the periphery of the city:

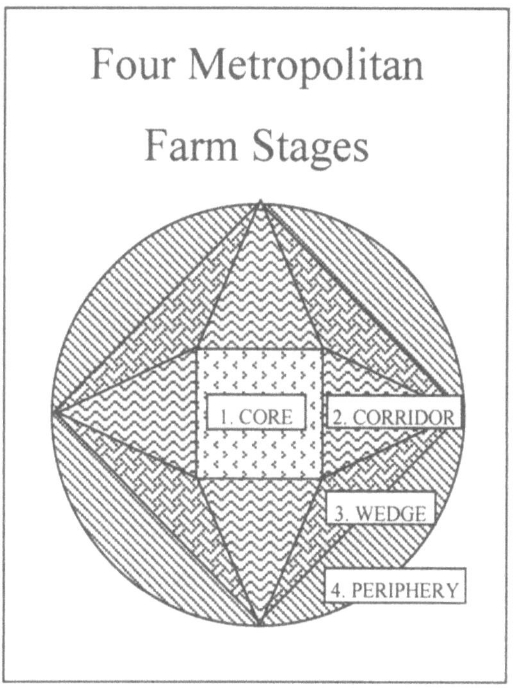

The sale of real estate from the core of the city can capitalise intensive agriculture in the corridors and likewise later to the wedges and the periphery. The main economic motivation for peri-urban agriculture is to produce the commodities for an ever-growing city demand under relatively low transport costs and at acceptable quality to those markets.

The following diagrams highlight the elements of required supporting systems for each of the four situations described. They show that the support services become more complex as agriculture progresses from subsistence to market-oriented specialised agriculture. The dynamics of this transition implies the application of agro-ecological and environmental precautions, ranging for land conservation, agroforestry, the recycling of marginal water, the reduction in the use of chemicals, and the installation of modern post-harvest and marketing infrastructure.

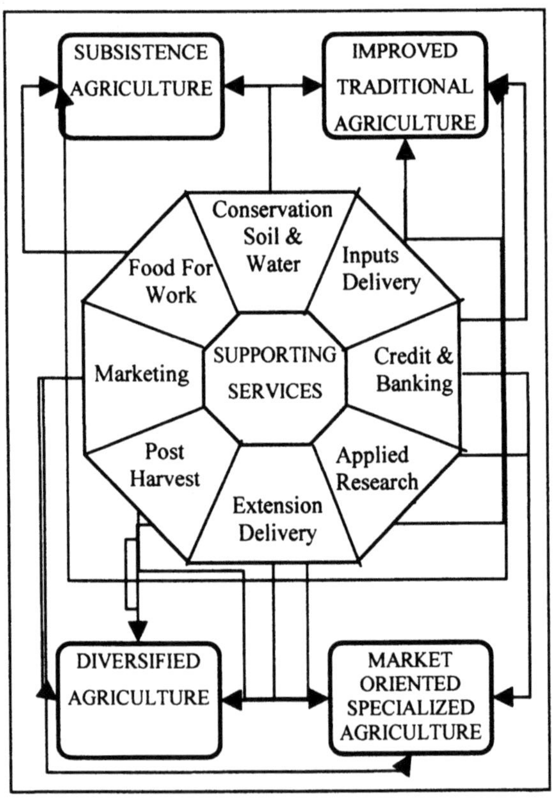

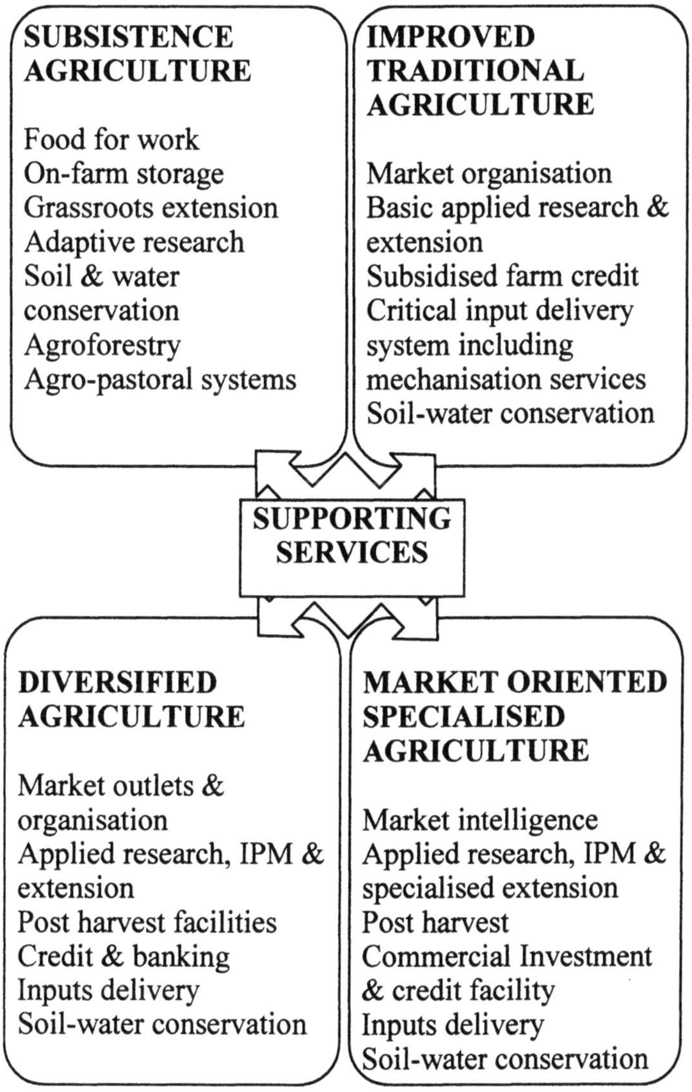

The next diagram summarises an overall policy aim to move farm populations out of subsistence to specialised agriculture over a time period of 2-3 generations.

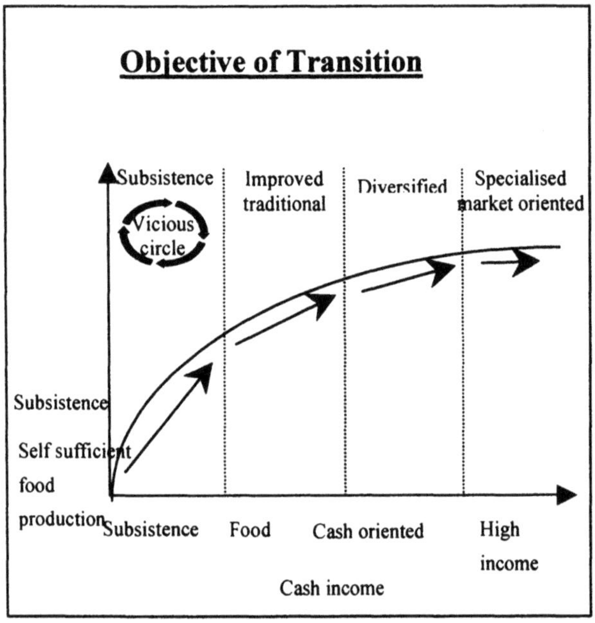

During this tedious transition it will be essential to conserve resources and prevent the degradation of farmlands by adopting the eleventh commandment of Walter Lowdermilk, one of the pioneers of resource conservation:

"The Eleventh Commandment"

Thou shalt inherit the holy earth as a faithful steward, conserving its resources and productivity from generation to generation

Thou shalt safeguard thy fields from soil erosion, thy living waters from drying up, thy forests from desolation and protect hills from overgrazing by the herds, that thy descendants may have abundance forever

If any shall in this stewardship of the land, thy fruitful fields shall become sterile stony ground of wasting gullies, and thy descendants shall decrease and live in poverty or perish from off the face of the earth.

Lowdermilk, 1939

References

1. Abt, Y. (1990) *Some Education and Training Aspects for Farm Managers in Developing Countries,* in Farm Management in Practice. Proceedings of the 7th International Farm Management Congress, Copenhagen. University of Reading, Farm Management Unit
2. Clark, Colin (1956) *Economics of Subsistence Agriculture*, Oxford Press.
3. Laudermilk, Walter (1946) *Palestine Land of Promise*
4. Schultz, Theodore (1964) *Transforming Traditional Agriculture*, Yale University Press
5. Seligman, Ruth (1998) *Constant Change and Farmer Creativity.* Shalom Magazine, issue no. 2
6. Weitz, Raanan (1965) *Rural Planning in Developing Countries* - proceedings of Rehovot Conference 1962. London Routledge & Kegan Paul
7. Weitz, Raanan (1971) *Compendium on Rural Development in a Changing World,* MIT Press

COMBATING POVERTY WITH PLANTS

Dov Pasternak
The Institutes for Applied Research, Ben-Gurion University of the Negev, P.O. Box 653, Beer-Sheva 84805, Israel, Tel: 972-7-6461972, fax: 972-7-6472984, email: dovp@bgumail.bgu.ac.il

DESERTIFICATION AND ITS CAUSES

Desertification, meaning the process of loss of soil fertility in the semiarid and in the dry sub-humid regions of the world, is not a new development. Rather it is a phenomenon which has been with us for hundreds and thousands of years. In the past, as is the case today, it was caused by both climatic changes and human interference.

An example of a climatic dependent desertification event is the extinction of Nabatean agriculture in the deserts of the Middle East. There is evidence to show that some 2000 years ago rainfall in the central and northern Negev desert of Israel was more plentiful than it is today, allowing the Israelites, the Nabateans and the Byzantines to cultivate extensive areas of land using water harvesting techniques (Issar, 1995).

Human interference with nature has also been responsible for desertification in the past. The best example is the salinization of soils in the irrigated areas of Mesopotamia, which resulted in the collapse of ancient empires (Jacobsen and Adams 1958).

However, in the twentieth century the process of desertification reached disastrous proportions. As in the past, the underlying causes can be divided into natural (climate changes) and man-made.

Combating Desertification with Plants, edited by Dov Pasternak and Arnold Schlissel.
Kluwer Academic/Plenum Publishers, New York, 2001.

Desertification caused by climate change

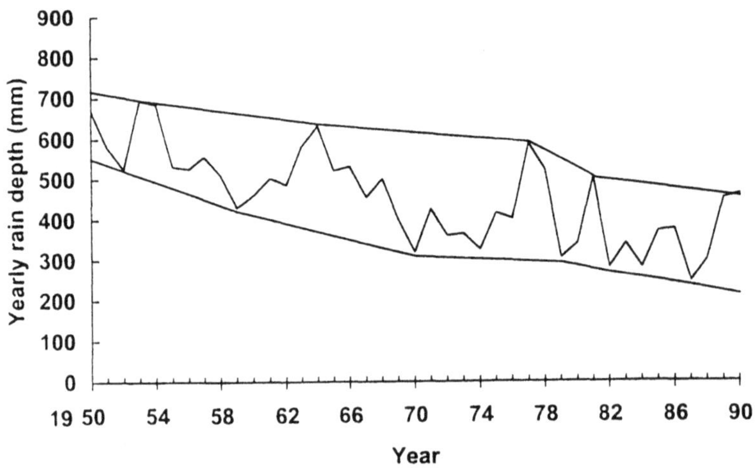

Figure 1. Rain Variation (mm) in Douendza, Mali Over a 40 Year Period

Figure 1 gives annual rainfall values for a period of 40 years in the Douendza region of northwest Mali. From 1950 to 1990 median rainfall figures dropped from 650 mm/year to 370 mm/year. This trend is typical of much of the Sahelian region of western Africa. However, since Morin (private communication) has recently observed a 50 year cyclical behavior in rainfall patterns in this region (Fig. 2), there are chances that this process will soon reverse itself.

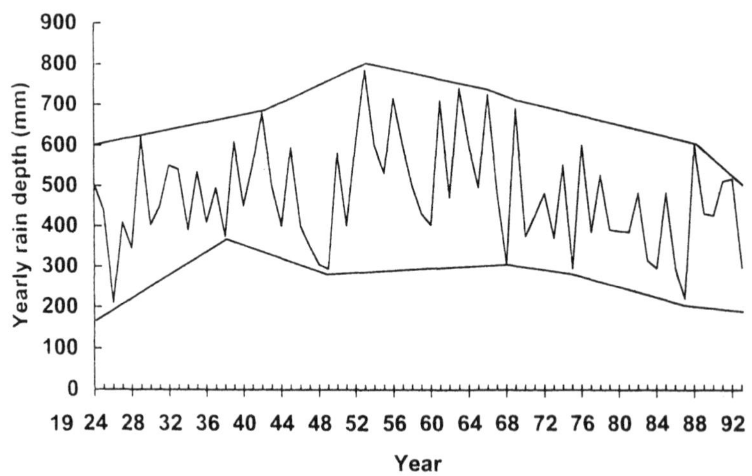

Figure 2. Rain Variation (mm) in Dori, Burkina-Faso, Over a 69 Year Period

The decline of rainfall in the Sahel has resulted in many places in desertification and land abandonment.

Desertification caused by human interference

Notwithstanding the importance of climatic changes on desertification, the present desertification crisis is to a large extent a process caused by human interference. The responsibility for desertification is shared among all the sectors of society: governments, the rich and middle income and the poor sectors. There are many examples of desertification due to government action. Here are just a few:

- The Ottoman Turkish Government introduced railway transportation to the Middle East at the beginning of the 20th century. The train engines were fuelled by wood. Within a few years the new demand for "firewood" resulted in the elimination of many forests in the Middle East and in massive soil erosion from the resulting bare hills.
- The governments of the former USSR, China, Pakistan and India developed impressive irrigation schemes in the semiarid regions of their countries. However, these governments did not take into account the need for appropriate drainage. The result was massive salinization of agricultural lands (Klaykov, 1991; El Swaify et al. 1983, Condom, 1999, Dwivedi, 1994).

Most cases of government induced desertification are the results of "good intentions". Governments wanted to promote fast development, but did not consider sustainability factors. On the other hand private companies have (and still do) spread desertification through exploitation for profit. The hills of Almeria in southern Spain are today totally bare due to the exploitation of the local forests by mining industries (Puigdefábregas and Mendizabal, 1998). Likewise, mining companies have denuded much of the Chilean semiarid regions by cutting timber to support mine shafts (Fuentes, 1978).

Timber explorers worldwide are about to complete a massive "denudation program" of the dry subhumid regions of the world with disastrous consequences to soil fertility (Dregne,1986).

The middle income sector has also had a hand in desertification. Examples are the destruction of useful vegetation (and the consequent reduction of soil fertility) in the Southwest of the USA by ranchers, the desertification of Patagonia by sheep breeders (Soriano 1986), and the salinization of the Imperial Valley in California or the irrigated areas of Santiago del Estero in Argentina by middle income farmers (Dregne, 1986).

Today, however, at the onset of the 21st century, the poor sector of society is by far the main contributor to desertification, particularly in the African continent.

Present day desertification and its consequences

Agenda 21, the document agreed upon at the Rio de Janeiro Earth Summit in 1992, says that:

• Desertification affects 70% (!) of all drylands, amounting to 3.6 billion hectares, or about one quarter of the world's land area.

• Almost one sixth of the world population is affected - some 900 million people.

The loss of arable land through desertification is occurring at a time when the world population is growing at a fast rate and the need for additional land to produce more food is becoming acute.

Desertification, particularly in Africa, has resulted in mass migration of the rural population to towns and cities and the creation of a huge mass of unemployed slum dwellers (World Bank, 1994a, 1994b). Furthermore, soil abandonment means that today there is less land to produce food for the local population (World Bank, 1989). Many of the semiarid countries of Africa, in the past net exporters of food, have become net food importers (Leisinger and Schmitt 1995). **Today some 60% of the world's food aid is directed to the populations of the arid lands of Africa** (Sasson, 1990).

This paper discusses the main reasons for poverty-induced desertification and highlights the importance of plants in the battle against desertification.

POVERTY AND DESERTIFICATION

Figure 3 is an illustrative demonstration of why and how poverty is causing desertification in Africa. I called it the "Vicious Triangle of Desertification".

There are many factors responsible for the current poverty of the rural sector in Africa and particularly the rural population of the dry regions. The main ones are listed below.

Negative government interference.

Examples of this are:

- Fixing of food prices by many African governments, and other measures favoring the urban community over the rural community (Dumont, 1986). The result of this policy is that the farmers stopped selling their products on the national markets.
- Emphasis on the development of "high potential areas" (this policy was also supported by many international aid agencies) at the expense of the dry areas, which were regarded as areas of "low potential"

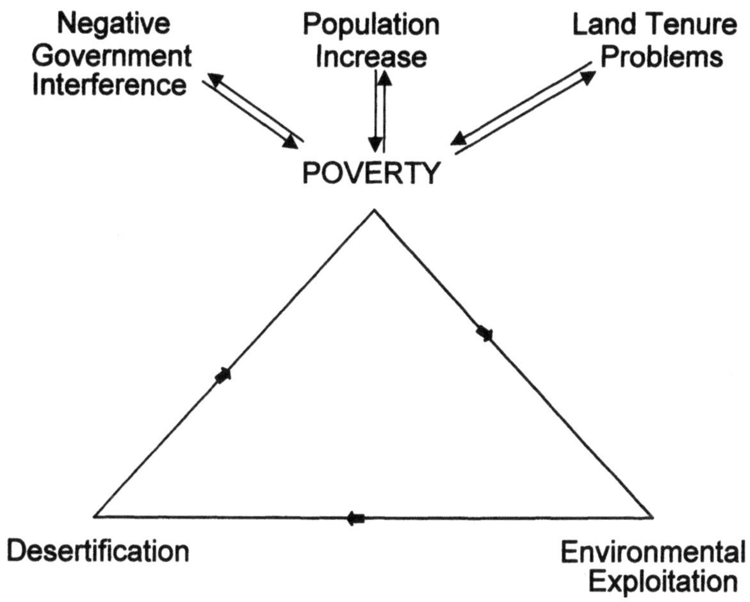

Figure 3. Poverty and Desertification: The Vicious Triangle

Uncontrolled population increase. At a rate of around 3%/year, population growth is probably the major reason for impoverishment of the rural population (World Bank 1994b). Increasing population pressure on the land has resulted in the cessation of "slash and burn" practices, in which the land was left fallow to recover fertility for several years after being used for crop production (Lallement, 1986). There just isn't any land left for this practice.

Problems of land tenure. Many African producers have no private land. Normally land is owned by the community, with farmers being given permission to work it. On the one hand lack of private ownership discourages investment by producers (especially long term investment, e.g. in tree planting), and on the other hand it encourages land exploitation (Werbner, 1982). Community ownership operates well in places of low population density where the "slash and burn" practice can be implemented. However, nowadays such places are very rare indeed.

Poor farmers lacking financial means can only ***survive*** *by exploiting the environment.* This exploitation takes many forms: Overgrazing, tree cutting to create new agricultural land and to produce firewood, and mining soil nutrients. Of all these exploitative activities, the practice of tree cutting probably has the most devastating and most irreparable effect on the soil.

Most soils of dry Africa have a light texture ranging from sandy to sandy loam (FAO-UNESCO, 1994). At the same time, in most places the monsoon rains fall in showers of very high intensity (*Fig. 4).* The result of these storms on the fragile soils of dry Africa is devastating erosion. In sandy soils with high water permeability, wind erosion is the primary agent of soil degradation (Bationo, 1993).

The situation is so acute that one can for practical purposes classify the soils of many dry African regions into two categories: soils that have already been degraded and soils that are in the process of being degraded.

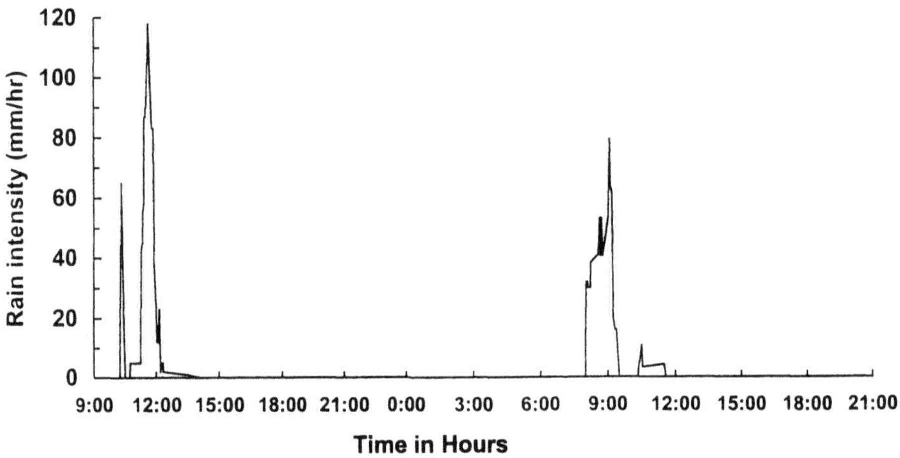

Figure 4. Rain Storm Analysis in Mokdo, Cameroon, 16-17 July 1988 (rain depth 148 mm)

So, how can one stop the progress of desertification in dry Africa? Poverty induced desertification is not an easy process to arrest. It is far easier for governments to stop desertification caused by private enterprises (mining, petroleum, lumber companies) or even by "well meaning" governmental or international developmental agencies. This can simply be done by legislation and law enforcement. *But what legislation will stop a hungry farmer from exploiting the soil or a poor mother from collecting firewood to cook her family's daily meal?*

Combating poverty can theoretically be done by attending to its main causes (negative government policies, population growth, feudal systems etc.). **But this will not work,** because poor people make more children to help the family earn more income while authorities discriminate against the rural poor and wealthy individuals exploit them because they are weak.

Furthermore, to enforce changes one would need to apply brute force, and such an approach, as the failed communist experience of Russia and Eastern Europe has demonstrated, is likely to have disastrous consequences.

Probably the best way to combat poverty-induced desertification is to attempt to improve the poor farmers' economic situation. *Their improved economic status will then become an effective tool which they themselves will use to fight the permeators of their poverty.* The introduction of new crops and of appropriate technologies that can increase farmer's income, improve soil fertility and increase crop yield seems the best way to improve the farmer's economic situation.

As illustrated in *Figure 5,* these crops and technologies should become the agent which will break the "vicious triangle of poverty".

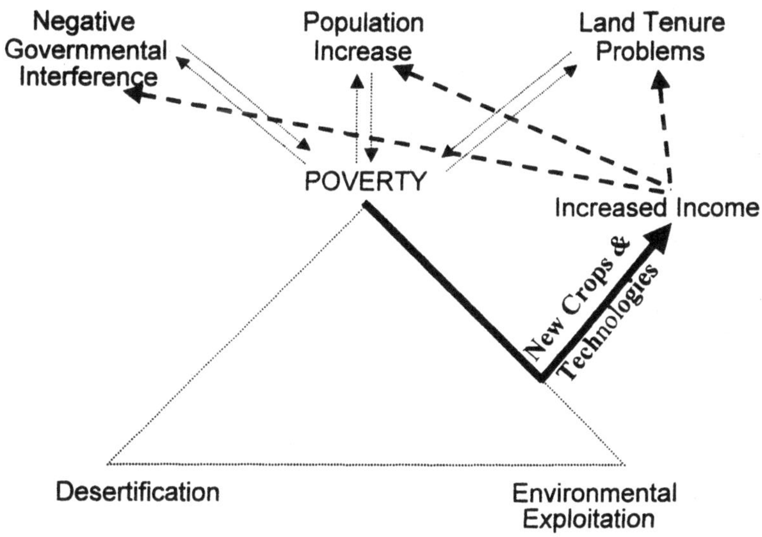

Figure 5. Breaking up the Vicious Triangle of Poverty

At this point the next question to be asked is: **how** is one going to provide the crops, technologies and other means of production that will be adopted by the poor farmers? We might have an answer.

IPALAC - The International Program for Arid Land Crops

Few people are aware of the obvious fact that the agricultural development of the temperate and tropical regions of the world is the direct result of a process of crop transfer from one region of the world to another.

The United States, which is the world's leading agricultural producer and exporter of agricultural products, has only five (!) native crops: cranberry

(*Vaccinium augustifolium*), pecan (*Caria illinoensis*), sunflower (*Helianthus annus*), blueberry (*Vaccinium augustifolium*), and Jerusalem artichoke (*Helianthus tuberosum*). All other crops came from abroad (IPGRI).

Wheat (*Triticum vulgare*) and potatoes (*Solanum esculentum*), the two main staple crops of Europe, were introduced from the Middle East and from Peru, respectively.

Brazilian native and wild plants such as cassava (*Manihot esculenta*), the rubber tree (*Hevea brasiliensis*) or the cocoa plant (*Theobroma cacao*), have become the staple and cash crops of Asia and tropical Africa but only after an intensive process of **domestication**. On the other hand, Brazil, the source of these important tropical crops, is a big exporter of coffee (*Coffea arabica)*, soybean (*Glycine max*) and sugarcane (*Saccharum officinale*), all respectively introduced to Brazil from Africa, China and Southeast Asia.

The process of crop transfer, domestication and dissemination has to a large extent bypassed the semiarid regions of the world, particularly those of Africa. IPALAC was created to correct this gap.

The objective of IPALAC is to become the catalyst of crop domestication, adaptation and transfer to the arid regions of the world with an emphasis on Africa.

The introduction of new crops and of affiliated and appropriate technologies to the farming community of dry Africa can and should become the lever that will lift up the population from its present state of deepening poverty.

From this point onwards, we will provide a very brief analysis of the constraints facing agricultural development in semiarid Africa and give some examples of how the IPALAC approach can provide solutions for these problems.

More than 90% of the farming activity of dry Africa is based on rainfed agriculture (Lallement, 1986). This situation must change because in dry regions rainfed agriculture is a very unreliable system, being too dependent on weather and climatic change. Even given an optimal scenario, rainfed agriculture would not be able to provide all the food needed by the fast growing local population in the long term.

The semiarid regions of Africa have vast resources of untapped water that can be used for large-scale irrigation. For example, Issar and Nativ (1988) quote French workers who estimate that under the semiarid and arid countries of western Africa lie deep aquifers holding enormous quantities of water. The main source of water presently used in most of dry Africa for irrigation is the phreatic aquifers in the vicinity of perennial and seasonal rivers, or ground water in natural depressions. These aquifers are relatively shallow and the water can be pumped by cheap and simple methods. The water of these aquifers is used for irrigation of small **market gardens** which supply vegetables and fruit to villages and towns.

The introduction of large-scale irrigated agriculture to semiarid Africa is a necessity, but irrigated agriculture normally requires large investments. Thus at the first stage, the fight against poverty should emphasize the improvement of these **market gardens**, through the introduction of new crops and appropriate technologies, together with improved varieties and management. The increased income of the market garden operators should automatically result in increased irrigation area.

Notwithstanding the above there is still much to be done to improve and stabilize the rainfed agricultural systems of Africa's drylands.

Rainfed systems

Agroforestry

The rainfed systems of semiarid Africa are based mainly on annual crops such as pearl millet (*Pennisetum glaucum*), sorghum (*Sorghum bicolor*) ground-nut (*Arachis hypogea*), and cowpea (*Vigna unguiculata*). Although trees are in many places a part of these systems, the food and the income presently generated from trees is in most instances limited relative to the contribution of annual crops.

By contrast, the Mediterranean rainfed agroforestry system, which is probably the oldest and most sustainable rainfed system in the world, is based mainly on fruit trees, particularly the olive tree.

In many so-called "drought years", total annual rainfall can actually be as high or even higher than in normal or in bountiful years. In most of these years crop failure results from long periods of dry spells between rains, rather than from lower mean annual rainfall. The annual crop is not built to withstand these dry periods but the perennial tree crops are. This is what makes trees a much more sustainable proposition for rainfed conditions as compared with annual crops (provided of course that long term income from trees is higher than that of annuals). The year 1999 was the rainiest year on record in much of Senegal. But for the farmers of this country it was a disastrous drought year due to the long dry spell that occurred immediately after the crops germinated.

So an equivalent of the olive tree should be found for the rainfed systems of semiarid Africa. Three of the most promising tree candidates for this system are:

1. *Zizyphus mauritiana*. This species, which is native to all the semiarid regions of Africa, was domesticated not long ago in India, where it is called Ber. The Ber tree is drought, heat, salt and flood tolerant. It produces a heavy yield of fruits the size (and the taste) of a small apple. The fruit, which

is rich in vitamins, can be eaten fresh or can be dried. It is a multipurpose tree. The leaves are both palatable and nutritious. The branches are used for firewood and as poles for roof support (Vashishta 1997).

2. *Balanites aegyptiaca.* This too is a very hardy small tree. It is a drought, salt and heat tolerant species with a very large ecological spread, somewhat similar to that of *Z. mauritiana. B. aegyptiaca* is a colonizing plant. It is among the first species to germinate and grow on disturbed or degraded soils, and therefore it can serve as an agent for rehabilitation of degraded wastelands. The kernels and pulp of *B. aegyptiaca* contain edible oils and valuable steroids (*Table 1*).

Table 1. Chemical composition of kernels and pulp of *Balanites aegyptiaca.* (Data from Abu-al-Futuch, 1983 and from Hosny et al, 1992).

Oil Composition (% of Fatty Acids)					
Fatty acid	Palmitic	Stearic	Oleic	Linoleic	Linolenic
	14-16	6-11	33-50	20-40	2-4

Kernel: Oil Content: 40-50%; Protein Content: 25%-30%.

Pulp: 70% carbohydrates of which 40% are sugars (glucose, fructose and sucrose) and 30% are glycosides.

Interesting glycosides are: balanitin 1, 2, 3 and 4, and spinostanol-type glycosides.

3. *Vitellaria paradoxa* (Shea or Karite) is a tree of the Sudano-Sahelian regions of Africa. The butter extracted from its seeds is widely used by the local population. Both *B. aegyptica* and *V. paradoxa* are now undergoing a process of domestication initiated by IPALAC.

Mulch plants

The torrential rains of semiarid Africa (*Fig. 3*) result in the creation of a thin crust on the soil which constitutes a barrier to water infiltration (Hoogmoed and Stroosnijder, 1984). The combined effect of heavy rain and crusts thus result in massive run-off of water from the field. *Figure 6* shows how mulching can prevent crust formation and reduce water run-off in sandy loam soils from 65% to 25% of fallen rains.

Mulching also increases the activity of white termites, which both decompose organic matter and increase soil porosity (Mando 1997). Recently, farmers in the central plains of Burkina Faso have started to mulch their fields (Critchley et al. 1992). Presently, however, the drylands do not have sufficient mulch, and therefore efforts must be made to introduce special mulch plants that will grow on marginal lands to provide this valuable element.

Irrigated agriculture

As mentioned earlier, the "irrigation revolution" of dry Africa should start through the improvement of existing irrigated market gardens.

Technologies. Drip irrigation is by far the most adequate system for irrigation in dry regions, particularly of horticultural crops. Until recently, the high cost of drip systems and the need to provide water pressure for irrigation and spare parts for expensive equipment prohibited its use by the owners of small market gardens in Africa. Recently, Israeli companies have developed low-pressure drip irrigation systems which overcome these limitations. IPALAC has developed a holistic approach for "African Market Gardens" which incorporates all the principles of modern fertigation (with low-pressure drip systems) and management practices. The cost of this system is affordable by many owners of market gardens.

The introduction of this new concept will provide the right conditions for increasing the yields and the quality of horticultural crops, will save water, and will protect the environment.

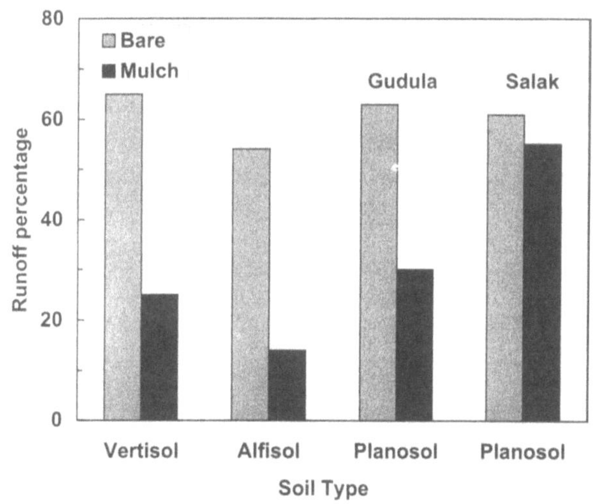

Figure 6. Average Runoff percent in North Cameroon in 1989 (rain 500-600 mm) as a Result of Mulching

Irrigated crops. The date palm is probably the most promising species for the market gardens of dry Africa for the following reasons:

- An adult date tree (six-years-old and more) produces between 50 and 100 kg of fruit per year. The price of 1 kg of date fruit in the markets of Sahelian countries is about US$1. Therefore a farmer with 10 female trees can earn $500-$1000 a year from his dates, many times more than present annual earnings

- Date fruit have a very high energy content (2,680 Kca/kg).
- Dates can be stored for at least one year. They are not perishable and can easily be transported to nearby or faraway markets.
- Dates planted inside existing market gardens provide a favorable microclimate for vegetable growth, thus further increasing the productivity of these gardens.
- Date yields are not dependent on rainfall and are solely a function of irrigation. The date fruit could therefore become a very important "food security" crop during years of crop failure.
- Date fronds and stems furnish raw material for a wide range of products.

All the above-mentioned crops and technologies are but a few examples of what can and should be applied to provide ammunition in the battle against poverty in the rural sector of dry Africa.

The IPALAC approach ("Combating Poverty with Plants") as presented in this contribution can help bring an end to poverty-induced desertification.

The battle against desertification will be won only when the "Program" part of IPALAC will be turned into a "Movement" encompassing everybody – the farmers, the NGOs, national and international agencies, governments and government organizations – all working to improve the performance of the rural sector through the use of appropriate crops and technologies.

In other words, to help arrest desertification, IPALAC (The International Program for Arid Land Crops) will need to be turned into IMALAC (The International Movement for Arid Land Crops).

REFERENCES

1. Abu-Al-Fatah, I.M. (1983). *Balanites aegyptiaca.: an unutilized raw material potential ready for agro-industrial exploitation*. UNDP Report TF/INT/77/021, Vienna.
2. Aubreville, A. *Climats, Forêts, et Desertification de L'Afrique Tropicale,*. Société des Editions Géographiques, Maritimes et Coloniales. Paris, 255 p.
3. Bationo, A. (1993). *The effect of crop residue and fertilizer use on pearl millet yields in Niger*. Fertilizer research, 34:251-258.
4. Condom, D. (1999). *Salinization, alkalinzation and sodification in Punjab (Pakistan): characterization of the geochemical and physical processes of degradation*. Land degradation and development. 10: 123-140.
5. Critchley, W.R.S., Reij, C.P., and Turner, S. (1992). *Soil and water conservation in sub-saharan Africa. Towards sustainable production by the rural poor*. A report prepared for IFAD by the Center for Development Cooperation Services, Free University, Amsterdam. Printed in the Netherlands by GCA, Amsterdam, 110 pp.

6. Dregne, H.E. (1986). *Desertification of arid lands.* In: Physics of Desertification. F. El-Baz and M.H.A. Hassan, editors. Martinus Nijhoff Publisers, Dordrecht, The Netherlands.
7. Dumont, R. (1986). *Pour L'Afrique, j'accuse.* Paris Plan, collection Terre humaine, 448 pp.
8. Dwivedi, R. S. (1994). *Study of salinity and waterlogging in Uttar Pradesh (India) using remote sensing data.* Land Degradation and rehabilitation 5: 191-199.
9. El Swaify, W., Arunin, S., and Abrol, I. (1983). *Development of salt affected soils.* In, Natural Systems for Development, What Planners Need to know. P.W. Jordan and R. Jensen, editors. MacMillan, New York p. 162-228.
10. FAO - UNESCO (1994). *Soil map of the world.* Revised legend with corrections. Technical paper No. 20, ISRIC, Wageningen, 139 pp.
11. Fuentes, E.R. (1978). *Interacciones hombre-clima en la desertificacion del Norte Chico chileno.* Ciencia e investigacion agraria 5: 137-142.
12. Hoogmoed, W.B. and Strossnijder, L. (1984). *Crust formation on sandy soils in the Sahel.* I. Rainfall and infiltration, Soil and Tillage Research 4: 5-23.
13. Hosny, M., Khalifa, T., Calis I., Wright, D.A. and Sticher, O. (1992). *Balanitoside, A furostanol glycoside, and 6-methyl-dioseginin from Balanites aegyptiaca.* Phytochemistry, 31: 3565-3569.
14. IPGRI - International Plant Genetic Resources Institute Regional Office for the Americas. Cali, Columbia. Map of "The Plant Diversity Belt".
15. Issar, S.A. (1995). *Impacts of climate variations on water management and related Socioeconomic Systems.* Technical Documents in Hydrology. IHP-UNESCO, Sc. 95/ws 26, 97 pp.
16. Issar, S.A. and Nativ, R. (1988). *Water beneath deserts; key to the past, resource for the present.* Episodes, 2: 256-261.
17. Jacobsen, T. and Adams, R.M. (1958). *Salt and silt in ancient Mesopotamian agriculture.* Science 128: 1251-1258.
18. Klotaykov, V.M. *The Aral Sea basin a critical environmental zone.* Environment 33: 4-9.
19. Lallement, D. (1986). *Prospects for rainfed agriculture in arid and semiarid areas of West Africa.* World Bank Symposium paper/dr 1/2. 14.81 to Washington, D.C.
20. Leisinger, K.M. and Schmitt, K. (1995). *Survival in the Sahel. An ecological and development challenge.* International Service for National Agricultural Research (ISNAR) ISBN 9 2-9118-020-3. 211 pp.
21. Mando, A. (1997). *The role of termites and mulch in the rehabilitation of crusted Sahelian soils.* Tropical Resource Management Papers ISSM 0926-9495 NO. 16, 100 pp.
22. Morin, J. (1993) *Rainfall analysis for tillage managements decisions.* Soil and Tillage Research 27: 241-252.
23. Puigdefábregas, J. and Mendizabal, T. (1998). *Perspectives on desertification:* western Mediterranean. J. Arid Zones 39:209-224.
24. Sasson, A. (1990) *Feeding Tomorrow's World.* UNESCO publication, Paris ISBN 92-3-102083-8 805 pp.
25. Soriano, A. (1986). *Erosion y desertizacion en la Patagonia.* Interciencia 11: 77-83.
26. Vashishta, B.B., (1997) *Ziziphus for drylands - a perennial crop solving perennial problems.* Agroforestry Today, 9:10-12.
27. Walsh, J. (1984) *Sahel will suffer even if rains come.* Science (Washington, D.C.,) 224, p. 1415.
28. Werbner, R. (1984). *Land Reform in the Making: Tradition, Public Policy and ideology in Botswana.* Rex Collins, London.

29. World Bank, (1989). *Sub Saharan Africa: from crisis to sustainable growth: a long-term perspective study.* Washington, D.C. U.S.A. 300 pp.
30. World Bank, (1994a) *World Tables*
31. World Bank (1994b) Social Indicators of Wealth.

Part I

POTENTIAL GERMPLASM FOR ARID LANDS

THE ROLE OF HORTICULTURE PLANTS IN COMBATING DESERTIFICATION

T.A. Mtaita, B.K. Manqwiro, A.N. Mphúru
Africa University, Faculty of Agriculture and Natural Resources, POB 1320, Mutare, Zimbabwe, Tel: 263-20-61611/60075, Fax: 263-20-66169/61785, email: dean@syscom.co.zw

1. INTRODUCTION

According to Rowland et al (1993), desertification is the extension of desert conditions into formerly productive areas resulting in loss of ecosystem productivity. He further observes that the causes of the process are complex and varied, involving both human activities and climatic changes. The definition adopted at the Nairobi Conference (1977) is that desertification is the diminution or destruction of ecological potential of land, which leads, ultimately to desert like conditions.

Certainly climatic changes in the form of droughts are often the trigger that sets off the desertification bomb. But it is as sensible to blame drought for desertification as it is to blame a rainstorm for a leaky roof. Desertification is caused by man: through such activities as overgrazing, over-cultivation, cutting too many trees and by unwise irrigation.

Various methods, for example agroforestry, have been employed in combating and controlling desertification. However their success and efficacy in hoarding the process have been low, possibly due to poor rates of adoption (Arnold, 1989). Planting horticulture crops such as fruits, vegetables, ornamental, medicinal and plantation crops, can be used to combat desertification.

Pareek, (1975), indicates that the arid, semi arid and sub humid environments have the potential to produce high quality horticultural produce. This is because many horticultural species can withstand the harsh conditions,

Combating Desertification with Plants, edited by Dov Pasternak and Arnold Schlissel.
Kluwer Academic/Plenum Publishers, New York, 2001.

which prevail in desert areas; in addition to harbouring the process of desertification, horticulture crops can also improve community nutrition and income.

The purpose of this paper is to identify some horticultural species which can be used in combating desertification, describe establishment strategies and to provide food for thought as to why it is difficult to combat desertification.

2. SCOPE OF HORTICULTURE IN DESERTIFICATION

Pareek and Mann, (1974) believe that many horticultural crops can exploit the harsh growing conditions of areas prone to desertification. A typical example is the Date palm (*Phoenix dactylifera*) which produces high quality fruit on saline soils and high atmospheric water deficiency. Indigenous horticultural species like *Strychnos*, *Zizyphus* and *Uiapaca* in Zimbabwe, are common to the *local indigenous* people and adapted to the conditions. Promoting a culture of these species is likely to be readily accepted since the produce increases food security and incomes.

Choundawat (1990) observed that deforestation spares such tree species because of the value attached to them. In Africa where desertification has resulted in malnutrition of about 16 million people (UNCOD, 1977) and nutritional disorders associated with low fruit and vegetable intake are common (ICAR, 1983; Chundawat, 1990), the use of edible horticultural plants in combating desertification will improve nutrition of affected communities.

3. HORTICULTURAL SPECIES SUITABLE FOR COMBATING DESERTIFICATION

Suitability of horticultural species in combating desertification varies with the local environment. Kadzere et al (1991) stated that indigenous fruit trees are capable of high yields on lands, which have lost their productivity. These species include: *Uiapaca kirkiana, Sclerocaria birrea, Azanza garkeana, Berchemia spp* and *Vangueria spp*. Tredgold (1986) observed that these species are not only used as a source of fruit, but also as firewood and timber.

During deforestation these indigenous fruit trees are normally spared (Campbell et al, 1991) as they are easy to establish and some of them are fast growing thus making them suitable to combat desertification. Kadzere et al

(1991) also noted that these indigenous trees have been adapted to thrive in the so called "invisible gardens" mostly found on marginal land.

Foliage from some of the fruit trees like *Strychnos spinnosa*, is an important fodder especially for small ruminants. Such species can sustain animals during drought periods and hence combat desertification through controlling the intensity of overgrazing. Despite the shining virtues of these indigenous species, the consistent lack of research has resulted in little being done to promote them, to improve food security and controlling desertification (Agritex – Zimbabwe, 1994). Some of the cited problems include:

- poor image and associated low demand for the products
- lack of improved planting material
- low economic yields and
- little edible flesh to seed ratio

In addition to the indigenous fruit trees cited above, a number of exotic fruit tree species are adapted to dry environments. In desert prone areas, these plants act as windbreaks and the extensive root network binds the soil thus reducing water erosion especially at the beginning of the rainy season when soil cover is at its lowest. Deciduous activity returns large amounts of organic matter to the soil in the form of leaf material which in turn support more vegetation biomass, and hence more soil cover and consequently erosion control. Eventually, ecosystems are reclaimed and desertification controlled. Some of the common horticultural crops adapted to harsh environmental conditions and their strong virtue in combating desertification are described below:

3.1 Olive (*Olea europeca*)

Olive is adapted to high temperatures and extreme water deficits. Under these conditions, olive can produce high quality fruit and oil (Flowers and Yeo, 1995). It can therefore be grown in the dry regions e.g. the Sahel, where its perennial cover can *slow* the process of wind erosion. The high market demand for olive oil could increase its adoption. This is also likely to improve the livelihood of affected people due to large returns.

3.2 Pineapple (*Ananas comosus*)

This perennial xerophyte is important for its fruit and fiber (silky pinna cloth). Some of the biological factors which makes it adapted to drought include thick, heavily cutinized, succulent leaves and sunken stomata. It can survive and produce high quality fruit in hot (29 – 25°C) and dry climates (< 600 mm yr^{-1} but > 250 mm yr^{-1}) (Bathlomew *et al,* 1977). It does well in seasonally humid climates where the dry season covers 7-8 months and the wet season covers the remaining 4-5 months. Pineapple fruit is quite nutritious, being fortified with vitamins and minerals. Therefore, promoting its cultivation would improve the nutrition of affected communities where health disorders associated with low fruit and vegetable intake are common. However, fruit and fiber production cannot be combined. Highly productive and commercial cultivars include Mulgoa, Neelum and Bombai (Nakasone, 1982).

3.3 Cashew Nut (*Anarcadium occidentale*)

Cashew seed is used as a confectionery and dessert. The oil is also edible and the seed juice is quite nutritious. Cashew is also ideal for improving food security in affected malnourished communities. It is a perennial evergreen tree, thus providing a persistent soil cover. This also reduces evaporation and consequentially soil desiccation, which exposes the soil to eroion. As a result, soil losses are reduced thus improving the productivity of these ecosystems.

3.4 Date Palm (*Phoenix dactylifera*)

The fruit is an important source of sugars and is equally fortified with vitamins and minerals. Leaves and stems can also be used as roofing material, while the fiber is used in manufacturing ropes. All of these products can be sold on a commercial scale, thus increasing the income base of the affected populations. This also makes the tree multipurpose and resourceful. Hot dry spells and huge atmospheric water deficits (< 25 mb) are vital for successful flower formation, pollination and fruiting. Irrigation might be necessary for development of large fruits. Therefore, it should be grown in orchards where supplemental irrigation can be provided. Brackish water can be used since the fruit trees are tolerant to salinity.

3.5 Fig (*Ficus* spp)

Ficus spp. are important for their edible fruit. *F. elastica* produces latex, which is comparable or similar to that of *Hevea braziliensis*. It is therefore an important economic species. Promoting its cultivation is likely to improve the economic significance of most areas prone to desertification. Inasmuch, *F. natalensis* is important as a source of bark cloth. Both fruit and cloth or rubber production can be combined. Ficus species are drought tolerant and profuse flowering is only possible after a dry spell of 4- 6 months. The deep root system and internal water binding mechanism, reduces water loss to the atmosphere. They also evade droughts through complete deciduous activity.

3.6 Guava (*Psidium guajava*)

Guavas are important as a source of nutritious fruit, jelly, pades and jams. They are adapted to areas of extreme heat and limited water supply, producing high quality fruit under these conditions. This species performs particularly well in high altitude areas, above 1500m. This makes it quite significant, since most upland deforestation has led to severe soil erosion, loss in ecosystem productivity and desertification. Increasing guava production in these areas is likely to increase plant cover, thus reducing soil erosion. The production of fruit and jelly can equally become commercialized. This increases the economic significance of the respective areas, hence triggers conservation activity against desertification.

3.7 Mango (*Mangifera indica*)

It is even more drought-hardy and adapted to high altitude areas. It is used to produce fruit, juices, jams and jellies. Its economic importance is comparable to that of guavas. However, it is not tolerant to acidity and does not perform well on sandy soils.

3.8 Tamarind *(Tamarindus indica)*

This legume tree is of much economic and ecological importance. Its pods are consumed in the immature stage, and serve as an important source of minerals. It is also used in meat sauces, beverages, sweet meats and curries. Therefore, commercial production is likely to improve the economic returns for the impoverished families as a result of desertification. The young leaves are also eaten as a source of protein. The tree is also a source of timber, and its high regeneration capacity ensures sustainability. Furthermore, through Biological Nitrogen Fixation, tamarind is quite important in restoring soil fertility. This would in the long run improve soil cover, reduce erosion and combat desertification. It is also drought hardy, capable of sustaining in areas less than 750 mm of rainfall.

3.9 Pomegranate (*Panicum granatum)*

Pomegranate is an important table fruit in most parts of the world, and it is also used in salads. High atmospheric water deficits do not affect the quality significantly, but the quantity of the final yield. However, it can be grown sustainably in these environments.

3.10 Papaya *(Carica papaya)*

Conditions in most arid lands prone to desertification are ideal for papaya production, since the fruit is susceptible to (powdery) mildews attack under humid conditions. In fact, in most papaya species, the best fruit is obtained in dry conditions. However, minimal irrigation with freshwater is necessary for a high yield per unit weight of fruit.

3.11 Lasoda *(Cordia myxa)*

This is extremely drought tolerant. It gives profuse flowering and high yields when exposed to dry conditions. More importantly, irrigation is not necessary after a year of establishment. This makes it quite ideal for dry climates. Furthermore, it is a good windbreak often planted alongside *Acacia albida* to prevent soil losses through deflation. Other species which are adapted to these environments include *Aonla* (*Emblica officinalis*), Acid lime (*Citrus aurantifoila*), Phalsa (*Grewia subinequalis*), Bael (*Aeglo marmelos*) and *Karonda* (*Carrissa grandifolia*)

3.12 Jojoba *(Simmondsia sinensis)*

The seeds of this native desert plant contain chemically unique oil, which can substitute for almost all the functions of sperm whale oil. The oil purity, lack of odour, resistance to rancidity has also made it a successful crop in the manufacturing of creams and ointments. It is therefore a prospective high value plantation crop, which could be adopted in desert fringes where the process of desertification is taking its toll. This could increase the income base and raise the standards of living for the affected communities. Native plant stands of Jojoba have been found to tolerate 460-610 mm per year of rainfall. Therefore, it could be successfully grown in semiarid areas, which are currently at risk or experiencing the process of desertification.

3.13 Species for Saline Environments

Several horticultural species can be used to control desertification in saline areas. The Olive tree is capable of producing high quality fruit under such conditions. Pomegranate can also be successfully grown in these environments without loss in yield quality and quantity. In fact, it has been observed to respond quite well to irrigation with brackish water. Bael and wood apple (*Feronia limona*) also possess the same characteristic. Citrus can also be grown, but a tolerant rootstock, e.g. rough lemon is needed. Jojoba is also salt tolerant, and could be successfully adopted for the arid and /or saline environments (Hogan and Bemis, 1983; Hogan et al, 1981; NRC, 1985).

3.14 Legumes

Leguminous plants like black gram (*Vigna mungo*), chickpea (*Cicer arictinum*), cowpea (*Vigna unguiculata*), green gram (*Phaseolus aureus*) and lentil (*Lens esculenta*) have been observed to be drought hardy, being capable of sustaining in areas receiving 500 – 750 mm of rainfall per year (Gibbon et al, 1985). Legumes are important in combating desertification and one should incorporate these legumes because:

- they are sources of food in these regions where the imminent desertification has resulted in reduced food production and starvation.
- they provide protein for the affected and malnourished people.
- they augment soil fertility through Biological Nitrogen Fixation. This is likely to reclaim the fertility of the soil and increase agricultural productivity of the soil in the long run.
- they are the source of fodder for animals, particularly during the dry season when fodder supply is limited. The legume forage can also be used as a supplement to cereal stovers during the dry season to increase their utilization and animal performance.

3.15 Oil crops

These can equally be intercropped with the orchard fruit trees during the first 3-4 years of establishment. Important species include: castor (*Ricinus communis*), groundnut (*Arachis hypogea*), niger seed (*Giuzotia abbysynica*), safflower (*Carthamus tinctorius*), sesame (*Sesame orientale*) and sunflower (*Helianthus anuus*). The oil produced can be used for different purposes depending on the crop including food, medication, lubrication, and as base for cosmetics. In addition, groundnut hauls can also be used as a supplement in residue based stockfeeds. These species can be incorporated into orchards at a planting space of 60-90 cm or 15-30 cm depending on the desired yield. In drier regions it is beneficial to have high population densities to reduce the losses due to desiccation from extensive evaporation.

4. PLANTING TECHNIQUES

Since the environments currently under desertification are hostile to plant establishment, the following horticultural strategies ought to be used.

(a) Promoting root growth – the root growth is the mirror image of its above ground portion. Therefore, promoting root growth through deep ploughing, mulching etc, increases the overall growth of the plant. This enables the plants to extract adequate nutrients and moisture from the remote regions of the soil.

(b) Small planting holes – the planting holes should be small, just enough to accommodate the root ball. The common practice of digging deep holes for planting is detrimental to plant growth. This is attributed to the subsidence of the root ball after planting, and as the root ball sinks, more soil covers the root ball, root collar and the trunk base. This inhibits absorption of water and mineral nutrients resulting in poor plant performance. Moreover, covering the root collar tissue with soil results in poor gaseous exchange, and also increases the susceptibility to several pathogens, e.g. *Phytophtora*. The net result is that of poor plant establishment. Planting holes should be dug no deeper than the size of the root ball. The idea is to place the root ball on undisturbed soil to prevent settling.

(c) The propagule should be planted in the same soil type in which it grew. This aids in its establishment since acclimation would not be necessary. Planting these tree propagules on a different soil may result in failure to penetrate the soil – root interface as the plant grows. The situation is worsened by addition of organic matter fortifications, which also add another interface. Failure to penetrate, which is common, results in the circling phenomena of roots and hence poor growth and establishment. In fact, the planting hole should be filled with soil from the nursery site.

(d) The tree propagules should be removed for planting in the field when they are well grown, say 50 – 100 cm. This increases its vigour and establishment in the field.

(e) The root ball and its soil cover can be planted into the hole within a polythene boundary. This reduces rapid shock on planting due to soil differences thereby improving plant establishment.

(f) Micro catchments or some form of crescent bunding is also necessary for water harvesting on sloping terrain and water conservation in drought prone areas. This reduces establishment losses due to poor water availability. On alkaline soils, an addition of 2 kg $CaSO_4$ and 1kg P_2O_5 is necessary to correct the pH abnormalities and aid in plant establishment. However, particular recommendations would depend on local conditions.

(g) Caliche hard pan layers in the soil should be removed at the time of planting to accommodate the root ball and further root growth. These indurations are usually found at 15-30 cm below the ground surface.

(h) Pruning trees before planting should be avoided. Removal of growing buds results in reduction of new growth hence plant vigor and growth. In addition, it has a dwarfing effect on the plant, especially if the apical bud is removed. Moreover, any removal of the top growth results in reduction of foliage, hence the capacity to primary productivity in the plant.

5. FOOD FOR THOUGHT

In theoretical terms, the action needed to halt desertification is obvious enough. Don't allow too many animals on grazing land, don't plough up soil where the rainfall is too low to support permanent agriculture and don't cut all the trees and shrubs for firewood. However, all these "don'ts" do not constitute afforestation no wonder that most arid and semi arid areas are still threatened with desertification.

Arnold (1987) points out that poor rates of adoption for some technologies is one major reason as to why combating desertification has not been successful. However adoption is not the only limiting factor.

The fundamental constraint is that the nations with the greatest biodivesity are frequently those without economic means to care for their environment. These nations utilize their biological resources to sustain the growing population. There is nothing wrong in using living things such as trees as a resource, provided it is done on a sustainable way. The trouble today is that poor nations are being forced to live on their environmental capital because of present needs and population pressures. As a result, they are destroying their biological diversity; which is their wealth for the future.

In our generation everyone must become part of a society that cares about conservation. It is a mistake to talk of a poor rural farmer who cuts down trees for fire wood and keep silent on the depletion of biodiversity caused by powerful global forces, primarily driven by markets. For any significant change, the major step may be the necessity of learning that we are all guilty; and therefore our primary object to effect and reverse desertification problems must start with our own societies.

International communities should contribute to the costs of combating desertification because biological diversity is a public good. Species and ecosystems in one part of the world can provide significant benefits to distant nations. Contributions can be done through the provision of economic incentives such as technical assistance, equipment, information, debt swaps, subsidies, loans and grants from the developed nations to the poor nations. However establishment, care and use of plants ought to be a responsibility of the affected community. To halt desertification we should dwell more in sharing a gift of useful knowledge and employ our wisdom and creative skills under the impulse of a strong sense of stewardship.

In Genesis 2:8-9 we are told that the Lord God planted a garden, made all kinds of beautiful trees grow there and produce good fruits. Verse 15 indicates further that Jehovah God proceeded to take the man and settle him in the Garden of Eden to CULTIVATE it and to TAKE CARE of it. This call from Jehovah God (the first horticulturalist) is a call for a caring attitude towards nature. A new life style that is based on stewardship and compassion in taking care of our life supporting system is a must for survival of all mankind.

It is indeed paradoxical that people are still falling down valuable trees and creating deserts. In the words of Buddha; "Because the cause was there, the consequences followed: because the course is there, the effects will follow."

New markets and villages from the ever-increasing population are escalating the demand for timber and firewood while arable farming is encroaching on the reserve lands of marginal productivity. Reversing any of these trends call for development agencies to take indigenous knowledge and stewardship seriously.

Within this should be an ethic with emphasis on the preservation of nature and developing moral principles for establishing duties and virtues with regard to non-human nature and future generations. Many indigenous people can provide traditional methods of living in harmony with the environment. Obviously we are not going to be satisfied with going back to live as they do, but we need to incorporate the best in their traditional wisdom. Mankind needs to be made aware that they are the trusties of nature – protecting and developing the natural world in accord with God's purpose and after the model of Christ's dominion (Philipians 2:5-11). It is indeed a moral obligation.

At present, the greatest challenge presented is to cultivate stewardship in all segments of society and create a population whose roots are in conservation, one which also realizes that, "the Earth is the Lord's and the fullness thereof". We are just the stewards!

REFERENCES

1. Agritex – Zimbabwe (1994). *The indigenous fruit tree resources of Zimbabwe: a survey carried out in association with the Horticultural Research Center of Zimbabwe*, University of Zimbabwe, Longman.
2. Arnold, J.C. (1989). *Towards a Solution for Desertification*: In Oshrakhi, P. (editor) The Impact of Desertification in the Sahel Region. Tristan Publishing Company.
3. Barthlomew, C. and Clarke, T.K. (1977). *Pineapple production in Israel.* Tata – McGraig Hill.
4. Bergeman, J.(1998). *Dispelling the common Horticultural myths: In Arid zone Trees.* Guest article No. 3.

5. Campbell B.M., Clarke, J.M. and Gumbo, D..J. (1991) *Traditional Agroforestry Practices in Zimbabwe*. Agroforestry Systems. 14,99-111.
6. Chundawat B.S (1990). *Arid Fruit Culture*. Oxford and IBN Publishing Co. Pvt. Ltd. New Delhi
7. Flowers, M.L and Yeo. O.P. (1995). *The Physiology of Salt tolerance in Olive* (Olea europeca): In Janick, J. Agricultural Reviews Vol.22. Prentice Hall.
8. Gibbon, D. and Pain, A. (1985). *Crops of the Drier Regions of the Tropics* ITAS.
9. The Good News Bible. Genesis 2: 8-9, Philippians 2: 5-11.
10. Hogan, L., Palzkilt, D.A. and Dennis, R.E. (1981). *Production of Jojoba in Arizona*. University of Arizona Agricultural Experiments Station Cooperation. Ext. Sero. Pub. 81132.
11. Hagan, L. and Bemis, W.P. (1983). *Buffalo Gourd and Jojoba: Potential for new and awplands*. University of Arizona, Agricultural Experiments Station Cooperation. Ext Sero pub 81132.
12. International council of Agricultural Research (1983). *The Nutritional Impact of Desertification in the indigenous communities if the Sahel* : UNEP UNO Pub.
13. Kadzere, I. and Jackson J.E. (1997). *Indigenous Fruit trees and Fruits in Zimbabwe : Some preliminary results from a survey in 1993 – 1994*: In Jackson, J.E, Turner, A.D and Matanda, M.L (editor) "Smallholder Horticulture in Zimbabwe". University of Zimbabwe Press.
14. Kiresiga, F. and Mwanza, S. (1995). *Under exploited wild genetic resources: the case of indigenous fruit trees of the Miombo Woodlands of Southern Africa*. Longman
15. Nakasone, T.C. (1982). *Pineapple production in Smallholder Agriculture*. ITAS Mcmillan.
16. National Research Council. (1985). *Jojoba: New crop for arid lands, new raw material for industry*. National Academy Press, Washington D.C.
17. Palgrave, K.C. (1993). *Trees of Southern Africa*. Capetown, Strurk Publishers.
18. Rowland J.R.J. (1993). *Dry land Farming in Africa*. CTA, Mcmillan.
19. Pareek, O.P. (1975). *Arid Horticulture*: In Desertification and its control : Released on the UNCOD. 29 – 08 to 9 – 09 – 1977, Nairobi, Kenya. Indian Council of Agricultural Research. New Delhi
20. Pareek, O.P and Mann, C.P. (1974). *Orcharding in the Indian Deserts*. Indian Council of Agricultural Research; New Delhi.
21. Tredgold, M.H. (1986). *Food Plants of Zimbabwe*. Gweru, Mambo Press.

TERFEZIAS, A FAMILY OF MYCORRHIZAL EDIBLE MUSHROOMS FOR ARID ZONES

Varda Kagan-Zur
Instutes for Applied Research, Ben-Gurion University, POB 653, Beer-Sheva 84105, Israel. Tel: 972-7-6461974, Fax: 972-7-6472984, email: zur@bgumail.bgu.ac.il

1. THE DESERT TRUFFLES

1.1 The Terfeziaceae family

The Terfeziaceae family of fungi belongs to the Tuberales, Ascomycetes, and includes several genera of hypogeaus mushrooms, of which the better known and culinary appreciated are the genera of the Terfezias and Tirmanias, on which this presentation focuses. Members of these genera are mycorrhizal, usually occupy the same habitat, and often share the same plant hosts. Fruit bodies may range from about 30 to 300g per truffle, the main body being about 100g. They are subglobous, and many times lobed *(Figures 1 and 2)*.

1.2 Nutritional value

Non exhaustive studies of their nutritional value have been performed, and it seems that of their about 20% dry matter 20-27% are proteins, some 85% of which is digestible by humans; 3-7.5% - fat, including unsaturated as well as saturated fatty acids; 7-13% of the dry matter is crude fiber; close to 60% carbohydrates; and appreciable amounts (2-5%) ascorbic acid (Ackerman et al., 1975; Al-Delaimy, 1977; Al-Shabibi et al., 1982; Sawaya et al., 1985; Bokhari et al, 1989; Saqri, 1989). High levels of potassium and

Combating Desertification with Plants, edited by Dov Pasternak and Arnold Schlissel.
Kluwer Academic/Plenum Publishers, New York, 2001.

phosphate, and fair amounts of iron have been found (Saqri, 1989). Non tested known toxic plant compounds were detected (Ahmed et al., 1981).

The white truffle is tirmania. The dark ones are terfezias.

Figure 1. Terfezia and Tirmania fruit-bodies
Photographed by: Mr. Y. De-Malach, Revivim

1.3 Distribution

Members of the Terfeziaceae family are found in arid and semi arid zones of the Mediterranean basin, Iraq and Kuwait, the Sahara and Saudi Arabia (e.g. Trappe, 1990), Hungary (Kiraly & Bratek, 1992) and Yugoslavia (Lawrynowicz at al., 1997), China (Zhang, 1992), and the Kalahari desert (Marasas & Trappe, 1973; Trappe, 1990). They are called desert truffles - to separate them from the highly praised European forest truffles. Trappe has defined a terfezia species in Australia, and several species in Texas and Mexico in the Americas.

Fruit-bodies of terfezias and tirmanias are collected and appreciated as food or a delicacy source in certain areas, such as Saudi Arabia, Kuwait, areas in the Magreb and the Kalahari, from ancient times to this day (Trappe, 1990). In other habitat regions, such as Egypt or China, they are little known.

Figure 2. A terfezia fruit-body on a "stalk"

2. GROWTH HABITS

2.1 Symbionts

Terfezias and Tirmanias form mycorrhizas mainly on roots of members of the *Cistaceae* family, such as different species of the genus *Helianthemum* (e.g. Awameh, 1981; Dexheimer et al. 1985; Morte et al., 1994; Roth-Bejerano et al., 1990) but have other symbionts as well. Both perennial (e.g. Roth-Bejerano et al., 1990) and annual (e.g. Awameh & Alsheikh, 1978) may support terfezia and tirmania mycorrhizas, though, possibly, with some specificity as to the truffle species. In Southern Africa, where no member of the *Cistaceae* family grows (Riley, 1963; Heywood, 1993), we were lately able to show that watermelon (*Citrullus vulgaris*), a domesticated plant, serves as a plant symbiont for *T. pfeilii* (Kagan-Zur et al. 1999). Watermelon is, probably, not the sole symbiont, as these Kalahari truffles are found where no watermelon exists.

2.2 Types of mycorrhizal symbiosis

The terfezias and tirmanyas exhibit a rather variable mode of mycorrhizal associations: *Terfezia arenaria* forms, under typical soil conditions, the specific terfezia-type ectomycorrhizal association (a developed Hartig net, but an underdeveloped, or no mantle, *Figure 3*), while *Terfezia pfeilii*, from the Kalahari, forms the terfezia type endomycorrhizal associations under its typical soil conditions (Kagan-Zur et al. 1999, *Figure 4*). Dexheimer et al. (1985) showed that *Terfezia leptoderma* forms the ectomycorrhiza type, while *T. claveryi* forms the endomycorrhizal association, under the same growth conditions. Furthermore, Fortas & Chevalier (1992) showed that the latter species along with *Terfezia arenaria* and *Tirmania pinoyi* could form the above described ectomycorrhizal associations in a phosphate rich medium (0.14% of total soil weight), but formed terfezia-type endomycorrhizal associations in phosphate poor (0.006% and below) media. So this family contains species which are ectomycorrhizal, facultative, and, possibly, endomycorrhizal.

Figure 3. Ectomycorrhizal associations between *Terfezia arenaria* (leonis) and *Helianthemum sessiliflorum.* Fungal cells (bead like cells) surround plant root cells

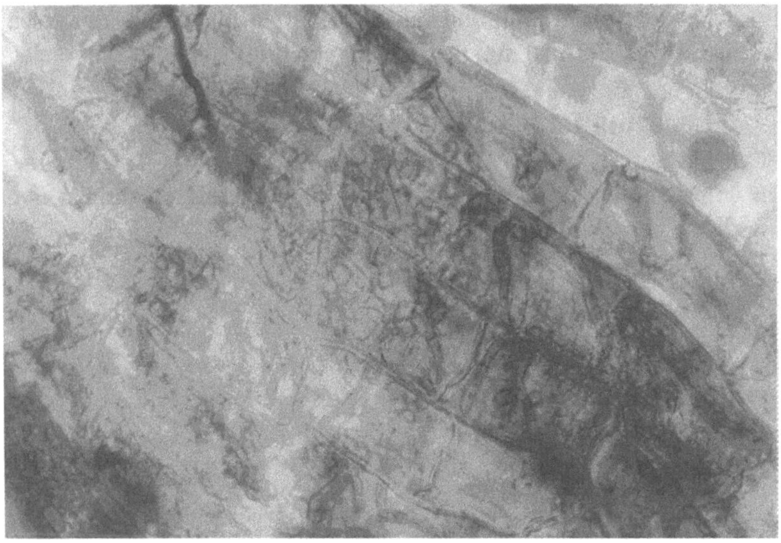

Figure 4. Endomycorrhizal associations between *Terfezia pfeilii* and watermelon roots. Fungal hyphae entangle within cells

2.3 Soil and climatic parameters

Most of the species develop in high pH calcareous soils (Fortas & Chevalier, 1992; Giovannetti et al. 1994), but some are found in soils with below neutral pH values (Kiraly & Bratek, 1992; Taylor et. al., 1995). All are found in rather sandy soils (*Table 1*).

Table 1. Range of variation of some characteristics of the top-soil of Terfezia productive fields.

	TERFEZIA SPP	*TERFEZIA PFEILII*
pH (water)	7.9-8.5	5.5-7.2
CaCO3 (total) %	4 - 54	0.5 - 3.1
Soil composition		
Sand %	80-95	94-97
Silt %	1-8	0.2-3
Clay %	4-9	2-3.5

In several countries there are legends connecting truffle appearance to thunder storms early in the season. The Beduins in the Israeli Negev tell the same story as truffle hunters in Morocco (Kagan-Zur, personal observations). The only proven fact is that desert truffles yields in the wild are highly dependent on rainfall and rain distribution during the rainy season, though as little as 200-250 mm per season may bring about a good yield.

2.4 Locating fruit-bodies

To collect desert truffles one does not have to rely on trained dogs or pigs, as is the case with European forest truffles.

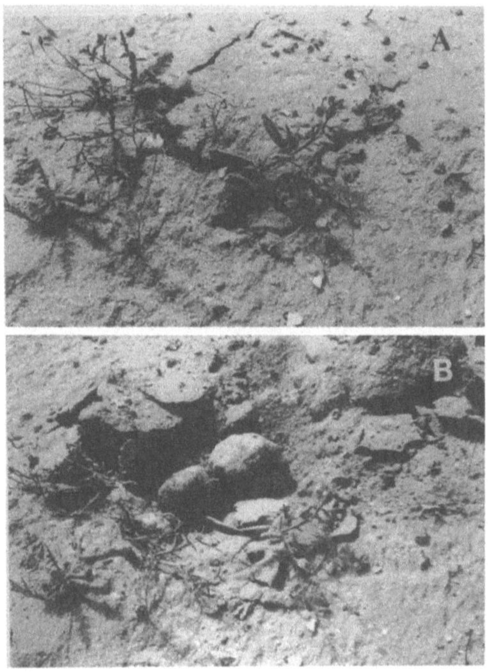

Figure 5. Truffle location. A cracked swallen soil surface near some *Helianthemum ledifolium* plants. B. The uncovered truffles. Photographed by Prof. A Danin of the Hebrew University, Jerusalem

Fruit-bodies are developing close to the soil surface, and as they swell they lift up the soil to form little cracked mounds, recognizable to the trained eye (Awameh & Alsheikh, 1978; Taylor et al. 1995, *Figure 5*). The fruit bodies are carried up on a 'stalk' (*Figure 6*) composed of entangled hyphae and sand particles. Some without (Taylor et al. 1995) some with (Kagan-Zur & Roth Bejerano, unpublished observation) parenchimateous tissue. This stalk may be between 2 and 10 cm in length, the bottom of which is connected to rhizomorphs (Taylor et al, 1995) and hyphae emanating from adjacent roots. Rhizomorphs may be detected to lead to a fruit body from up to 40 cm away.

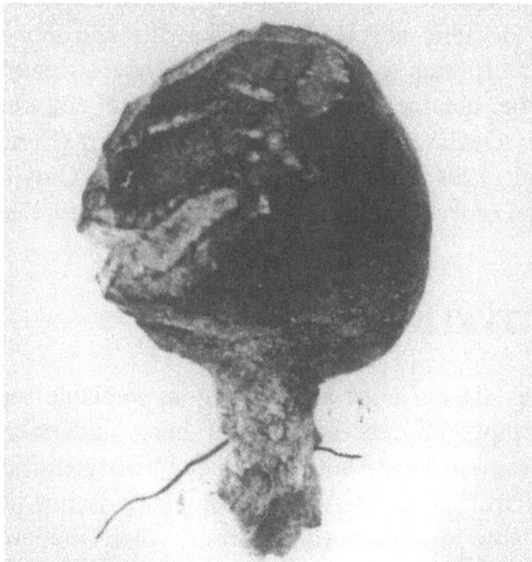

Figure 6. A truffle with its 'stalk'

2.5 Axenic cultures

Mycelial cultures have been obtained both from outgrowth of glebal hyphae (Roth-Bejerano et al., 1990; Morte et al., 1994, Figure 7) and from spore germination (Awameh & Alsheikh, 1979; 1980a; 1980b; Fortas & Chevalier, 1992.

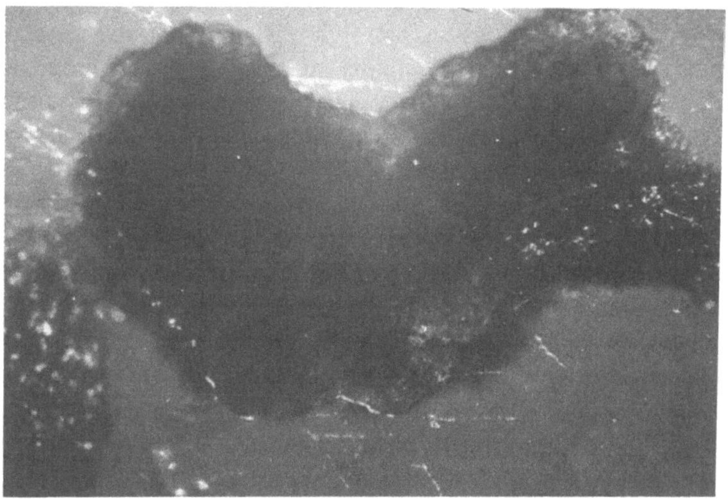

Figure 7. Terfezia hyphae growing out of a piece of a fresh, immature truffle, lain on a synthetic, sterile growth medium.

As opposed to Tuber species, both dikayotic and monosporic *Terfezia* cultures can infect (Fortas & Chevalier, 1992). Mycorrhizas of *Terfezia* and *Tirmania* may be obtained both in semi axenic cultures (sterile soils) (Awameh, 1981; Dexheimer et al. 1985; or *in vitro* (Fortas & Chevalier, 1992: Morte et al., 1994; Roth-Bejerano et al., 1990). Only one report exists showing initiation of fruit bodies *in vitro* (Fortas & Chevalier, 1992).

3. CULTIVATION?

Desert truffles are not cultivated on any appreciable scale anywhere in the world. Attempts at cultivation have been undertaken decades ago (Awameh & Alsheikh, 1978) but stopped for non-scientific reasons. Cultivation of desert truffles on perennial host plants is not likely to be very commercially viable in developed countries, where modern agriculture on fertile soils is practiced - as desert truffles fetch far lower prices than forest truffles, yields are not high (and will remain low for decades before research will bring about an appreciable improvement) and the soil will have to be occupied by a non yielding crop for several years. This kind of symbiosis may, however, be domesticated for deserts in which these truffles are natural and appreciated by the locals, and no new introductions need to take place, provided a sufficient water supply can be guaranteed in dry years.

However, two other avenues are open to research:

1. The availability of annual couples makes it a desirable aim to try and domesticate these for an eventual more sophisticated agriculture.
2. The finding that Terfezia pfeilii forms mycorrhiza with roots of watermelon leads to a possibility to develop a dual crop yielding couple, which may prove economically viable even at early stages of domestication. Further more, with this couple no domestication of watermelon needs to be done.

Traditionally, desert truffles are sold at local rural markets around the places where they are found. Already, a wider market seems to open up in Europe, especially Germany - both in specialty markets, and canned, in supermarkets. Other possibilities for introducing desert truffle as a common place commodity may open up once these truffles are regularly available. The European forest truffles are far too expensive to contemplate such an approach.

REFERENCES

1. Ackerman LGF, PJ Van Wyk and LM du Plessis. 1975. S.A. Food Rev. pp. 145-146.

2. Ahmed AA, MA Mohamed and MA Hami. 1981. Mushroom Sci. XI 833-842
3. Awameh MS. 1981. Mushroom Sci. 11:843-853.
4. Awameh MS and A Alsheikh. (1978). Mushroom Sci. 10:507-517.
5. Awameh MS and A Alsheikh. (1979).Ann. Phytopathol. 11:223-229.
6. Awameh MS and A Alsheikh. (1980a). Mycologia 72:50-54.
7. Awameh MS and A Alsheikh. (1980b). Mycologia 72:494-499.
8. Bokhari, Y.H.A., Suleiman, A.A.A., and Basalah, M.O. (1989). J. Food Prot. 52:668-669.
9. Dexheimer J, J Gererd, J.P. Leduc and G. Chevalier. (1985). Can. J. Bot. 63:582-591.
10. Fortas, Z and G Chevalier. (1992). Can. J. Bot. 70:2453-2460.
11. Giovannetti G, N Roth-Bejerano, E Zanini and V Kagan-Zur. (1994). Hort. Rev. 17:71-107.
12. Heywood VH ed. (1993). *Flowering Plants of the World.* BT Batsford Ltd Publ. London. ISBN 0-7134-7422.
13. Kagan-Zur, V., Kuang, J. Tabak,S. Taylor, F.W. and Roth-Bejerano, N. (1999). Myc. Res. 103:1270-1274
14. Kiraly, I. and Z Bratek. (1992). Micol. Veget. Medit. 7:303-304.
15. Lawrynowicz, M., Markovic, M., Milenkovic, M. and Ivancevicm, B. (1997).
16. Acta Mycol. 32 (2) 233-238.
17. Marasas WFO and JM Trappe. (1973). Bothalia 11:139-141.
18. Morte MA, A Cano, M Honrubia and P Torres. (1994). Agri. Sci. Fin. 3:309-314.
19. Riley, H. P. (1963). *Families of flowering Plants of Southern Africa.* University of Kentuckey Press.
20. Roth-Bejerano N, D Livne and V Kagan-Zur. (1990). New Phytol. 114: 235-238.
21. Saqri FA. 1989. Iraqi J. Sci. 30:421-424
22. Sawaya WN, A. Al-Shalhat, A. Al-Sogar and M. Al-Mohamed. (1985). J. Food Sci. 50:450-453.
23. Taylor, F.W., Thamage, D.M., Baker, N., Roth-Bejerano, N. and Kagan-Zur, V. (1995). Mycol. Res. 99:874-878
24. Trappe, JM. (1990). In: M. Bencivenga and B. Granetti, (eds.) Atti del secondo congresso internazionale sul tartufo. Spoleto, Italy (1988). Com. Mont. dei Martini, Italy. pp. 19-30.
25. Zhang B-C. (1992). Mycol. Veget. Mediter. 7:39-42.

THE ROLE AND POTENTIAL OF TRADITIONAL TUBER CROPS IN MALAWI

Moses F.A. Maliro
University of Malawi, Bunda College of Agriculture, P.O.Box 219, Lilongwe, Malawi, Tel.: 265 277 222/ 420; Fax: 265 277 364/ 420,E-mail: mmaliro@unima.wn.apc.org or mbkwapata@malawi.net

1. INTRODUCTION

A wide range of traditional tuber crops exist in Malawi. These include several domesticated and wild yam species (*Dioscorea species*), African potato (*Plectranthus esculentus*), *Disa species*, *Habenaria walleri* and *Satyrium species* (Williamson, 1975). Most of these are semi domesticated and are used to supplement food crops among smallholder farmers(Malawi Government, 1990 and 1994). The nutritional composition of these tuber is, in most cases, superior over the commonly grown roots and tuber crops like cassava (*Manihot esculenta*), potato (*Solanum tuberosome*) and sweet potato (*Ipomea batatas*). They are often rich in starch/carbohydrates, proteins, fats and minerals (*Table 1* and *2*). Therefore these traditional root and tuber crops play very important roles in the nutrition and economy of the rural families. These crops have a great potential of supplementing in the people's nutrition through yield improvement and economy if research attention is given to them. There is great potential of developing or selecting for crop varieties or clones adaptable to a wide range of environments.

Combating Desertification with Plants, edited by Dov Pasternak and Arnold Schlissel.
Kluwer Academic/Plenum Publishers, New York, 2001.

2. ROLE OF ROOT AND TUBER CROPS IN FOOD SECURITY

Roots and tuber crops are among the important staple foods not only in Malawi, but through out tropical Africa. They account for well over 50% of the total staple foods produced in Sub-Saharan Africa and thus the backbone of the region's economy (Hahn, 1994, Malawi Government, 1994). In Malawi the current agricultural production policy has moved from emphasis on maize production to crop diversification incorporating tuber crops in the systems. These tubers play a vital role in the country's food security because they are tolerant of environmental stresses and give reasonable yields under marginal soil conditions (Malawi Government, 1994; Sauti et al., 1994).

Table 1: Nutritional Composition of Traditional/semi-domesticated tuber crops

	Plectranthus esculentus	*Coleus rotundifolius*	*Disa spp.*	*Dioscorea spp* (yam)
Moisture (%)	70	75	-	60-70
Protein (%)	6.35-7.71	1.4	-	2.0
Fat (%)	0.27-2.63	0.5	-	0.2
Carbohydrate (%)	29.32-85.40	21	-	24
Fibre (%)	5.99	0.7	-	0.4-0.8
Ash (%)	4.48-5.56	1	-	0.7-2.6
Calcium (Ca) (mg/100g)	55-90	17	-	10
Iron (Fe) (mg/100g)	100-155	6	-	1.2
Vitamin A. (mg/100g)	-	-	-	0.8
Vit. B1 (mg/100g)	-	-	-	0.09
Vit B2 (mg/100g)		-	-	0.03
Ascorbic acid (mg/100g)	5	1	-	10
Food energy (Cal./100g)	-	-	-	104

Sources: Alleman, 1996; Gila, 1984; Kay, 1987; Kyesmu & Akueshi, 1989; Latham, 1978; Onwueme, 1982.

Root and tuber crops are important as a source of food and cash. They are efficient producers of calories and play an important role in meeting food needs in the country. Even in areas where maize is a staple food, root and tuber crops are important food supplement. They are basic sources of low cost calories because of their adaptability to a wide range of ecology and their inherent efficiency in the conversion and storage of solar energy. Calories from cassava, for instance are only 25 to 50% as costly as calories from grains and pulses (Hahn, 1994). The current Malawi national agricultural policy is to increase their production in areas where they are a staple food and encourage every farmer to grow them as security and cash crops (Malawi Government, 1994; Sauti et al, 1994).

Table 2: Nutritional Composition of Commonly cultivated root and tuber crops

	Cassava (*Manihot esculentus*)	Potato (*Solanum tuberosome*)	Sweet potato (*Ipomea batatas*)
Moisture (%)	60	82	70
Protein (%)	0.7	2.10	1.65
Fat (%)	0.2	-	-
Carbohydrate (%)	37	17.10	25.30
Fibre (%)	1.0	-	-
Ash (%)	-	-	-
Calcium (Ca) (mg/100g)	25	13.00	21.10
Iron (Fe) (mg/100g)	1.0	1.10	2.61
Vitamin A. (mg/100g)	Trace	Trace	100
Vit. B1 (mg/100g)	-	-	-
Vit B2 (mg/100g)	-	-	-
Ascorbic acid (mg/100g)	30	-	30
Food energy (Cal./100g)	153	70	114

Sources: Alleman, 1996; Gila, 1984; Kay, 1987; Kyesmu & Akueshi, 1989; Latham, 1978; Onwueme, 1982.

3. DIVERSITY OF TRADITIONAL TUBER CROPS IN MALAWI

While researchers and agricultural extension workers in Malawi have been emphasizing on three commonly cultivated root and tuber crops, thus cassava, potato and sweet potato, there remain several species of tuber crops that have served the rural communities for a long time but remain, if at all, with little appreciation. Studies on food use and nutritional composition of these have shown that there is a great potential to promote their usage and production. Most of them have been reported to have adaptation over a wide range of climates, tolerant to diseases and pests and capable to produce under poor soils (Allemann, 1996; Kay, 1987). The following are selected traditional tuber crops with potential of becoming of economical importance if promoted.

3.1 *Plectranthus esculentus* (African Potato)

African potato is a traditional tuber crop belonging to Lamiaceae (Labiatae) family. Its origin is believed to be from two centres of dispersal, one in the south central Africa (Malawi or Zambia) and the other, Ethiopia (Greenway, 1944). The crop is almost semi-domesticated in Malawi. It is a herb growing to about 60 cm with opposite leaves and yellow flowers. It bears underground tubers which grow finger-like from the central point. The crop is important because of its tubers which are very rich in starch, vitamins A and B, minerals and protein (Alleman, 1996) (*Table 1*). When harvested the tubers are first washed and eaten raw, baked, or cooked as a vegetable. They are liked by many local people who use them as a substitute to potatoes (*S. tuberosome*) (Alleman, 1996; Maliro and Woodward, 1998; Williamson, 1975;).

Use of *P. esculentus* has declined over the years mainly due to loss of planting material because of emphasis on the introduction of crop hybrids or exotic crop varieties and the social stigma which is attached to the use of semi-domesticated crops. With the current policy of promoting tuber crops for crop diversification and food security, the African potato is now being evaluated as an important tuber crop for increased production in areas where it is adapted, for use in areas of low agricultural potential which are generally not suitable for production of exotic crops (Allemann 1996; Maliro and Woodward, 1998). The Malawi Plant genetic Resources Centre and the University of Malawi have planned a country wide germplasm collection of *Plectranthus esculentus* for evaluation (Maliro, 1999).

Preliminary survey by Maliro (1999) and available literature (Williamson, 1975) show that the crop is widely adapted and has a wide variability in Malawi. Distribution of the African potato in Malawi shows that it is widely adapted ranging from the warm low rainfall areas of Salima to cool high rainfall areas of the Kirk range mountains. Similar studies in South Africa and Zimbabwe have also revealed high variability and adaptation of the crop to a wide range of environments (Allemann, 1996; Dlihwayo, 1999). The crop has been found to thrive in almost all climates, growing as successfully in the humid areas as in hot dry climatic conditions (Kay, 1987; Tindall 1983).

Agronomical studies by Allemann (1996) have shown that *P. esculentus* has a yield potential in excess of 60 tonnes/ha. under good growing conditions. Under proper cultivation in central Africa this species is estimated to yield 13 to 25 tonnes/ha from 2 to 6 tonnes/ha under poor growing conditions (Burkill, 1995; Tindall 1983).

Most of the pests and diseases attacking *P. esculentus* rarely cause serious yield losses (Tindall 1983) and no serious attacks by either insects or diseases have been noted in field trials done in south Africa (Allemann and Coertze,1996). The *P. esculentus* tubers have also been found to store well in underground silos, or kept indoors in baskets that have been packed with straw. The tubers can also be stored successfully in dry sand or in cool, well ventilated conditions (Allemann, 1996).

It is all these attributes that now make the *P. esculentus* a potential important tuber crop that can add to the diversity of tuber crops for farmers in Malawi.

3.2 *Solenostemon rotundifolious* (Hausa Potato)

Solenostemon rotundifolius (Poir) J .K. Morton syn. *Coleus rotundifolius* Chev. and Perrol, locally known as Hausa potato, is another semi-domesticated tuber crop in Malawi. It grows into a small herbaceous plant of 15 to 30 cm high, prostrate or ascending, with a succulent stem and thick leaves. It produces dark-brown round tubers (like those of *S. tuberosme)* but smaller in size, in clusters at the base of the stem. It is believed to have originated in central or east Africa but later spread throughout tropical Africa. The crop is also known to have spread to South East Asia where it is cultivated on a small scale. The distribution of this crop in Malawi is the same as that of *P esculentus.*

The *S. rotundifolius* tubers are used as a substitute for potato and are usually cooked in curry and eaten with rice. They can be boiled, baked or fried in a similar way to potato chips. The tubers are also reputed for their aromatic flavour. Extracts from the plant leaves are reported to be used for treatment of dysentery and some eye disorders. The tubers are rich in carbohydrates, protein and minerals like calcium and iron (*Table 1*).

S. rotundifolius has shown adaptation to a wide range of climates from the preliminary survey in Malawi and those done in South Africa (Allemann 1996; Maliro, 1999). However for optimum yields it favours high rainfall and low night temperatures for development of tubers. It is suited to well drained, sandy loam soils. Like most other tuber crops it does well on marginal/poor sandy soils. Allemann, et al (1996) reported that *S. rotundifolius* can be cultivated on almost all soils as long as they are well drained, however optimum yields are obtained on sandy loam soils.

Attempts to improve the yield through agronomic practices elsewhere have shown that there is a great potential to improve tuber yield of the crop. Preliminary research work on plant spacing, planting on ridges, fertilization and weed control in South Africa and else where show that yields of *S. rotundifolius* can be greatly improved. Tuber yields from the range of 7 to 15 tonnes/ha to 18 to 20 tonnes/ha under very favourable conditions have been reported (Kay, 1987). Experimental results in South Africa showed that this plant has a yield potential of more than 45 tonnes/ha if the correct selection and good agronomic practices are used (Allemann, 1996). *S. rotundifolius* is susceptible to attack by nematodes while other pests and diseases are rarely serious

The available information shows that the crop can become a valuable tuber crop particularly in low potential areas to attain household food security if promoted through research and extension. The crop is able to grow and produce yields under marginal conditions, and with few pests and diseases it can be produced in sustainable agricultural systems. However there is still a need for a great deal of research on this crop before it can be considered for commercial production.

3.3 *Dioscorea* Species (Yams)

There exist several species of yams (*Dioscorea spp.*) in Malawi that have a potential of becoming a valuable food security crop. These yam species differ widely in shape and colour of the tubers but the commonest are *D. rotundata* (white yam) and *D. alata* (Greater yam). Other yam species grown in Malawi include *D. bulbifera* (Potato yam), *D. cayenensis* (yellow yam), and *D. esculenta* (lesser yam).

Most of the yam species in Malawi are domesticated but their wild relatives (like *D. dumetorum* (bitter yam)) are also used as food source in times of famine. Although they do not form part of staple food crops in the country they are a rich source of carbohydrates, protein and ascorbic acid, and are commonly eaten as a vegetable, either boiled and baked/roasted. They are often found for sale at markets ready cooked (Asiedu, 1986; Onwueme, 1982; Okoli and Onwueme, 1987; Williamson, 1975). The tuber is liked by many people but its consumption is limited by its production quantities.

Distribution of yams in Malawi shows that they are grown by farmers throughout the country but vary in the species from place to place. Most of these are commonly seen planted to grow over fences and refuse pits just enough for consumption as snacks. Yams in general do very well in areas with temperatures of about 30°C with a wide range of annual rainfall of 600 to 3000mm on deep well-drained sandy loam soils for optimum yields. Most yams can be grown successfully at low or medium altitude, however *D. alata* is reported to be grown at altitudes up to 2700 m however yields are considerably reduced above 900m (Kay, 1987). With this adaptation to a wide range of environments yam crops can grow well as one of major tuber crops in the country as a food security crop. However yam production in Malawi is constrained by lack of suitable varieties, planting materials and agronomic packages for farmers (Malawi Government, 1994).

In major yam growing areas such as West Africa yams are usually intercropped with maize and vegetables, such as cucurbits, pumpkins, peppers and okra. Intercropping is a traditional practice among Malawian farmers and if yam production can be promoted in the country it can be easier for farmers to integrate within their farming systems. Gross yields of yam tubers range from 7.5 - 18 t/ha to 20 - 30 tonnes/ha. Under optimal conditions yams can yield up to 70 tonnes/ha for *D. esculenta* and 58 tonnes/ha for *D. alata* (Kay, 1987; Onwueme, 1982).

To promote production, its use and elevate it to an important food crop there is a need for research to carefully evaluate growth and yield performance of the existing yam species and landraces for recommendation to farmers. With research and agricultural extension support yams can become one of the important tuber crops in Malawi.

3.4 *Disa* species (Orchid) (Known as 'Chikande' in Vernacular)

In Malawi there are some *Disa* species, *Habenaria walleri* and *Satyrium species* that are in a group of ground orchids (Williamson, 1975). They are still wild and are commonly found in the hilly areas of Ntcheu, Dedza (Kirk Range area) and Nkhotakota Districts. They bear tubers that are variously shaped some growing like fingers from a central point while others are unbranched, some are red in colour and other are white.

Although these tuber plants are still not domesticated they are heavily harvested by local people in the areas where they are found for consumption and sale. When harvested, the tubers are washed, then pounded to remove the skin, dried in the sun and repounded. They are then boiled in a solution of potashes ('Chidulo'), the surplus liquid is squeezed out and the product left to dry. At this stage it becomes a firm jelly, reddish in colour and is usually sold in this form.

The texture of the jelly varies with the species used. The jelly is used as relish taken together with a thick porridge prepared from maize flour. As relish, the jelly is cut into strips and boiled with salt and, with groundnut flour added to it. The dish is liked apparently because of its resemblance to meat, being red and having a texture like gristle (Williamson, 1975).

There is very little research work or none that has been done on the nutritional composition of these orchid tubers let alone on its diversity and conservation. Due to population pressure on natural resources these tuber plants continue to be depleted. There is an urgent need for conservation activities and research to look into its domestication to promote its production and use.

With the gelling ability of these tubers there is also a need for research to investigate on possibilities of developing a gelling agent for pathological and plant tissue culture media in biological research laboratories.

Currently the National Herbarium and Botanic Gardens of Malawi has been mandated to take lead on conservation of these important orchids while the Malawi National Plant genetic Resources Centre is planning for germplasm collection of the same.

4. FUTURE PROSPECTS OF TRADITIONAL TUBER CROPS IN MALAWI

The Malawi national policy on agriculture production has been encouraging crop diversification with emphasis on roots and tubers as food security crops since early 1990s. As the result tubers have since then received much attention by researchers and Agricultural extension workers which has resulted into increased production of roots and tubers.

The Government is also currently supporting research work on agrobiodiversity, which is concerned with conservation and research on traditional crops including tubers.

With economic crisis affecting the country smallholder farmers can not afford inorganic fertilizer inputs for the cereal production. Frequent droughts for the past decade has made majority if not all smallholder farmers food insecure.

As a result agricultural researchers have been persuaded to investigate on food crops that smallholder farmers can grow with minimum inputs and be able to harvest some in years of droughts. Consequently root and tuber crops have become obvious candidates as food security crops because most of them do well on marginal soils even with low annual rainfall.

In addition to doing well under poor agricultural conditions, traditional tubers have been found to have high nutritive values and in some cases even higher than the exotic or popular grown tuber crops. Therefore prospects of *P. esculentus*, *S rotundifolius*, *Dioscorea species* (yams) and the *Disa species* being promoted as important tuber crops in Malawi are high.

REFERENCES

1. Allemann, J. and Coertze, A.F. (1996). *Plectranthus. Indigenous Root Crops*, ARC-Roodeplaat Vegetable and Ornamental Plant Institute, Private Bag X293, Pretoria 0001, South Africa.
2. Allemann, J, Viljoen, J. and Coertze, A.F. (1996). *The Zulu round potato- an endangered minor root crop.* ARC-Roodeplaat Vegetable and Ornamental Plant Institute , Private Bag X293, Pretoria 0001, South Africa.
3. Allemann, J. (1996). *Two indigenous tuber species being evaluated as possible food crops.* ARC-Roodeplaat Vegetable and Ornamental Plant Institute, Private Bag X293, Pretoria 0001, South Africa.
4. Asiedu, J.J. (1986). *Processing and Physical/chemical properties of Tropical Products.* Centaurus - Verlagsgesellschaft, PfaffenWeiler.
5. Burkill, H.M. (1995). *The useful plants of West Tropical Africa*, Vol 3. Royal Botanic Gradens, Kew.
6. Dlhwayo, P. (1999). *Status of Plectranthus esculentus crop in Zimbabwe.*Country Report at the First meeting of Plectranthus Research Network, held in March 1999 at ARC- Roodeplaat Vegetable and Ornamental Plant Institute, Private Bag X293, Pretoria 0001, South Africa.
7. Gila, M.A. (1984). *Morphogenetic variation in 'Rizga' Plectranthus esculentus.* N.E. Br. M.Sc. Thesis, Department of Botany, University of Jos, Jos, Nigeria.
8. Greenway, P.J. (1944). *Origins of some East African food plants.* Part 1. East African Agric. Journal, 10:34-39.
9. Hahn, S.K. (1994). *Root crops for food security in sub saharan Africa* in Akoroda, M.O. (Ed.). Root crops for food security in Africa. International Society for Tropical Crops - Africa Branch (ISTRC-AB), Ibadan, Nigeria. pp
10. Kay, D.E., 1987. Root crops (Revised). *Crop and Product Digest* No. 2. (Second Edition). Tropical Development and Research Institute. London. 380pp
11. Kyesmu, P.M. and Akueshi, C.O. (1989). *Effects of relative humidity variation on crude protein, crude fibre, fats and ash contents of Coleus Potato (Plectranthus esculentus N.E. Br.) under storage.* Nigerian J. Bot. 2, 1-7.
12. Latham, M.C. (1978). *Human Nutrition in Tropical Africa.* Food and Agriculture Organization (FAO) Rome. 268 pp
13. Malawi Government (1990). *Aguide to Agriculture production in Malawi* 1990 - 91 and 1992 - 93. Ministry of Agriculture, Lilongwe.pp122
14. Malawi Government, 1994. *Aguide to Agriculture production in Malawi* 1994/95 - 1995/96. Ministry of Agriculture and Livestock Development, Lilongwe.
15. Maliro, M.F.A. and Woodward, B. (1998). *Effect of photoperiod, night temperature and sucrose level on In vitro Growth of Plectranthus esculentus shoots* (unpublished). University of Malawi, Bunda College of Agriculture, Lilongwe, Malawi. 22 pp
16. Maliro, M.F.A. (1999). *Status of Plectranthus crop in Malawi.* Country Report at the First meeting of *Plectranthus* Research Network, held in March 1999 at ARC-Roodeplaat Vegetable and Ornamental Plant Institute, Private Bag X293, Pretoria 0001, South Africa.
17. Onwueme, I.C. (1982). *The Tropical Tuber Crops - Yams, Cassava, Sweet Potato, and Cocoyams.* ELBS and John Wiley & Sons, Chichester. 234 pages.
18. Okoli, O.O. and Onwueme, I.C. (1987). *Yam and the African food crisis.* In Terry, E.R., Akoroda, M.O. and Arene, O.B. (Eds.). Tropical Root Crops and the African food crisis. Proceedings of the third Triennial Symposium of The International Society for

Tropical Root Crops - Africa Branch held in Owerri, Nigeria, 17 - 23 August 1986. Pages 38 - 42

19. Sauti, R.F.N., Khonje, E.M.H., Thaulo, G., Chibambo, M. And Bulla, G.M. (1994). *The importance and production constraints of cassava in Malawi.* In Terry, E.R., Akoroda, M.O. and Arene, O.B. (Eds.). Tropical Root Crops and the African food crisis. Proceedings of the third Triennial Symposium of The International Society for Tropical Root Crops - Africa Branch held in Owerri, Nigeria, 17 - 23 August 1986. pages 93-96
20. Tindall, H.D. (1983). *Vegetables in the tropics.* Macmillan Press, London.
21. Williamson, J. (1975). *Useful Plants of Malawi.* Government Printer, Zomba, Malawi.

HONEYBEES AND NECTARIFEROUS PLANTS IN THE DESERT ARE AN IMPORTANT FACTOR TO SUSTAIN MODERN AGRICULTURE

D. Eisikowitch
Department of Plant Sciences and The Institute for Nature Conservation,The George S. Wise Faculty of Life Sciences, Tel-Aviv, Israel, Tel: 972-3-6409849, Fax: 972-3-6409380, email: eisik@post.tau.ac.il

1. BACKGROUND

Bees are insects that are dependent, exclusively, on a current supply of nectar and pollen derived from the flowers. All honeybees are classified in the genus Apis that includes five species. The common honeybee, *Apis mellifera* is one of the most investigated insects and is the subject of this paper.

2. INTRODUCTION

Apis mellifera probably originated in the African tropics, emigrated to Western Asia and later spread up to Central Europe. These well-adapted species include several subspecies and various local ecotypes (Ruttner 1975a). The natural distribution of the honeybees lies between a very wide range of climates and habitats, beginning with the tropical forests, through the African Savannas, Mediterranean regions in South Africa and Asia minor, ending in the colder European climates (Winston 1987).

The honeybee is a social insect that lives in a colony including 30-80 thousand female worker bees, scores of males and one egg laying queen that

Combating Desertification with Plants, edited by Dov Pasternak and Arnold Schlissel.
Kluwer Academic/Plenum Publishers, New York, 2001.

controls the colony. Nest sites are variable according to race and the habitat with honeybees can be found in hollow stems, rock clefts or within man made shelters - the hive. As with other insects, bees are tremendously affected by the ambient temperature, and are usually not active outside the hives at temperatures below 10° C. They do not fly in rain or under high wind velocity. Certainly if all the proper physical conditions are provided but no suitable food source is available, the bees will still not be active. The blooming period of the suitable flowers is one of the key factors for bee's existence. If we examine the Israeli blooming chart (*Fig. 1*), (Zohary 1962) we can see that in Israel, the blooming period lies mainly between February and May. This blooming period coincides with the local bee's activities (Bodenheimer and Ben-Neria 1948). During the cold days in Israel, bees remain in their hives, controlling their nests at a fixed temperature. The stored honey is their source of energy. In warmer seasons, when flowers appear, the bees are fully active, collecting and storing pollen and nectar, laying eggs and raising the next generation. In the mid-summer, the bees reduce activity, due to the shortage of flowers. During this period, bees use stored honey. If the foraging period is rich and long, and honey production rich enough to cover the dearth period, the honeybees are able to survive under these conditions.

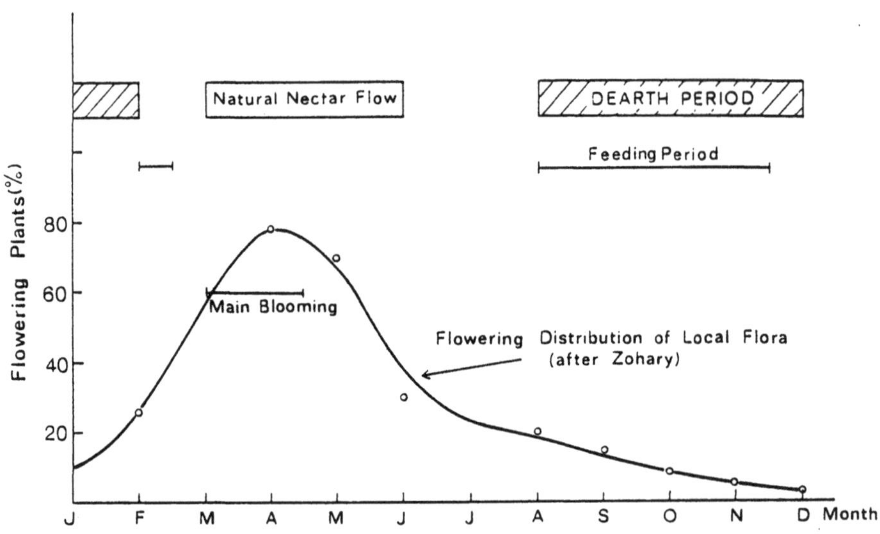

Figure 1. The Israeli Blooming Chart (Zohary 1962)

3. BEES AGAINST DESERTIFICATION

Colonization of the desert by modern man created among other problems also the demand for fresh food, which was followed by establishing modern agriculture in the desert. Sixty years of modern agriculture in Israel revealed that many of the cultivated plants are partially or fully in need of the service of pollinating insects. Almonds, certain leguminous, cucurbits and several solanum cultivars are totally dependent on bees, mainly honeybees. (Zohary and Hopf 1988, Free 1993, Orr and Eisikowitch 1988).

The transfer of beehives into the desert is a common practice in Israel (as in many other countries) (McGregor 1976); however, this practice is just for a restricted period. Artificially feeding honeybees for a long period creates many problems and does not enable the beekeeper to produce valuable honey. Due to the fact that keeping bees with the aid of natural food is not as yet solved, permanent colonies are still outside the desert cycle.

As far as we can see today, shifting honeybees towards the desert must be accomplished by establishing nectariferous and pollen supply plants, which will fulfill the need of pollen and nectar along the dearth period of the desert.

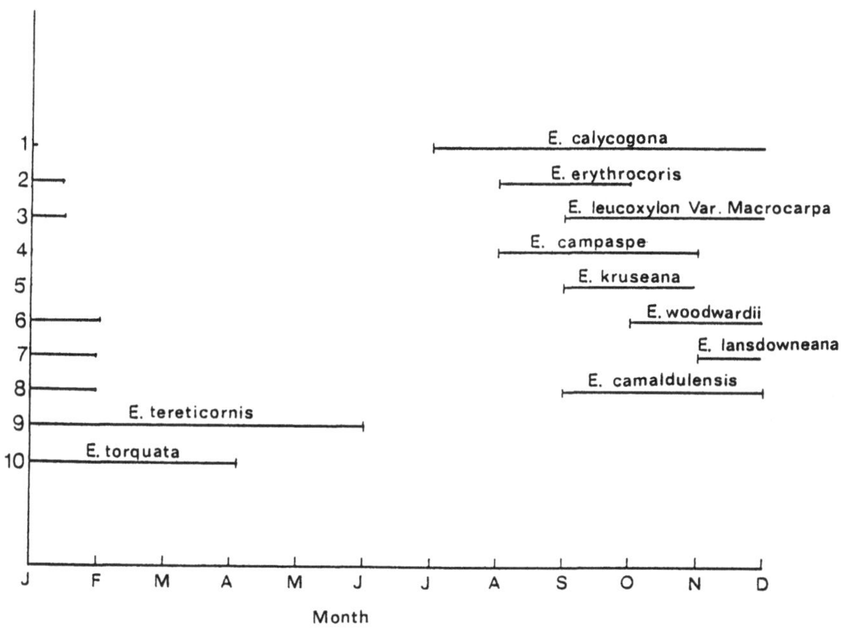

Figure 2. Flowering Calendar of Eucalyptus

From the beginning of the present century, many attempts to accomplish afforestation have been made in our desert, most of them carried out by the K.K.L. and also by the University of Ben-Gurion. In spite of many failures there were places, even under extreme conditions, where plants survived. Most of the plantations were Eucalypts, which are known to be a good nectar and pollen source (Eisikowitch and Reves 1979, Chipendale, 1973, Blake and Roff 1972, Gindel 1956, Lupo and Eisikowitch 1990). Some of the introduced eucalyptus coincide perfectly with the dearth period in Israel (*Fig 2*) and might be excellent candidates for nectariferous plants.

The introduction of both honeybees and nectariferous plants must be carried out very carefully and I would suggest the following steps:

- First and foremost, select the local honeybees, if they exist.
- If there are no local honeybees, bring them from the closest location.
- There is a need to recognize the local flora and its blooming period in order to define the dearth period.
- Introduce suitable nectariferous plants to close the gap of the dearth period.
- Bear in mind that any introduction of plants or animals to a new area can interfere with the local delicate balance of interaction between various organisms. Although this is a long and painstaking task, any short cuts will just complicate matters.

REFERENCES

1. Blake, S.T. and C. Roff. (1972). *The honey flora of Queensland.* Dep. Of Primary industries. Queensland, Australia.
2. Bodenheimer, S.P. and, Ben Neria, N. (1948). *One year of honeybee investigation in Palestine*. Beeskeeper board. Tel-Aviv.
3. Chipendale, G.M. (1973). *Eucalyptus of the Western Australian Gold fields*. Dep. of Primary Industry. Forestry and timber bureau. Canbera, Australia.
4. Eisikowitch, D. and Reves, J. (1979). *Acclimatization of Eucalypts as melliferous flora in Israel*. In: 27th International Congress of apiculture of Apimondia. Athens. Greece. Apimondia publishing house.
5. Free, J. B. (1993). *Insect pollination of Crops*. Academic Press.
6. Gindel, I. (1956). *Acclimatization of Plants*. Am Oved (Hebrew)
7. Lupo, A. and D. Eisikowitch, (1990). *Eucalyptus erythrocoris: a source of nectar and pollen for honeybees in Israel*. Apidologie 21: 25-33.
8. McGregor, S.E. (1976). *Insect pollination of cultivated crop plants*. Agriculture Handbook No 496. U.S.D.A. USA
9. Orr, N. and Eisikowitch, D. (1988). *Interaction between melon (cucumis melo) flowers and honeybees under extreme desert conditions in Israel*. Apidologie 19: 85-96.
10. Ruttner, F. (1975). *Races of bees*. In: The hive and honeybee. Dadant & Sons, Hamilton IL.
11. Winston, M. (1987). *The biology of the honeybee*. Harvard University Press.
12. Zohary, D. and Hopf, M. (1988). *Domestication of plants in the Old World*. Oxford.
13. Zohary, M. (1962). *Plant life of Palestine*.The Ronald Press.New York.

IMPORTANCE OF NON-WOOD FOREST PRODUCTS AS SOURCE OF RURAL WOMEN'S INCOME IN WESTERN BURKINA FASO

Niéyidouba Lamien and Gaspard Vognan
INERA/CRREA du Nord-Ouest, BP: 49 Tougan, Burkina Faso, Tel: 226-534006, email: nlamien@hotmail.com

1. INTRODUCTION

Burkina Faso is one of the African countries affected by the phenomenon of desertification. Since the 70s, it committed itself, with the support of its partners, to struggle against desertification using various actions. Among those, biological technology and tree planting, have been used a lot as an approach to control desertification.

However, as the PNUE (1994) points it out, all these actions have not been able to curtail the effects of desertification. Yaméogo (1994) esteemed that the main limiting factors of the struggle have been the insufficiency of financial resources and the weak adherence of local populations. To fill these gaps, the PNUE (op. cit.) recommends participative approach in its principles mentioned in the United Nations convention against desertification. However the participation of populations remains subordinate to the consideration of their vital needs. Rochette (1989) reported that meeting women's forest products needs is the most powerful lever of rural agroforestry development.

The International Council for Research in Agroforestry (ICRAF) based its domestication program on local species which were considered very useful by local populations (Simons, 1996). The infatuation of rural populations for the plantation of citrus and mangoe trees, whose economic

Combating Desertification with Plants, edited by Dov Pasternak and Arnold Schlissel.
Kluwer Academic/Plenum Publishers, New York, 2001.

importance is proven, is an expression of this logic. When these populations clear new land, local species with edible products such as shea nut tree, Nere tree, tamarind tree, Baobab tree, etc., are selected and preserved on the farming plots.

The farming system in Burkina Faso is characterized by the omnipresence of trees owing to their Non-Wood Forest Products (Lamien et al., 1996).

Proving the paramount importance of Non-Wood Forest Products of the local species, as source of income in the rural area, is expected to arouse awareness among decision-makers for a better valorization of these products. This valorization could lead to better husbandry of the potential of production of these products that is the forest and the establishment of home garden or woodlot with local species as people practice so for citrus and mangoe trees.

In Burkina Faso, the economy depends, largely, on low input agriculture systems practiced on soils undergoing severe degradation and farmers therefore, have no other choice than to undertake diversification of their income sources. In this respect, Non-Wood Forest Products, generally considered as minor products, need to be seriously considered. FAO (1995) reports that it is socially disadvantaged groups, mainly rural women, which rely more on these products for their subsistence. In Burkina Faso, more than 80% of the population, mainly women, are in this situation.

Nevertheless, the decision-makers and Non Governmental Organisms which are interested in promoting women's lucrative activities, hesitate, up to now, to invest in this sector. All have doubts on the real importance of these products and the profitability of an investment in this sector.

The present study aims at contributing to arouse awareness of these actors, by generating more information on the economic and social importance of these products in the life of socially disadvantaged groups such as women and children.

Taking in account previous studies and the daily living, we set the following hypotheses:

(i) Non-Wood Forest Products contribute in a significant manner to the strategy of women's financial self sufficiency development and,

(ii) the processing activities of these products are economically profitable.

During a former study, 10 products were identified by Lamien (1996) as a very important source of income in the rural area. For the present study, we assessed the gross earnings and budgeted the processing activities of the main agricultural and Non-Wood Forest Products, in order to assess the contribution of Non-Wood Forest Products in the rural woman's income.

2. METHODOLOGY

2.1 Study Area

The study area is the western part of Burkina Faso, located between 10-13° North and 3-6° West. The climate is tropical semiarid type with an annual rainfall ranging from 700 to 1000 mm. The vegetation is woody savanna, woodland savanna and clear forests. The farming landscape is featured by the presence of local species providing edible products such as *Vitellaria paradoxa* Gaertn, *Parkia biglobosa* (Jacq.) Benth, *Tamarindus indica* L. and *Borassus aethiopum* Mart.

The villages of Pompoï and Kawara representative of two sud-zones: (i) cotton production and (ii) rice production were visited.

2.2 Assessment of the Gross Productions of Agricultural and Non-Wood Forest Products

The assessment focused on the main agricultural and Non-Wood Forest Products that rural women usually gather for their cash and food needs. The evaluation consisted in a survey of a sample of 50 women to identify and quantify the different products they mobilize throughout the year. Quantities were obtained from direct stock measurements in households and estimation of quantities which were already used according to declarations of women. The monetary values were obtained using the market value of the goods during the period of the survey.

2.3 Assessment of the Economic Efficiency of Agricultural and Non-Wood Forest Products Processing

Evaluation of the economic efficiency focused on shea nut, Nere seeds and sorghum processing. A sub-sample of 12 women processing their products for sale were followed for data recording.

The budget methods were used. The index of economic efficiency that was calculated for each processing budget were: (i) gross earning, (ii) gross margin, (iii) monetary costs, (iv) family labor cost, (v) the work remuneration, (vi) the equipment depreciation, (vii) the profitability

The cost of the labor was estimated from an opportunity cost that is the remuneration of the agricultural work. It is about 57 F CFA according to Vognan (1998). Concerning the cost of the equipment depreciation, the method of linear depreciation was adopted assuming that the equipment is exclusively used for these activities. Profitability was calculated by making a ratio between gross earnings and total cost.

3. RESULTS AND DISCUSSIONS

3.1 Agricultural and Non-Wood Forest Products Gathering by Women

The sample of women we met comes from households whose main activity was agriculture. In such households, women's work time is shared between domestic job, family farm job and lucrative activities. Saul (1989) indicated that the main constraint to women's economic activities is the fact that they have no time for themselves. Nevertheless, as the woman advance in age and earns daughter-in-laws, she could liberate herself of the family agricultural activities to devote to her own activities (Beck-Engelberg, 1994).

To achieve a certain financial autonomy, the rural woman develops several strategies to generate incomes. Agricultural production and Non-Wood Forest Products they harvested play a considerable role in these strategies.

During the cropping season, after their domestic tasks and work in family fields, women release a time for the exploitation of their individual fields. Sorgho (1992) reports that women used to go very early in their individual field and return around 9h, to cook. After which they join the family field. Around 16 h they return to their individual fields. Those who don't go to their individual field devote these hollow time to search products such as shea nuts, vegetables for sauces or fuelwood.

During the harvest period, a woman can offer her working force, when she has time, to her brothers and uncles in return for a reward in nature of the product harvested. After a day of harvesting millet for example, she can receive at least the content of the container she used for the work.

The individual field exploitation and somebody else's assistance with the crop allow the woman to constitute a stock of supplies. Such supplies, generally made up of cereals, oil seeds and vegetables for sauces, are the exclusive property of the woman responsible of their management. These supplies, that can be qualified as stocks of security, are generally saved to alleviate food unforeseen crisis and cash needs of the household.

The same strategy of stock constitution is developed for Non-Wood Forest Products. As Schreckenberg (1996) underlines, salable products such as shea nuts and seeds of Nere tree, have special harvest seasons where women compete for trees. It is necessary to be the first under the tree, especially on areas with free access. That is the reason why every woman gets up very early in the mornings for a 5 to 6 hour trip. On farm land and fallows, trees are managed by the land-owner.

Table 1 shows agricultural and Non-Wood Forest Products that were observed in the households. The monetary values of the products vary from 1800 to 73 440 F CFA. In term of contribution to the potential income of the woman, agricultural products appear to be the most important contributor, 73 to 83% against 17 to 27 % for Non-Wood Forest Products. The multiplicity of agricultural products justifies their great contribution to the gross income. Indeed, one counts 4 to 6 agricultural products, according to the site, against only 2 forest products. The seasonal appearance of forest products justifies the low number of recorded products.

Table 1: Contribution of forest and crop products in the potential income of rural women

Village sites	Pompoï		Kawara	
	Monetary Value (FCFA)	Contribution (%)	Monetary Value (FCFA)	Contribution (%)
Crop products				
Maize	3600	5		
Sorghum	11375	15		
Millet	9414	13		
Rice			73440	62
Fonio			4500	4
Bambara groundnut	1826	2		
Cowpea	2075	3		
Peanut	26400	35	14025	12
Sesame			6400	5
Partial gross income (1)	**54690**	**73**	**98365**	**84**
Forest products				
Shea nut	14875	20	13905	12
Nere seeds	5600	7	5525	5
Partial gross income (2)	**20475**	**27**	**19430**	**16**
Total gross income (1)+(2)	**75165**	**100**	**117795**	**100**

NB: 1USD = 600 F CFA

Lamien et.al. (1996) found out, after a whole year investigation, that about 30 raw and processed products were used in the households, in the same area. Shea nuts and Nere seeds were the permanent products observed during the year. Peanuts and rice constitute the greatest contribution ranging from 35 to 65% respectively. Shea nuts occupy the second position with 12 and 20% of contribution.

According to the results of the survey, Non-Wood Forest Products, especially shea nut and pods of Nere are access limited for women. In the rural landscape, one can generally distinguish two areas of tree products picking: (i) the area with free access that corresponds to spaces not exploitable for agriculture and government domains, (ii) the private area, represented by fields and fallow. According to rules of trees tenure, trees existing on an area belong to the land owner. In general only men are land owners. Only women coming from the land owner family are authorized to pick the products for food needs.

Stemming products of the family area are first devoted to family consumption. The salable products come generally from the free access area and the surplus of the family area products. This type of tree tenure limits considerably the access possibilities of women whose husbands are not landowners and hardly allows the other women to make money.

3.2 Economic Efficiency of Agricultural and Non-Wood Forest Products Proccessing

Besides the gross productions of agricultural and Non-Wood Forest Products, women process these products for their cash and food needs. The results of the survey show that 90% of women process Non-Wood Forest Products for the family consumption, against 10% of women who sell them. According to Serpantié et al. (1996), the harvest and processing of shea nuts and even others products such as seeds of Nere tree, aim first to avoid buying these products, because they don't have enough money, rather than to generate income. The social organization imposes to woman the responsibility of providing sauce ingredients, while the man supplies the staple. This forces her to produce enough shea butter and fermented seeds of nere tree for her family food needs. Only surplus of products can be sold. The women whose processed products are first devoted for sales are age advanced (45 to 65 years) ones and sometimes widows. Younger women prefer to sell their products in the raw state. On the other hand, they are more active in the processing of agricultural products, especially in the manufacture of local beer from sorghum. The selling facility of the local beer is their main motivation. Within 12 hours or at worst in 24 hours, a woman sells the totality of her local beer, while the flow of a production of shea butter or fermented seeds of nere tree requires one to two weeks with random and weak daily receipts. Lamien (1996) found about 150 F CFA for ordinary day receipts and 856 F CFA for market days for the butter, 70 and 752 F CFA for the fermented seeds of nere tree. Concerning local beer the daily receipt is about 2500 FCFA (table 2). Woman can, at least, produce local beer once a week and throughout the year, which is not conceivable in the current context of the shea butter and fermented seeds of nere tree manufacturing.

3.3 Processing Budgets

Table 2 presents budgets of the processing of two Non-Wood Forest Products (the shea nut and the seeds of Nere tree) and one crop product (the grains of sorghum) that women used to practice in the study area. The gross margin, the remuneration of the work and the profitability are economic efficiency index that have been focused for analysis.

Table 2: Budgets (F CFA) of processing of non-wood forest and crop products

	Non-Wood Forest Products		Crop product
	Shea nuts	Nere seeds	Sorghum
Gross earning	2985	6392	2500
Monetary cost	1725	3975	1550
Gross margin	1260	2417	950
Labor cost	412	1139	480
Hourly remuneration of the job	152	75	73
Equipment depreciation	39	28	82
Profitability	137%	124%	118%

NB: 1USD = 600 F CFA

Table 2 shows that processing of products generates a highest gross margin for Non-Wood Forest Products. The gross margin of sorghum processing corresponds to 75% of shea nuts processing gross margin and 39% of Nere seeds one.

In terms of work remuneration, it appears more advantageous for women to devote their time to the processing of Non-Wood Forest Products rather than to sorghum. Manufacturing of the shea butter remunerates the hour of work twice more than fermented seeds of nere tree and local beer (*Table 2).* During the cropping season, women should devote their time to those processing activities rather than to offer their labor for farm works, whose remuneration by hour would be 57 F CFA according to Vognan (1998).

The ratios between gross earnings and total costs are superior to 100% for all processing activities. Which means that for each invested ''Franc'', each activity allows to recuperate the investment and to release a profit. Non-wood forest product activities had, however, a higher ratio than sorghum's. One can therefore conclude that processing activities of these products, especially shea nuts, remunerates better the invested capital than sorghum. Manufacturing of fermented seeds of nere tree would have been more profitable if the process was not very labor consumer. It costs two time more expensive than the one of shea butter and local beer.

If we focus on the remuneration of the capital, woman would earn more by investing in the production of shea butter and fermented seeds of nere tree rather than local beer.

Depending on the evaluation method, highest ratios can be obtained. Sidibé (1994) who evaluated an annual remuneration rate by making the ratio between the annual net margin and the sum of the value of the equipment and the moving capital, found 1,96 for the shea butter and 4,61 for the fermented seeds of nere tree. De Boer (1992) even found a ratio equal to 20 for this kind of activities. If one considers only the values of the ratios, one can conclude with these authors that Non-Wood Forest Products processing activities appear to be economically profitable.

However an activity can appear very profitable through this index without being easy to develop because of many constraints. According to FAO (1995) the existence of markets, the quality demands, the food habit evolution and the existence of others competitive products are notable constraints to the development of economic activities based on Non-Wood Forest Products. Forest products which are concerned in this paper do not escape these constraints.

Results of survey revealed that the major part of rural women (90%) make a self production of these products to avoid to buy them and this limits local demand for these products. In this context, only an external demand can enhance the development of a market for these products. In fact there are possibilities for external market but facility policy has to be implemented before. Such policy should include the establishment of good infrastructures such as roads to favor products access.

The recent decision of the European Union to allow the utilization of others vegetable oil, to a rate of 5%, in replacement of cocoa butter, will increase the demand for shea butter. This demand increase will arouse more women's enthusiasm for shea nuts gathering as it has been the case in the course of 80s (CSPPA, 1987). Private shea trees tenure would have emphasized from these years.

Concerning the fermented seeds of nere tree precursory signs of market development also exist. The growing interest for the natural product consumption may be auspicious for this product that was being replace by industrial products in the flavoring of sauces.

Taking into account, the non quantifiable advantages, in particular the attenuation of drought effects, owing to the presence of a vegetative cover, made up with multipurpose species, the economic efficiency index of a forest based activities would not have to be the main criterion determining the choice of investments.

4. CONCLUSION

It appears, through this study that Non-Wood Forest Products contribute significantly to the potential income of women in rural area. Shea nuts occupy the second position after rice or peanut, in term of quantitative contribution to woman's income. It appears more advantageous to manufacture and sell the shea butter and fermented seeds of nere tree than local beer.

However, constraints to the implementation of these activities exist. The traditional form of processes requires much labor and therefore is very time consuming. The weak level of local demand brakes the products flow. That is the reason why young women prefer to manufacture products which are easy to sale such as local beer, even if their profitability level appears weak in comparison with shea butter and fermented seeds of nere tree.

The scientific community underlines that the development of forest based activities stumbles to multiples constraints that it is necessary to have in mind, when one wants to implement a program of development of these products. Among these constraints, the development of the product market, the competition of substitution products, the sporadic aspects of products, etc. have to be mentioned.

Nevertheless, it remains that all investment in the sector of forestry whose indirect or direct effects are to favor the development of the biodiversity by conserving or enriching the vegetative cover, will be always profitable to all viewpoints. In that, the policy of the government of Burkina Faso to promote shea butter extractors in order to improve the efficiency of the process and to give small grants to support women's lucrative activities, have to be strengthened in general, and especially for Non-Wood Forest Products.

ACKNOWLEDGEMENTS

The financial support of International Foundation for Science (IFS) has permitted to carry out this work. We express our gratitude to this institution and to our colleagues in particular Mr. San Traore, Dr. F. Hien and Dr. A. Mando who have accepted to devote part of their time to reading the paper and make suggestions.

REFERENCES

1. Rochette R. M. (1989). *Le Sahel en lutte contre la désertification, leçons d'expériences*, CILSS/PAC GTZ, 592 p.
2. Yaméogo M. (1994). *8000 villages 8000 forêts, un vaste programme de reforestation,* dans Arbre et Développement, bulletin technique d'information et d'échanges N° 9 MEE, Burkina Faso 4-6 p.
3. PNUE , (1994). *Convention des nations unies sur la lutte contre la désertification dans les pays gravement touchés par la sécheresse et/ou la désertification, en particulier en Afrique,* brochure 71p.
4. Vognan G., (1998). *Evolution des revenus des producteurs en zone cotonière ouest,* Rapport de recherche, INERA, Bobo-Dioulasso 20p.
5. Lamien N., (1996). *Importance des produits forestiers non ligneux comme source de revenus des femmes en milieu rural.* Rapport de recherche à mi parcours, INERA/FIS, Bobo-Dioulasso, Burkina Faso.
6. Sidibé A, (1994). *Analyse économique de quelques activités de femmes liées à l'utilisation de produits non ligneux des essences forestières locales dans le sud-ouest du Burkina,* in Recherche intégrée en production agricole et en gestion des ressources naturelles, Rapport technique, INERA/projet ARTS, Université de Perdue, 266-274p.
7. De Boer L; (1992). *Compte rendu des activités de recherche socio-économique à Thiougou du 11 au 13/08/99,* INERA/RSP, Ouagadougou Burkina Faso.
8. FAO, (1995). Report of the International Expert Consultation on Non-Wood Forest Products, Non-Wood Forest Products 3, Rome Italy, 465 p.
9. Simons A. J. (1996). *ICRAF's strategy for domestication of Non-Wood Tree Products,* in Domestication and Commercialization of Non-Timber Forest Products In Agroforestry Systems, FAO, Non-Wood Forest Products, (9), 8-22 p.
10. Saul M., (1989). *Expenditure and Intrahousehold patterns among the southern Bobo of Burkina Faso.* In H.j. Rutz & B.S.Orlove (Eds). The social economy of consumption, Lanham, pp 349-377.
11. Sorgho M-C. (1992). *Dimension socio-économique de la femme en milieu rural dans le plateau central du Burkina Faso.* Mémoire de D.E.A, Département de Sociologie, Université Nationale de Côte d'Ivoire.
12. Lamien N. Sidibé A. et Bayala J., (1995). *Use and commercialisation of non timber forest product in western Burkina.* Communication présentée à la conférence Internationale sur la domestication et la commercialisation des produits forestiers non ligneux des sytèmes agroforestiers, tenue à Nairobi au Kenya du 19 au 23 Fev. 1996, 17 p.
13. Beck-Engelberg J. (1994). *A qui ce baobab? Les systèmes de gestion des ressources naturelles et leur impact sur la situation socio-économique des femmes au Sahel,* Etude

de cas de 6 villages au Burkina Faso, Projet d'économie familiale DED, AUS DER PRAXIS 1/94, 80p.

14. Schreckenberg K., (1996). *Forest, Fields and Markets: a study of indigenous tree products in the woody savannas of the Bassila region, BENIN.* Thesis for Degree of Ph.D., Departement of Geography, University of London, U.K. 325 p.

15. Serpantie G., Bayala J., L., Devineau J. L., Helmfrid S., Lamien N. et Ouégraogo J. S., (1996). *Rôle des jachères dans la production arborée non ligneuse en savane: cas du karité dans l'ouest du Burkina Faso.* Communication présentée lors de l'atelier CNRST/ORSTOM/FED, ''La jachère, lieu de production'' Bobo - Dioulasso, 2 au 4 octobre 33 p.

ETHNOBOTANY AND IMPORTANCE OF THREE LOCAL SPECIES IN NORTHERN CAMEROON

Anogeissus leiocarpus, Vitellaria paradoxa and Balanites aegyptiaca

J.M. Fondoun and J. Onana
IRAD, B.P. 2067, Yaounde, Cameroon, Tel: 237-225924, Fax: 237-233538,
email: iradpnva@iccnet.cm

1. INTRODUCTION

The northern and far northern provinces in Cameroon are part of the Sudan-Zambesian Region in Africa. They are characterised by a semi arid climate, with rainfall limited to 1200 mm per year and temperatures up to 40°-45° C. The hydrology distribution is characterised by river flow in two main drainage basins associated with the river Niger and lake Chad. The Logone river is the major one. The mayos of the Maroua region are seasonal streams running down from Mandara, which during dry season disappear into their sandy alluvial beds. Soils in the middle Benoue and the Logone plain are mainly hydromorphic (F. Loung).

Vegetation is typical savannah, traditional parklands, wooded grasslands, and steppes in the far north announcing desert (R. Letouzey, 1973). Major species found in this particular zone are *Acacia* and related species, *Brachystria spp.* and *Hyparrhemia* grasses.

There are many other important species of particular interest to the local population but they are highly threatened. Among these threatened species are: *Anogeissus leiocarpus, Butirospermum pakii/Vitellaria paradoxa, and Balanites aegyptiaca.*

The presumed threats include the following: over exploitation by local population, overgrazing by domestic animals, bush fire, drought etc. In order to stop slash and burn activities and put to good use the potential tree/shrub species while fighting against desertification, an investigation on the ethno-

Combating Desertification with Plants, edited by Dov Pasternak and Arnold Schlissel.
Kluwer Academic/Plenum Publishers, New York, 2001.

botanical value of the above mentioned three key species was carried out in 1998. At the same time, genetic material for field genebanks establishment and domestication was collected.

A semi structured questionnaire was used to interview farmers in order to understand the importance and uses of the trees in their community e. g. wood/fuelwood, fodder, charcoal production, fruit/food production, medicine, posts, beekeeping, fencing, etc. The other aim of the interview was to assess the level of threats and make well-founded recommendations for further management, conservation, and domestication of the available biodiversity of the chosen species to stop desertification.

2. MATERIALS AND METHODS

A preliminary survey for key species identification was carried out at the market level in the two provinces, northern and far northern Cameroon. Three target species, *Anogeissus leiocarpus, Balanites aegyptiaca* and *Vitellaria paradoxa,* were identified for further investigation on the basis of their multipurpose uses by local population.

Twenty villages were randomly selected in two zones for surveys coupled with germplasm collection (*Fig. 1*):

Zone I representing 10 villages around Maroua in the far North.

Zone II representing 10 villages around Garoua in the North.

In each village 8 to 10 households were randomly sampled and interviewed using a questionnaire focussing on farmer awareness of plant species, their use, and the exploitation regime. Farmer compounds and farmlands were also visited to assess the distribution of the target species.

Trees were counted within quadrats of 100 x 100 m and the average percentage of existing stand per village was calculated as follows:

$$\frac{\Sigma \text{ Counted trees/quadrat/H.}}{\text{Total Number of H}}$$ (where H refers to households).

The following symbols were used to classify different groups according to the number of trees found in a quadrat in an average household per village:

(-) for 0 to 1% tree stands found per village household on average

(+) for 2 to 19% tree stands found per village household on average

(++) for 20% to 59% tree stands found per village household on average

(+++) for 60% to 99% tree stands found per village household on average.

The questionnaire was focussed on: wood/fuelwood, charcoal, fruit/food, medicinal uses, fodder, posts, beekeeping, fencing etc.

Response frequencies of local populations on any type of use identified for the three studied species were recorded per zone and compiled in tables.

Table 1: Farmer knowledge of plant species in zone 1 and 2

ZONE	**I (Villages around Maroua)**										Total
Villages	1	2	3	4	5	6	7	8	9	10	
Ethnic group	a	b	c	d	e	f	g	h	i	j	
A. leocarpus	WU	WU	WU	WU	WU	WU	WU	WU	WU	WU	
V. paradoxa	O	O	O	O	WU	WU	WU	O	O	O	
B. aegyptiaca	WU	WU	WU	WU	WU	WU	WU	WU	WU	WU	
Other species	WU	WU	WU	WU	WN	WN	WU	WU	WU	WU	
Number of Households	10	10	9	10	8	9	10	8	8	9	91
ZONE	**II (Villages around Garoua)**										Total
Villages	11	12	13	14	15	16	17	18	19	20	
Ethnic group	k	l	m	n	o	p	q	r	s	t	
A. leocarpus	WU	WU	WU	WU	WU	WU	WU	WU	WU	WU	
V. paradoxa	WU	WU	WU	WU	WU	WU	WU	WU	WU	WU	
B. aegyptiaca	WU	WU	WU	WU	WU	WU	WU	WU	WU	WU	
Other species	WU	WU	WN	WN	WU	WU	WU	WU	WU	WU	
Number of Households	10	10	10	9	8	8	8	10	10	10	93

Total of households: 184

1 = Mountourwa a = Guizigua / Foulbé	6 = Doukoula f = Daba	11 = Gashiga k = Foufouldé	16 = Tcholiré p = Dourou/Foulbé
2 = Mouda b = Foufouldé	7 = Yagoua g = Banana	12 = Dourbey l = Daba	17 = Tcheboa q = Sara laka
3 = Mindif c = Foufouldé	8 = Barguine h = Massa Toupouri	13 = Louggéré m = Foulbé	18 = Tongo r = Baya
4 = Laf d = Guizigua	9 = Kalfou i = Zébé	14 = Sangéré Paul n = Fali	19 = Na'ari s =
5 = Kaélé e = Moundan	10 = Zébé - Marao j = Massa Toupouri	15 = Ngouf o = Foulbé	20 = Figil t = Binoua

O = Not known WU = Well known and used WN = Well known and not used

3. RESULTS AND DISCUSSION

3.1 Farmer Knowledge of the Species

In zone I in three villages out of 10 farmers were fully aware of the existence of natural stands of *V. paradoxa,* while the other two species were well known to the local population in all the villages.

In zone II all the three tree species were well known to the local population (*Table 1*).

3.2 Exploitation/Uses

In zone I the calculated and compiled exploitation percentage for the 8 main uses of the three tree species varies from 0 to 100% (*Table 2*). *V. paradoxa* emerges as the most frequently used species with 62.5% average rate of utiliation. *Balanites* was mainly used as fodder (70%) and vegetable for human consumption (100%), while the main uses of *Anogeissus* were as wood/fuelwood and charcoal.

Table 2: Global response of farmers regarding frequency (%) of use of study species in zone I

Species Uses	*Anogeissus*	*Balanites*	*Vitellaria*
Wood / fuel wood	100	50	100
Fodder	25	100	50
Charcoal	75	0	100
Fruit/Food	0	100	100
Medicinal	25	0	75
Fence	25	75	0
Bee keeping	50	25	75
Posts	50	0	0
Average over different uses	43.75	43.75	62.5

Results in zone II (*Table 3*) were more or less the same as in zone I, but *Anogeissus* ranked second after *V. paradoxa.*

Table 3 Global response of farmers regarding frequency (%) of use of study species in zone II

Species Uses	*Anogeissus*	*Balanites*	*Vitellaria*
Wood / fuel wood	100	50	100
Fodder	25	75	40
Charcoal	75	0	100
Fruit/Food	0	100	100
Medicinal	25	0	80
Fence	25	70	0
Bee keeping	40	30	75
Posts	60	0	0
Average over different uses	43.75	40.625	61.87

3.3 Estimated Occurence of the Three Tree Species in Villages

Table 4 clearly shows that only the villages of Kaele, Doukoula, and Yagoua, still have natural populations of *V. paradoxa* in zone I, confirming farmers' awareness of the species. The occurence of the three tree species was higher in zone II than in zone I.

The results also indicate that farmers or local population in surveyed villages are familiar with the key species. In zone I negative responses were just a sign of the absence of natural stands, which almost completely disappeared because of over exploitation. The main causes of the over exploitation scored in zone I and II are:

1- Exploitation as wood/fuel wood, which for *Anogeissus* and *Vitellaria* reached 100%.

2- Exploitation as fodder, which for *Balamites* reached 100% in zone I and 75% in zone II.

3- Exploitation as charcoal, which for *Vitellaria* and *Anogeissus* reached 100% and 75%, respectively, in both zones.

4- Exploitation for fruit/food, which for *Balanites* and *Vitellaria* reached 100% in both zone I and II.

5- Exploitation for medicinal uses, which for *Vitellaria* reached a high 80% in zone II and 75% in zone I.

6- Exploitation for bee keeping, for which *Vitellaria* and *Anogeissus* scored respectively 75% and 50% in zone 1 and 75% and 40% in zone 2.

7- Exploitation for fencing, which for *Balanites* scored a high 75% in both zones.

8- Exploitation as posts. Only *Anogeissus* scored high in both zones for this use.

To be more precise, it was mentioned that *Anogeissus* was used by traditional healers to kill worms and cure syphilis. *Vitellaria* products or sheanut butter were used to stop hair loss. *Anogeissus* was also reported to cure skin diseases when added to *Guierra senegalensis* and *Pilostigma reticulata.*

Table 4: Estimated occurence of study species in villages

ZONE I VILLAGES AROUND MAROUA			
Villages / *Ethnic group*	*Anogeissus Leiocarpus*	*Balanites aegyptiaca*	*Vitellaria paradoxa*
Moutourwa / *Foufouldé*	+	+	-
Mouda	+	+	-
Mindif / *Mouskoum Foulbé*	+	+	-
Laf	+	+	-
Kaélé / *Moundang Toupouri*	+	+	+
Doukoula / *Toupouri*	+	+	+
Yagoua / *Banana Toupouri*	+	+	+
Kalfou / *Foulbé*	+	+	-
Barguine / *Massa Toupouri*	+	+	-
Zébé-Marao / *Massa Toupouri*	+	+	-
ZONE II VILLAGES AROUND GAROUA			
Gashiga / *Foulbé*	++	++	+
Dourbey / *Daba*	++	++	+
Louggéré / *Foulbé*	++	++	+
Sanguéré Paul / *Fabi*	++	++	+
Ngouf / *Foulbé*	++	++	++
Tcholiré / *Foulbé dourou*	++	+++	++
Tcheboa / *Sara-laka*	++	++	+++
Tongo / *Baya*	++	+++	+++
Na'ari	++	++	+
Figuil / *Binoua*	++	+	+

From the distribution *Table 4* it appears that *Vitellaria* populations in zone 1 are limited to three villages out of 10 and the frequency is low. The symbol (+) indicates that natural stands are lower than 20%, that is, are seriously threatened. Also notable is the finding that *Anogeissus* populations in all villages of zone I are lower than 20%, as are *Balanites* populations. Although these species are of great importance, the danger of extinction is very high in zone I, where the population of *Vitellaria* needs to be reintroduced by establishing new plantations. It is obvious that extinction is due to drought combined with over exploitation and grazing. Fortunately, our investigation has led to identification of areas where there are still aboundant populations of the threatened species, e.g. villages in zone II where more collection of *Vitellaria, Anogeissus* and *Balanites* needs to be carried out for genetic diversity tapping and reintroduction in zone I. Vegetative propagation techniques need to be developed, and collection of the superior genotype material must be achieved before it disappears.

Vegetative propagation of other useful tree / shrub species present in the savannah dry land of Cameroon also needs to be developed in order to combat desertification, which is gradually progressing from north to south.

4. CONCLUSION

From the present investigation it is clear that useful plant species of high value occur naturally in the Sahelian zone of the north and far north of Cameroon. The ethnobotanical survey coupled with germplasm collection of *A. leiocarpus, V. paradoxa and B. aegyptiaca* carried out in 20 villages has revealed the many uses and critical importance of these species in the day-to-day life of the populations in the Sudano-Sahelian zone of North Cameroon. The threat to *Vitellaria* species and the depleted populations of *Anogeisus* and *Balanites* identified and mentioned in the results reflect an alarming situation, which, however, may be overcome by re-introducing genetic material from areas that still possess potential for arresting the progress of desert in that particular region of Cameroon. Meeting this challenge will involve further surveys to identify and collect suitable material, such as seed and vegetative propagules.

REFERENCES

1. J.M. Fondoun, (1998). Atelier régional sur la conservation, la gestion, l'utilisation et le renforcement des Resources génétiques forestières de la zone sahelienne. Rapport National du Cameroun. 29 p.

2. J.M. Fondoun, J. Kengue et M. Tchatat, (1991), *Conservation des Resources Phytogénétiques forestières au Cameroun.* Acte du Seminaire Régional: Gestion des Resources et des Reserves de la biosphère et éducation relative à l'environnement (projet pilote du Dja) Sangmélima 6-10 Mai 1991.
3. J.F. Loung, (1980), Atlas of the United Republic of Cameroon. p. 5-8.
4. R. Letourzey, (1973), Vegetation Atlas of the United Republic of Cameroon, p. 20-24

POTENTIALS OF NON-WOOD FOREST PRODUCTS OF THE SEMIARID REGIONS OF UGANDA

J. Paul Elokaokich
Forest Department, P.O. Box 7124, Kampala, Uganda, Fax: 256-41-3470086, email: catchaff@infocom.co.ug

1. INTRODUCTION

Uganda is a small country with a total area of 241,540 km^2 and a population of about 20 million people. It has an agricultural based economy, hence depends on agriculture and livestock production for economic well being.

Uganda is located in East Africa astride the equator within 4°N - 1°S and 29°E - 35°E. This location bridges the ecological zone of dry East African savanna grassland and the West African rain forests. The convergence of major geographically distinct vegetation formations and ecologically important Tropical High Forests (THF) provides Uganda with modified tropical climate of 1000mm – 1500mm as annual rainfall. The combination of this and general relief status of 900m – 1500m above sea level makes Uganda conducive to the prolific growth of diverse ecosystem including forests.

2. BACKGROUND

Forests of Uganda fall in two categories, Natural forests and man made forests (plantations). Forest cover of Uganda is small about 63,660 km^2. Despite this relatively small size Uganda forests contain immense and exceptional biodiversity. The forests host a number of species of plants,

Combating Desertification with Plants, edited by Dov Pasternak and Arnold Schlissel.
Kluwer Academic/Plenum Publishers, New York, 2001.

birds and animals all living interdependent with each other. It is estimated that Uganda may have between 250,000 to 500,000 species altogether. About half of these are estimated to be insects. Birds total to 100 species of which 33% are forest dependent while about 16% are water dependent. Flowering plants are about 4000 *(Biodiversity Inventory Report, Forest Department, 1996)*. Therefore loss or fragmentation of natural forest threatens the very survival of these flora and fauna.

About 26.5% of the total forested area of Uganda (c16, 920 km^2) are under gazetted ownership and management by the Government. The gazetted forest resources in Uganda inclusive of national parks stand as follows:

- Tropical forest *(fully stocked)* 4,200
- Tropical forest *(degraded)* 900
- Woodland 8,600
- Bush 3,000
- Deciduous plantations *(Eucalyptus)* 60
- Conifer plantations *(pines, cypress)* 160

(Source: National Biomass study, Forest Department 1996).

The bulk of the remaining natural forest resources are found on public lands or privately owned lands.

3. MANAGEMENT OF FOREST RESOURCES IN RESERVED AREAS

Only the reserved areas mentioned above are under government control. Prior to the 1970's management of natural forests for production was well planned. Before harvesting the potential yield of the forest in question would be determined on the basis of in-depth and systematic appraisal. However due to the socio-economic problems that prevailed in the country thereafter, forest management declined with resultant agricultural encroachment and uncontrolled harvesting that threatened the structure and composition of most THF. Today this trend has since reversed.

Management of forests in Uganda has been based on forest policy that emphasized sustainability, optimization of economic benefit and promotion of people's participation in management of the forest resources. In line with the current forest policy, forest management has focused on natural forests, coniferous plantations, poles and fuel plantations with a view to ensuring sustained yield in perpetuity, maximum economic and social benefits, while taking into account environment and biodiversity conservation.

4. MANAGEMENT OF FORESTS OUTSIDE RESERVED AREAS

Little data currently exist on natural forest outside reserves. So far natural forests outside reserves have neither been reserved nor received any kind of planned management system. As this is mainly public land, only species defined as reserved trees belong to the government in these areas. Local people freely collect fuelwood and building materials. Commercial utilization is subject to licensing and the reserved tree species must not be cut unless authorized.

Freehold land areas which the ownership includes tree vegetation represent a very small proportion of the total land area of Uganda. However private forests still are very important as considerable areas on the shores of lake Victoria fall under this category.

Planned forest management practices in the past were directed mainly to Tropical High Forests and plantations mainly for the production of timber with no attention paid to savanna woodland and the significance of Non-Wood Forest Products (NWFP). The only planned management system directed to savanna woodland has been clearance of some areas of savanna woodland for the establishment of plantations *(pine, cypress and eucalyptus)*. There has not been any specific policy or management for NWFP. Non-Wood Forest Products were looked at as "minor" forest products and therefore no much concern was paid to their contribution to the forest ecosystem or national economy.

5. NON-WOOD FOREST PRODUCTS

Rural communities of Uganda have for time immemorial depended upon NWFP as a source of subsistence economy and employment. Despite this there has been negligible attention on the management systems that enhance NWFP growth and sustainable utilization. The present forest policy just makes a passing mention over NWFP as the focus has been mainly on timber as a principal forest product.

There are numerous ways in which NWFP have been used in Uganda. NWFP have been used in the cultural and symbolic functions *(FAO No.6, 1990)*. NWFP resources have also been used to support production sectors such as agriculture, fishing and livestock production.

Significant links to culture and belief: In many rural communities forest trees are believed to provide links to their ancestral past. Forest areas and specific trees are protected and revered for particular cultural occasions.

Rituals and ceremonies which draw on forest symbols often serve to link people with their cultural heritage as well as their ancestral past *(FAO No.30. 1995)*. Each community has its own traditions associated with sacred forest areas and species.

In symbolic functions forest resources are believed to intervene in situations where man is not able to solve a conflict. For example in a traditional rural communities in Uganda, if there was too much rain in the rainy season or rain delayed or failed to return timely in the rainy season, some rituals would be performed in the forest under big trees to appease their "gods". These big trees are believed to act as mediators between them and their angered "gods".

Forest resources also serve a judicial function. They are used as boundary markers to define property and provide evidence of user-rights in judicial disputes.

Ash of some leguminous plants like beans and peas are used *(instead of sodium bicarbonate)* to soften and shorten the cooking time of some vegetables that would take long to cook well.

Stories are that Non-Wood Forest Products like some epiphytes are used by rural teenagers to win love from girls that prove difficult in accepting advances from boys. *(charm girls to accepting advances).*

In agriculture NWFP resources play a role in improving agricultural productivity by:

- Helping to maintain and restore soil fertility and stability
- Helping to protect water supplies
- Providing raw materials for fabrication of implements like hoe handles
- Providing harvesting, transporting and marketing equipment like baskets and sacks
- Protecting crops i.e. fencing
- Food storing i.e. wood ash placed in storage bins.

These products may be rudimentary by modern agricultural standard, but they may be the only resources cheaply and easily available to the rural communities particularly those living in the arid and semiarid lands (ASAL).

In fishing and hunting activities NWFP resources play a role as raw material for the production of nets, trap, gum *(for gumming spears)*, snares, bee hives, fodder *(nectars for bees)*, boats, bows, arrows and fuelwood for fish preservation.

In livestock production NWP resources are a source of fodder and also widely used as livestock medicines.

6. SEMIARID REGIONS OF UGANDA

The arid and semiarid lands (ASAL) are found in the north eastern part of Uganda (Karamoja region) and the neighbouring districts of Kitgum, Lira Soroti and Kumi. Similar climatic conditions and vegetation exist in Nakasongola district in the northern part of the central region and Ankole – Masaka corridor.

Knowledge base for such area is woefully lacking as already mentioned that management practices in Uganda have concentrated mainly on THF and plantations. Emphasis in these areas like similar areas in Africa is placed on technological fixes to what are more often than not social issues. Land conversion to seemingly more productive land uses like irrigation and dry land agriculture is preferred to improving natural resources (Barrow 1997). The changes may be externally imposed, thereby ignoring the wealth of local knowledge and management experience of the local inhabitants.

The inhabitants of the ASAL in Uganda are pastoralists and livestock has an obvious important value to their livelihood. The environment in which they live is hash. So these people have to (make use) survive and thrive on the limited resource base they have at their disposal – a resource base limited by the unforgiving climate. The uses and perceptions of trees by these people in the ASAL show that there is an extensive ethnobotanical knowledge, with keen appreciation of species' properties.

Semiarid regions of Uganda, particularly the Karamoja region is very dry. The climate is relatively severe with erratic rainfalls. Land–surface temperatures are high and evaporation is between two to four times that of annual precipitation *(East Africa 1965)*. Under such conditions any mismanagement of rangeland can aggravate and distort the deleterious effects of such regime. The consequences are rapid deterioration in vegetation cover with resultant soil erosion.

The ASAL regions of Uganda, despite the hash environment, have potential and can provide wide range of NWFP such as gum arabic, resins, fodder, nuts, roots, medicinal plants, fibres, fruits, fuel, bush meat, mushrooms, honey etc. The collection and primary processing of these NWFP could play important role in providing rural communities with opportunities for employment and income generating. Currently individual or a few groups of craftsmen are engaged on production and processing of NWFP for local consumption.

There are several tree species in the semiarid areas of Uganda which can produce NWFP with commercial value, examples:

Acacia species particularly *Acacia senegal* that produces high quality gum arabic grows well in the region. Reasonable stock of *Acacia senegal* exists in Karamoja region and surrounding districts. In Karamoja in dry seasons the tree naturally produces exudates which are collected by local pastoral Karamojong for local consumption. Before civil unrest in the country, the commercial collection of gum arabic from natural exudates had reached about 3000kgs to 5000kgs by the early 1970's. The *acacia* plants can also act as windbreak. Their canopies intercept rain drops while the root systems are effective in reducing soil erosion, thereby stabilising soil. The species also improve soil by nitrogen fixation. The foliage and pods are valuable dry season fodder while the stem has wide application in fencing, energy and construction. The environment benefits of these plant resources in the region are therefore significant.

Between 1987 and mid 1991some work was carried out on gum producing trees/shrub in Karamoja region. Due to insecurity at the time, not all areas were accessible. However the areas covered, revealed that ample stock of *acacia* species exit in the region growing in association with other species. Species like *Balanites aegyptiaca, Euphorbia candilbrum, Ziziphus abyssinica, Camiphora africana* and *Combretum species*, which also are capable of producing gum *(Kityo 1991)*. The study also showed that *Acacia senegal* seeds have a viability of above 70% and regenerates well. If regeneration is undertaken with proper management can add to the exiting natural stands.

Butyrospermum paradoxum, a deciduous tree which grows 7-25m with much branches. It is an important oil producing plant. Its fruit is fleshy pulp and is edible. The fruit tastes sweet and perfumed likened to the over ripe pear *(FAO No.19, 1988)*. The fruits are useful in the "hanger periods" in agricultural cycle when stores are depleted and crop harvests are still time off. The tree produces shea butter, which traditionally in Uganda is used as cooking fat. Shea butter is also important as medicinal and cosmetic ointment. It is also used in soap production.

7. CONCLUSION

The semiarid regions of Uganda have not received necessary management attention. As a result the local communities that live within these areas utilize the resources without much concern on their sustainability. The semiarid regions are still sources of a range of NWFP which in turn sustain the socio-economic and cultural well being of the communities that live in. There is urgent need to carry out survey of Non-Wood Forest Products particularly gum arabic which is of great economic potential, and to revive its commercial production for the home and international markets.

REFERENCES

1. Barrow. E. (1997) *Indigenous knowledge and its application in resolving conservation and utilization problems* (paper presented to the Regional conference for Africa on Gum).
2. Biodiversity Inventory Report – Forest Department (1996).
3. East African agricultural and Forestry Journal Vol. 30 No. 3, (1965).
4. FAO No.6, (1990) *The major significance of "minor" forest products.*
5. FAO No.19, (1988) *Non-timber uses of selected arid zone trees and shrubs in Africa.*
6. FAO No.30, (1995) *Income generation from Non-Wood Forest Products in Upland conservation.*
7. Kityo.P.W. (1991) *Natural gums and resins* (Uganda) Technical Report.
8. National Biomass Study Forest department report 1996.

BIODIVERSITY PROSPECTING IN DRYLANDS OF LATIN AMERICA

Barbara N. Timmermann
Department of Pharmacology & Toxicology, College of Pharmacy, The University of Arizona, Tucson, AZ 85721, U.S.A., Tel: 1-520-626-2841, Fax: 1-520-626-4063, email: btimmer@pharmacy.arizona.edu

1. INTRODUCTION

The "Bioactive Agents from Dryland Biodiversity of Latin America" begun in 1993 with funding from the International Cooperative Biodiversity Group (ICBG) program which is jointly funded by the National Institutes of Health (NIH), the National Science Foundation (NSF) and the U.S. Department of Agriculture (USDA).

Headed by U.S.-based group leader, Barbara N. Timmermann, Ph.D. from The University of Arizona, the project involves collaborations in the U.S. with the G.W.L. Hansen's Disease Center; American Home Products Corporation's Wyeth-Ayerst Research Laboratories and American Cyanamid Company; and other units at the University of Arizona (Office of Arid Lands Studies and Department of Management Information Systems). Partners in Latin America are scientists from the Pontificia Universidad Catolica de Chile, Santiago, Chile; and from Argentina, the Instituto de Recursos Biologicos, Buenos Aires, the Universidad Nacional de la Patagonia, and the Centro Nacional Patagonico, Chubut; and in Mexico the Universidad Nacional Autonoma de Mexico.

The Latin American ICBG program consists of three Associate Programs headed by a Principal Investigator. Associate Program 1 (Biodiversity Procurement and Inventory) is located in each of the foreign country institutions including Argentina, Chile and Mexico. Associate Programs 2

Combating Desertification with Plants, edited by Dov Pasternak and Arnold Schlissel.
Kluwer Academic/Plenum Publishers, New York, 2001.

(Drug Discovery) and 3 (Information Management and Dissemination) are based at The University of Arizona in Tucson.

This international project seeks to discover and develop pharmaceuticals and crop-protection agents from plants of arid and semiarid ecosystems in Argentina, Chile, and Mexico. While plants from arid ecosystems are well known to produce a variety of natural products as defensive agents and poisons, they have received much less attention than those from the tropical rainforests as potential sources of useful biological agents. Despite the commonly held perception that they are unproductive and desolate, dryland environments are diverse in character and support one-fifth of the world's population (Middleton & Tomas, 1997). More than 80 percent of people living in tropical and subtropical drylands are rural and are largely dependent on local plant resources for food, fuel, shelter, medicines, and many other products.

In addition to scientific study, the work promotes economic growth in areas where the plants are collected, involves local populations wherever possible, collects indigenous knowledge about the plants and their uses, and works to conserve biological resources through educational programs.

2. BOTANICAL COLLECTIONS AND INVENTORY

Many xerophytic plants are known for their medicinal properties and for the complex arrays of natural products they manufacture as adaptations to extreme conditions of heat, desiccation, ultraviolet radiation, and herbivory to which they are exposed. Arid-adapted plants are noteworthy by providing host defenses against infectious disease, parasitism or predation (Timmermann & Hoffmann, 1985; Hoffmann et al., 1993; Maatooq et al., 1996).

This ICBG program focuses on plants from different regions in Latin America and allows for the systematic screening of medicinal, endemic and local plants with a battery of high throughput biological assays. The feasibility of a plant-screening program depends on effective procurement strategies. Our ICBG uses a combination of random and biorational strategies for plant collections in arid and semiarid ecosystems of the three countries. Ethnobotanical approaches are practiced whenever possible to identify particular species reported to induce physiological effects.

In order to incorporate national priorities into the site selection process, regions of high interest for conservation and community development are considered as recommended by governmental and non-governmental conservation organizations in each host country. In each area, we attempt to incorporate local participants through a local non-governmental organization or the local government. We also invite researchers from local institutions or universities to collaborate in fieldwork as part of botanical cataloguing of regional floras.

Plants collections were performed in diverse ecosystems such as the cold deserts and steppes of Patagonia and Tierra del Fuego, as well as the phytogeographical provinces of the Monte and the Chaco in Argentina; the hyper-arid desert of Atacama and the semiarid and arid central Chile, and the drylands of central and western Mexico including deserts, the seasonally dry tropical forests, and the pine-oak forests.

This ICBG was designed so that plant collections, inventories, and other activities are in agreement with the appropriate domestic and international laws, such as laws on endangered species (CITES) and plant conservation. Import licenses and phytosanitary certificates are obtained for each sample.

When working in the natural areas of origin of the plants, permission is always sought at the national and local levels. The appropriate collecting permit requests in Mexico are filed with the Instituto Nacional de Ecologia (of the Secretaria de Medio Ambiente, Recursos Naturales y Pesca, SEMARNAP). In Argentina, permission is sought from Direccion de Parques Nacionales, the Secretaria de Medio Ambiente y Desarrollo Sustentable and from local authorities in each provincial state of collection. In Chile, Corporacion Nacional del Medio Ambiente (CONAMA), Corporacion Nacional Forestal (CONAF) and Servicio Agricola Ganadero (SAG) are the organizations that grant permission for collecting. In addition, permits are obtained from individual landowners to collect on private lands.

On the local level, the ICBG project is presented to the landowner, president of the municipio or equivalent authority where collections will be made. A signed letter of agreement is made in which the purpose of the project is stated and the terms of royalty benefit sharing are specified. This letter of intent is backed by copies of publications on the ICBG project and a copy of the agreement between The University of Arizona and the collaborating institutions in the source countries.

In some cases, agreements are made with local non-governmental organizations and provincial governments in source countries. Because of the common heritage of the medicinal plants, we established these working relationships with organizations rather than with traditional medicine specialists.

Once a plant species is located in the wild, a necessary amount of above-ground biomass is collected to yield approximately 1 kg of dry weight of material. The collections are made from areas in which the species occurs in relative abundance and using environmentally sound methods that leave the plant and the population alive and able to reproduce either vegetatively or generatively.

When available, local collaborators in the collection areas are contacted and interviewed about plant remedies from the local flora. This knowledge is gathered primarily to help insure preservation of this cultural knowledge, to increase the chance of drug discovery, and to maximize the potential of rewarding the local community with financial benefits.

Standard herbarium voucher specimens are made and a corroborative specimen may be obtained if needed. Standard habitat and ethnobotanical data also are collected. Noted on specially designed two-page plant collection data forms is a variety of information, including the longitude, latitude, and elevation of the collection site, the collected plants' phenology and growth, other species that grow in the population, type of soil, any ethnomedical uses, etc. In addition, the location of each plant population is documented using a hand-held geopositioning system. This information is entered into the BIOD database, along with related bibliographic references.

Standard herbarium specimens have been prepared for classification by taxonomists and for routine distribution to participating herbaria (Chile, Argentina, Mexico, Arizona) as well as major herbaria in the world (U.S. National Herbarium, the New York Botanical Gardens, the Missouri Botanical Garden and the Royal Botanic Gardens at Kew in the United Kingdom).

Our plant collections have been made by botanists in each source country with taxonomic identification by scientists in the herbaria of the same country. Herbarium specimens are routinely provided for regional cataloguing projects such as the Patagonian Flora under production by our collaborators at the Instituto de Recursos Biologicos in Argentina. Since the initiation of this ICBG, over 2,000 plant samples have been collected in the three countries of study.

3. EXTRACTION AND SCREENING ACTIVITIES

Plants are usually dried at room temperature and ground to 3mm particle size before extraction. We routinely test the organic-soluble plant extracts (dichloromethane-methanol, 1:1). Since the beginning of the project, over 7,000 extract samples have been generated from the collected species. Most of the extracts have undergone primary and secondary assays and selected ones in animal models.

This ICBG involves three institutions in the screening process in different therapeutic areas to maximize efficiency and avoid potential conflicts of interests. Plant extracts, fractions, and pure compounds are submitted to a battery of bioassays using several automated, high throughput enzyme assays developed by American Home Products Corporation (formerly

American Cyanamid Company). The biomedical assays are performed at Wyeth-Ayerst Research Laboratories (Pearl River, New York) while the agrichemical and veterinary tests are performed at American Cyanamid Company (Princeton, New Jersey). A wide assortment of mechanism-based, whole organism, and enzyme-induction assays are employed in the primary screening program to detect lead extracts with interesting mechanistic properties. An extensive battery of organisms resistant to a wide variety of clinically used antimicrobial agents is employed for secondary testing of active leads. Because of the proprietary nature of the therapeutic areas of potential target applications, the details of the bioassays performed at American Home Products Corporation will not be presented here.

This ICBG also includes the search for therapeutics for developing country diseases such as tuberculosis. Similar tests are not usually available in early screening methodologies of major pharmaceutical companies. Bioassays against *Mycobacterium tuberculosis* and *Mycobacterium avium* (responsible for causing tuberculosis and opportunistic infections in immuno-compromised patients, respectively) are performed at the Department of Health and Human Service's GWL Hansen's Disease Center located at Louisiana State University (Baton Rouge).

During the first four years of the project, all *in vitro* evaluations were conducted using the BACTEC 460 radiorespirometric assay, the clinical TB drug susceptibility assay of choice. We currently use rapid, inexpensive, microplate-based assays such as the Microplate Alamar Blue Assay (MABA; Collins & Franzblau, 1997). MABA has shown good correlation with BACTEC in testing crude plant extracts, fractions and pure phytochemicals (Collins et al., 1997; Shawar et al., 1997).

4. BIOASSAY-GUIDED ISOLATION

When biological activity is detected and confirmed for a sample in at least one screen it is considered a positive lead. Bulk plant collections in the order of one-to-three kilograms of dry biomass are obtained for these positive samples or hits following the initial screens. The prioritization of active leads is carried out at regular drug discovery meetings or through correspondence of the senior investigators a few times a year. As they become available, all test results obtained from the plant collections are provided in a confidential manner to the collaborating source country botanists and chemists.

Fractionation and bioassay-guided isolation and characterization of the active natural products is done using state of the art techniques in natural products chemistry laboratories both at The University of Arizona and

Universidad Nacional Autonoma de Mexico. The University of Arizona is involved in the chemical investigations of active plants of Argentina and Chile, whereas Mexico is dedicated to the exclusive study of Mexican plants.

The bioassay-guided approach requires the determination of biological activities prior to detailed chemical investigations of plant extracts. The results of the screening programs at Wyeth-Ayerst Research Laboratories, American Cyanamid Company and the GWL Hansen's Disease Center are the basis for any decision to investigate the chemistry of a plant extract.

Once the decision to continue an investigation has been made by the drug discovery group, we typically extract the total amount of plant material collected, usually 1 kilogram, with methanol-dichloromethane (1:1). The two principal methods used in our laboratories for an initial fractionation of plant extracts obtained in this manner are solvent extraction with polar and non-polar solvents followed by column chromatography. A combination of both methods was found to be suitable for most extracts. Liquid chromatography is our most general method for isolating and purifying natural plant products in sub-milligram to gram quantities. The methods used include HPLC, medium pressure systems, low pressure glass columns ("flash chromatography"), vacuum liquid chromatography and preparative TLC. For the detection of compounds we use TLC with various spray reagents, single wavelength UV, diode array UV and refractive index detectors. More important than the size of the column, the pressure used and the method of detection are the types of adsorbents available for each method. A large variety of chemically modified silica gel phases, such as C18, C8, C4, diol, cyano-, amino, or phenyl- phases, are available either as pre-packed HPLC columns or as bulk material. However, initial fractionations of large amounts of complex mixtures are mostly performed with inexpensive silica gel or reusable reversed phase C18 silica gel. A useful phase not based on silica gel is lipophilic Sephadex LH20, which is very effective for removing chlorophylls or fat and for the separation of phenolic compounds. Typically, it is sufficient to purify a few milligrams of a compound to determine its structure and perform a series of biological tests.

Sometimes larger amounts of a pure compound are required for further testing in animal models. In these cases, we obtain additional plant material from the source countries in larger quantities (5–10 kg). The isolation procedures can usually be optimized at this stage since the structure and physical properties of the compound are already known. To date, we have isolated pure compounds in sufficient amounts for two independent animal studies. In one case, we have also synthesized a natural product from a more abundant analog.

Among the modern techniques in mass spectrometry, LC/MS and LC/MS-MS have gained wide popularity in natural product chemistry. These methods which combine the separation power of HPLC with the sensitivity and selectivity of a mass spectrometer have given us a great advantage to obtain a mass spectrum of a compound without having to isolate it from a complex mixture. In most cases, the structure of small organic molecules can be determined if one milligram of the sample is available. Structure elucidation relies largely on NMR and two-dimensional NMR has become a common technique. Inverse detected experiments, like HMBC and HMQC, are now routinely used by our ICBG for the identification of unknown compounds. The instruments currently used for our research are a Varian Unity 300 and a Bruker AMX 500. The data generated by these instruments can be transferred through an Internet connection directly to our laboratories where powerful software is available for manipulating and displaying the results without using instrument time. So far we have isolated and identified more than 200 pure biologically active compounds of which about one-fifth are novel chemical entities. In addition to one- and two-dimensional NMR, we routinely characterize isolated compounds by IR, UV, high resolution MS, specific optical rotation and melting points. Capabilities for x-ray crystallography, which is not used as routinely, are also available to us.

Although the largest part of our drug discovery research remains confidential, we have been able to publish chemistry results in peer reviewed journals following the filing of a provisional patent application (Wæchter et al., 1998, 1999 a, b; Valcic et al., 1998, 1999; Flagg et al., 1999).

5. INFORMATION AND DISSEMINATION COMPONENT

The goal of the Information Management, Dissemination component (Associate Program 3) has been to support the research, conservation, and economic growth efforts of the overall project by building information handling capabilities at all project sites and by promoting the exchange of information between cooperating institutions. Based on a survey of all project participants in each country at the beginning of the project, specific objectives identified were to:

1) develop general project-related communications products;
2) build an arid lands plant database catalog integrated with bibliographic and geographic information systems (GIS) functions;
3) provide technology transfer and training in the use of these information systems. Throughout the course of the project, these objectives have remained largely the same, although specific activities have changed and evolved according to newly identified needs and interests.

5.1 BIOD Project Web Site

Originally the project's web site was used to provide general information about the project, the ICBG program, and to offer wider distribution of the *BIOD-Prospects Newsletter*. Besides the full-text of each issue of the Newsletter, the site also contained a continuously updated annotated list of hypertext links to other biodiversity-related resources on the Internet. Recently, however, the web site has undergone a complete revision and expansion. The new site provides a complete historical record of the project including lists of publications and presentations resulting from the project, the full-text of selected documents, slide shows of conservation activities, and special features such as links to web-versions of the two project databases (http://ag.arizona.edu/OALS/ICBG/bio-home.html)

5.2 BIOD Database

The copyrighted database program is a relational database designed specifically for this ICBG project to manage, retrieve, and analyze plant data in a simple and efficient manner. The total BIOD database management system actually consists of a number of distinct parts: the data file, an interface, an interface to a geographic information system (GIS), the actual GIS system, and a bibliographic file. These parts have been integrated into a seamless system providing access to the plant data collected, bibliographic references, and additional geographic data. The complete database is available only to project participants due to the sensitive nature of some of the data collected.

One of the strengths of the BIOD database system is that it provides users with the opportunity to see plant biodiversity in a different manner than just a textual sense. The capabilities of the BIOD system go far beyond placing point data on a map. By linking the database to a GIS system, users can discover new relationships between plants and their collection areas by overlaying point data and then classifying it according to the database fields.

5.3 Flora of Chile Database

The Flora of Chile web database project began as a supplementary initiative to this ICBG project. Its purpose is to organize and disseminate pictorial information about plants, providing easy access to photographs of vegetation based on background information and objects contained in the pictures. This prototype database system is available via the web

(http://misdb.bpa.arizona.edu/~guoxiang/chile/biod/webdesc.html).

6. BIODIVERSITY CONSERVATION, RESTORATION AND DEMONSTRATION PROJECTS

One of the project's central goals is to address and promote biodiversity conservation and sustainable economic activity, including development of strategies for minimizing negative environmental impacts while ensuring that equitable economic and social benefits from discoveries accrue to the country, community, and organization which facilitated the discovery of the natural product.

In the area of biodiversity conservation and sustainable development, the first year of the project focused on establishing cooperative relationships among academic, governmental, and non-governmental agencies in each of the three source countries. Team members gathered background materials on biodiversity, sustainability, and traditional uses of plants. They also briefed the research team on the principles of the Biodiversity Treaty and its implications for this project. Finally, a background paper covering treaties, intellectual property rights, and conservation strategies was prepared to provide the framework for workshops in Latin America in 1996 and 1997. The purpose of these workshops was to insure that biodiversity, intellectual property rights, and cultural issues are considered in the process of prospecting for plant resources. Results of the Chilean and Argentinean workshops have been published (Timmermann & Montenegro, 1997; Suarez et al., 1999).

Conservation and development goals are closely linked to this ICBG's academic research process. The plants collected for drug discovery purposes form only a part of the information gathering process. Research to support biodiversity management is integrated into independent research projects such as the study of adaptations of native plants to the local arid environment; facilitating plant regeneration following harvesting for medicinal purposes; as well as growth dynamics, interactions with pollinators, defensive mechanisms against predation, and other relevant fields.

Propagation and regeneration programs for native plants were initiated in several sites by recruiting local municipalities, citizens, teachers, and rural communities to participate in conservation strategies. The ICBG also collaborates in conservation programs including reforestation, restoration of polluted habitats, propagation of native species, establishment of medicinal plant gardens, seed banks, and city and school parks. These projects focus on community participation in "reforesting" with native plants, which not only conserve plant communities, but soil and water, as well.

As part of the *ex situ* conservation component, several hundred accessions of propagation materials (e.g., seeds, cuttings, etc.) have been obtained for subsequent processing and cultivation.

7. CAPACITY BUILDING AND TRAINING

A considerable effort in this ICBG is directed towards capacity-building and professional training in the source countries to develop long-term collaborative and sustainable relationships between the institutions involved. Among the measures that have been implemented in this ICBG is the training of source country scientists in several laboratories in the U.S. On the other hand, U.S. researchers have offered on-site training in database management and natural products chemistry in each host country and have lectured on biodiversity prospecting and drug discovery at several academic institutions as well as governmental and non-governmental organizations.

U.S. graduate students have also had the chance to conduct field work in Chile and Mexico in support of their Ph.D. dissertation and M.S. thesis studies. Students as well as post-doctoral researchers have worked in the areas of systematics, ecology, reproductive biology, conservation, anthropology, ethnobotany, chemistry and microbiology. This active exchange of researchers for the past six years is expected to continue in a similar manner in the future.

With respect to infrastructure development, this ICBG allowed for the purchase of modern equipment and supplies for natural product chemistry laboratories and herbaria, computers, geographic information systems and other database technologies. This project also purchased two vehicles for field trips in Mexico and Chile, and helped to complete the construction of a germplasm facility in Argentina.

8. COLLABORATION AGREEMENTS

The successful collaboration of the members of this ICBG required detailed agreements among the various participants, which defined work and funding commitments, ownership of materials, licensing rights and distribution of future financial benefits, if any. This design of the agreements resembles a wheel in which The University of Arizona serves as the hub of the wheel and each of the collaborators as a spoke (Rosenthal, 1997). This meant that seven agreements had to be negotiated. The challenge of this construction was the necessity to assure that all agreements were consistent with the others. The advantage was our ability to address specific concerns on an individual collaborator basis.

Each two-way agreement defined the scope of work obligations of The University of Arizona and of the collaborator, responsibility for permits and for obtaining informed consent, the collection and preservation of data, ownership of inventions, confidentiality, funding support, bioassay screening, reports, responsibility to establish a sustainable agricultural source of the bioactive plant in the region of its collection, and collection and distribution of royalties to the participating parties. Royalties, should there be any, will be divided into a "collector's share", an "inventor's share" and a "conservation share". The employing institutions of all named inventors of a patent will equally divide the "inventors share" (45% of all royalties). Further distribution by the institutions to the inventors will be dictated by existing policies. Since collectors are not generally recognized as inventors, the collector will receive a separate "collector's share" (5%) and the remaining (the largest) share (50%) will be distributed to a conservation fund in the area of the collection of the country of collection.

It is important to realize that the probability that any one of the ICGBs will discover and develop a commercially viable drug is quite small. For example, it is estimated that it may require the evaluation of 50,000-100,000 compounds in order to obtain a single marketable drug (Kuhlmann, 1997). Not all leads will produce a drug; nearly all (49 or 50) of the compounds that show promise at an early stage in the development process will fail when put into more advanced animal models. Therefore, the real benefits from the ICBG are in the collaborative interactions established among the participating countries, the databases developed as a result of the project and the training of students and faculty through active exchange programs.

The 1998-2003 ICBG has been expanded to include work on endophytic microbes that grow in association with the plants in the areas of collection. To date, we have limited materials sent to American Home Products Corporation for assay to non-reproducing plant extract materials. That is, any seeds, roots, tubers, etc., by which the plants reproduce were reduced to extracts before they were forwarded for testing. This allowed us to alleviate concern by some source country parties that someone in the chain could acquire large amounts of plant material without accounting to the source country. Unfortunately this is not feasible for microbial material and a new process of collection and material control, with a corresponding legal instrument, must be negotiated. We intend to use the NIH Universal Biological Materials Transfer Agreement (BMTA) [Federal Register, March 5, 1995] as the starting basis for the instrument.

We believe we have established a working model that will become a foundation for further collaboration both among members of The University of Arizona ICBG and among others interested in pursuing similar collaborations. A successful outcome of this ICBG has been the use of our structure and agreements by collaborating source countries in drafting their own national and regional regulations on conservation and biodiversity prospecting.

9. COMMUNITY DEVELOPMENT AND BENEFIT SHARING

An important objective of the ICBG program is to promote local responsibility for the conservation of biological diversity. In particular, we are interested in deriving products from ecologically healthy, diverse habitats that will enhance the well being of the local people, as well as benefiting humanity. If an industrialized product is developed from plant material, the community from which the plant sample originated will have priority in producing raw material from *in situ* populations, if management allows a sustainable production. Cultivation of commercial crops in the priority community will be necessary if the natural population cannot sustainably produce sufficient material, as is very likely. Hence, the local community will benefit economically by additional jobs and taxes as well as by the conserving habitat.

Intensive agricultural plots in several indigenous communities and schools have been established with success in Mexico. Native plants raised in greenhouses were planted on the grounds of elementary and high schools as well as in local parks in Argentina and Chile.

This ICBG is actively involved in community work at the request of source country schools, local communities, agricultural extension stations, and non-governmental organizations. Regular training workshops are provided on conservation of native plants to elementary and high school teachers and students as well as farmers and medicinal plant collectors and processors.

In collaboration with the Jardín Botánico of UNAM and Organización de las Nacional Unidas para la Agricultura y la Alimentación (FAO), the ICBG Mexico group (along with collaborators from the ICBG Argentina group in collaboration with Bosque Modelo Chihuahua), surveyed the literature on conventional and biotechnological propagation techniques for arid and semi-arid land plants. The results were published in a manual (Zárate et al., 1996). The information summarized in this publication is available to those interested in propagating plants for ecological restoration programs and cultivation; it also indicates existing deficiencies in current academic programs which need to be remedied.

Many communities are interested in formalizing the use and production of herbal remedies. The Mexican and Chilean Secretarias de Salud will soon enforce national regulations requiring the registry of these products. Such registration includes a monographic study of the botany, ethnobotany, chemistry, pharmacology and toxicology of each ingredient present in the phytomedicine. The first effort in this direction was recently published in two monographs by our project and others are to follow (Pena et al., 1998; Montenegro et al., 1999).

As part of the collaborative agreements, an environmental conservation trust fund (fideicomiso) is to be set up to administer any royalty that could be generated by products derived from this ICBG. Environmental trust funds have not been established at this time as such funds have been put on hold in the source countries pending legislation on access and benefit sharing on biological resources. This ICBG has made contact with advisors of the Argentinean, Chilean and Mexican Senate's commissions that are drafting these laws to offer assistance. We also have participated in public consultation forums to offer this ICBG as a pilot project upon which to base future legislation.

10. CONCLUSIONS

This chapter used one of the ICBG programs to illustrate a particular framework developed for research into drug discovery from natural products from plants of xeric environments, biodiversity cataloging and conservation as well as economic development. As such, this ICBG has been specifically designed for the existing scientific, technological, cultural, biogeographical, legal and technological situations in Argentina, Chile and Mexico.

In the long term, this project has built institutional and international relationships between the U.S. and developing countries that will continue to grow beyond the life of the project and will serve as an effective model for others who seek to develop similar relationships.

ACKNOWLEDGEMENTS

This program is supported by a grant 2 U01 TW00316-07 from the National Institutes of Health (NIH), the National Science Foundation (NSF), and the U.S. Department of Agriculture (USDA). The author is very grateful for the excellent contributions of the collaborating scientists as well as the laboratory contributions of the postdoctorals, technicians, graduate and undergraduate students.

REFERENCES

1. Collins LA, Biswas AN, Cantrell CL, Fischer NH, Franzblau SG (1997): *Microplate Alamar Blue Assay vs. BACTEC 460 for Screening Natural Products against Mycobacterium tuberculosis and Mycobacterium avium*, Interscience Conference on Antimicrobial Agents and Chemotherapy, 37th, Toronto.
2. Collins LS, Franzblau SG (1997): *Microplate Alamar Blue Assay versus BACTEC 460 system for high throughput screening of compounds against Mycobacterium tuberculosis and Mycobacterium avium*. Antimicrob Agents Chemother 41: 1004-1009.
3. Flagg M, Valcic S, Montenegro G, Gomez M, Timmermann BN. (1999). Pentacyclic Triterpenes from Chuquiraga ulicina. *Phytochemistry 52*(7): 1345-1350.
4. Hoffmann JJ, Timmermann BN, McLaughlin SP, Punnapayak H (1993): *Potential antimicrobial activity of plants from the Southwestern United States*. Int J Pharmacognosy 31: 101-115.
5. Kuhlmann, J. (1997): *Drug Research: From the Idea to the Product*. International Journal of Clinical Pharmacology and Therapeutics 35: 541.
6. Maatooq GT, Stumpf DK, Hoffmann JJ, Hutter LK, Timmermann BN (1996). *Antifungal eudesmanoids from Parthenium argentatum x P. tomentosa*. Phytochemistry 41: 519-524.
7. Middleton N, Thomas D (1997): *World Atlas of Desertification-UNEP*, London, Edward Arnold Press, p. viii.
8. Montenegro G, Pena RC, Mujica A, Iturriaga L, Gonzalez L, Timmermann BN. (1999). *Posibilidades de un control botanico analitico de la Hierba de San Juan Hypericum perforatum L.* Rev. Academia Colombiana de Ciencias Exactas, Fisicas y Naturales 23(88):455-460.
9. Pena RC, Timmermann BN, Iturriaga L, Gonzalez L, Montenegro G (1998): *Posibilidades de un control botanico analitico de la una de gato Uncaria tomentosa (Willd.) DC.* Rev Academia Colombiana, 22, 595-600.
10. Rosenthal JP (1997): *Equitable sharing of biodiversity benefit agreements on genetic resources* In: Proceedings of the OECD International Conference on Biodiversity Incentive Measures for Conservation and Sustainable Use of Biological Diversity, Cairns, Organization for Economic Cooperation and Development (OECD), pp. 253-274.
11. Shawar RM, Humble DJ, Van Dalfsen JM, Stover CK, Hickey MJ, Steele S, Mitscher LA, Baker W (1997): *Rapid screening of natural products for antimycobacterial activity by using luciferase-expressing strains of Mycobacterium bovis BCG and Mycobacterium intracellulare*. Antimicrob Agents Chemother 41: 570-574.
12. Suárez EY, Fortunato RH, Elechosa MA, Casamiquela R, Saavedra E, Timmermann BN (1999): *Aspectos Técnicos, Culturales, Políticos y Legales de la Bioprospección en Argentina*, Proceedings of an International Workshop, Buenos Aires, San Lorenzo Press, pp. 1-174.
13. Timmermann BN, Hoffmann JJ (1985): *Resins from Grindelia: a model for renewable resources in arid environments*. In: Wickens GE, Goodin JR, Field DV, eds., Plants for Arid Lands, London, George Allen & Unwin Press, pp. 357-368.
14. Timmermann BN, Montenegro G (1997): *Proceedings Taller Internacional Aspectos Ambientales, Eticos, Ideologicos, y Politicos en el Debate sobre Bioprospeccion y Uso de Recursos Geneticos en Chile*, Santiago, Sociedad de Biologia de Chile, vol. 5, pp. 1-119.
15. Valcic S, Montenegro G, Timmermann BN (1998): *Lignans from Chilean propolis*. J Nat Prod 61: 771-775.

16. Valcic S, Montenegro G, Mujica AM, Avila G, Franzblau S, Singh MP, Maiese WM, Timmermann BN (1999): *Phytochemical, morphological, and biological investigations of propolis from Central Chile*. Zeitschr f Naturforsch 54c: 406-416.
17. Wächter GA, Franzblau SG, Montenegro G, Suarez E, Fortunato RH, Saavedra E, Timmermann BN (1998): *A new antitubercular mulinane diterpenoid from Azorella madreporica Clos*. J Nat Prod 61: 965-968.
18. Wächter G, Matooq G, Hoffmann JJ, Maiese WM, Singh MP, Montenegro G, Timmermann BN. (1999a). *Antibacterial Diterpenoid Acids from Azorella compacta*. J. Natural Products 62(9): 1319-1321.
19. Wächter G, Valcic, Flagg M, Franzblau SG, Montenegro G, Suarez E, Timmermann BN. (1999b). *Antitubercular activity of pentacyclic triterpenoids from plants of Argentina and Chile*. Phytomedicine 6(5): 341-345.
20. Zarate-Pedroche S, Bye R, Elechosa M, Monroy de la Rosa MA, Morales G, Mendoza Cruz M (1996): *Metodos de propagacion convencionales y biotecnologicos de la especies medicinales en zonas aridas*. In: Izquierdo J, Palomino G, eds., Tecnicas Convencionales y Biotecnologicas para la Propagacion de Plantas de Zonas Aridas, Serie Zonas Aridas y Semiaridas, No. 9, Santiago, Organizacion de las Naciones Unidas para la Agricultura y la Alimentacion (FAO) pp. 305-347.

PROVENANCE TRIAL FOR INDIGENOUS TREE SPECIES

"Schinziophyton rautanenii schinz"

Lovemore Simwanda
Zambia National Farmer's Union, P.O. Box 30395, Lusaka, Zambia, Tel: 260-1-252649, Fax: 260-1-252648, email: znfu@zamnet.zm

1. INTRODUCTION

Schinziophton Muell. - Arg. has three tropical south, west African species (1) and *S. rautanenii* schinz (mungongo - Lozi) is confined to Africa South of the equator in the Democratic Republic of Congo, Angola, Namibia, Botswana, South Africa (Northern part), Mozambique, Zimbabwe and Zambia. In Zambia, it occurs almost in all provinces except the copper belt. It is a typical tree of the open *Baikiaea plurijuga, Guibourtia coleosperma, Brachystergia spiciformis* woodland on Kalahari sands in the southern and western provinces. It grows in woodlands near dambo heads on sandy alluvium in the eastern province. It appears occasionally in *Acacia chipya*, escarpment woodlands and secondary thickets.

The economic importance and desertification control potential of *S. rautanenii* lies in its high oil yielding fruit, light easy-to-carve wood, medicinal properties, and remarkable adaptation for flourishing under harsh conditions in Kalahari sands. Divergent research reports regarding the oil content range from 38 to 60 percent, suggesting a vast variation of species.

In view of the species high economic commercial value, desertification control potential and high adaptability levels under harsh conditions, there is need for domestication of this plant species. However, the threat of ever increasing deforestation leading to possible desertification in the country, due to increasing human population use of forest resources and its wood products, desertification control programs need to take centre stage. A

Combating Desertification with Plants, edited by Dov Pasternak and Arnold Schlissel.
Kluwer Academic/Plenum Publishers, New York, 2001.

moderate program has been initiated to study the suitability of various provenances for establishment of *Schinziophyton* local plantations in Zambia. This project further intends to provide data on the performance of the species under different silvicultural management regimes, its adaptation to an assortment of soil types, various rainfall and climatic patterns.

2. MATERIALS AND METHODS

2.1 Study area

Two plots of 2.02 ha each were stumped and cleared on already deforested land using hand tools. A tractor after which the area was marked and squared for planting did ploughing and discing.

Two areas had each six provenances randomly selected and planted. Total number of observation plots were 35 i.e., 8 at 3.0 m x 3m and 27 at 4m x 4m.

2.2 Study Site

The experiment site has sandy soils, which are darker, finer, pale, coarse, loose, infertile white or grey sands. This being the typical Kalahari soil types, the pale sands are normally associated with depressions, flatter areas where drainage is likely to be less free and at lower altitudes. In dense thicket with dense surface rooting, the upper 60cm of soil could remain at or below wilting point from April to the first rains in late October. At 2m depth the soil dries out and later wetted with the rains achieves standard soil type water holding capacity. It is generally agreed that under such conditions *Schinziophyton* survives on the current season's ephemeral rainfall and roots within the annually wetted soil zones. *Soil moisture is a critical primary factor in the ecology of the natural and planted forests.*

2.3 Rainfall, High Temperatures and Wind

Rainfall undoubtedly exercises a marked effect upon flowering and fruiting, especially in deeper-rooting species such as *Schinziophyto rautanenii*, less so in shallower-rooting species. Severe storms in the mid and late rains can greatly reduce flowering and fruiting potential especially of *Schinziophyton*. Irregular early season rainfall affects germination and seedling survival, as obviously does poor rainfall.

High temperature plus low humidity levels contribute greatly to high fire hazard and damage severity in the late dry season. Temperatures on the soil surface under less than 1.0 cm of litter can approach a potentially lethal level of about 59 degrees centigrade (Fargular, 1970).

2.4 Temperature

The study area (in Sesheke) is a place with climatic extremes. The mean temperature during the hottest month is 25.8°C and the temperature during the coolest months is 14.2°C giving the largest temperature range 11.6°C in the country. Mbala (Northern Province) and Kawambwa (Luapula Province) have the lowest (4.1°C and 4.7°C respectively) annual temperature range; Sesheke has the highest maximum temperature ever recorded 40.6°C during the hottest month.

2.5 Rainfall Climatology

The rainfall decade amounts (70% probability) in November at Sesheke are clearly insufficient for sowing. Only in decade I of December can Sesheke expect 23mm, which just equals 0.5 PET. However, the risk of dry spells soon after sowing is very high at Sesheke as the probability of receiving more than 30mm rainfall is obvious. Rainfall at Sesheke does not exceed PET during the rainy season (*Figure 1*). The total growing season at Sesheke is relatively short as compared to other areas in the country.

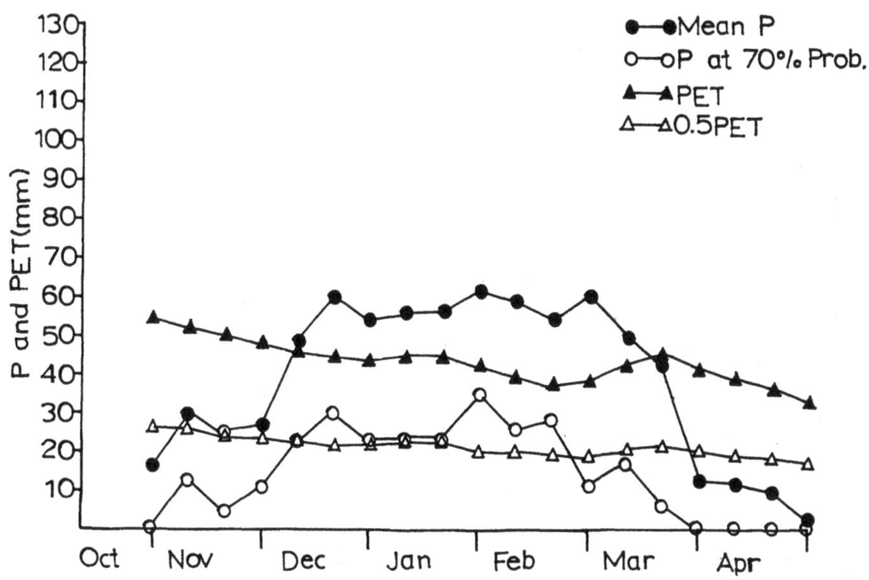

Figure 1. Water balance diagram for Sesheke

Sesheke is known for its frost - proneness during the cool dry season and for its drought proneness during the rainy season, rendering crop production at this location rather risky.

3. MANAGEMENT OF THE EXPERIMENT

3.1 The Experimental Design

The design was a complete randomized design in which area one (1) had 17 observation plots for six (6) provenances and 18 for area two (2) respectively. Planting for 3.0m x 3.0 m plots was 7 plants 3m apart on one line by seven (7), forming a square of forty-nine (49) plants; of these, twenty-five (25) forming the *"central core"* were for assessments. The plots where planting was 4m x 4m, each line had six (6) plants, 4m apart and this gave a number of thirty-six (36) plants. The *"central core"* was of sixteen (16) plants. Protection of the plants in the *"central core"* was by the *"surrounding plants"*. Every plant in the *"central core"* was tagged and numbered. Weeding was only done manually after planting.

3.2 Assessment Procedures

3.2.1 Height, Survival and Diameter

Assessments for height and diameter growth were by a measuring ruler 1.0m and a caliper respectively. Height was in Centimeters (cm) while diameter was in Millimeters (mm). Diameter of the root (Lignotuber) collar was taken at points 10 cm from the ground. Assessment for survival was by physical count. One observation was made quarterly.

3.2.2 Results and Discussion

Main results are presented in *Table 1*. Mean height growth for provenance planted at 4.0m x 4.0m and 3.0m x 3.0m spacing was 36.1cm and 33.5cm respectively. Root collar diameter was 8.2mm for 4m spacing and 7.7mm for 3m spacing. Provenance 1 (Kazembe) showed best height growth and root collar diameter at 4m spacing. The least was Provenance 5 (Mongu) which showed 25.26 cm and 6.34 mm, for height and root collar growth respectively. For the 3m spacing, Provenance 11 (Zambezi) also showed better growth of height and root collar diameter.

However, root collar diameter, at both spacing showed no significant difference between provenance means at 5 percent level but height growth showed a significant difference between provenances at 5 percent level in the 4m spacing. The results from measurements done in the last quarter showed Provenance 1 (Kazembe) to have superior growth rate for both height and diameter. This possibly indicates that despite this provenance being from the high rainfall region, it could still perform better in drier regions as compared to those from drier areas.

There is a need for research to continue and confirm these results in much harsher conditions.

Table 1 showing growth performance of 6 months old *Schinziophyton rautanenii* (Schinz)

Provenance	Number of plots	Spacing	Mean Height	Root Collar diameter
1	2	4 m x 4 m	50.24	13.15
2	1		38.92	8.48
3	2		34.49	6.16
4	2		37.91	11.16
5	3		25.26	6.34
6	2		25.84	7.87
7	2		34.56	7.54
8	1		34.14	9.69
9	3		46.34	12.72
10	3		37.60	8.98
11	3		37.61	7.90
12	3		30.52	7.27
		3 m x 3 m		
6	3		30.74	7.87
7	3		32.19	7.54
11	2		37.71	7.90

Key:

1 = Kazembe; 3 = Magoye; 5 = Mongu; 7 = L/stone; 9 = Nchelenge; 11 = Zambezi
2 = Kabompo; 4 = Mbala; 6 = Mporokoso; 8 = Senanga; 10 = Amwala; 12 = Mumbwa

The variation in height growth for provenances planted at 4m spacing is a possible indication of greater competition among plants for light. The results are indicating also better establishment and adaptability to the site conditions at this spacing. The growth of Provenance 6 (Mporokoso) shows better performance at 3m spacing where as 7 (Livingstone) shows poor performance at the same 3m spacing. This may indicate wider spacing for Provenance 7 (Livingstone) and narrow spacing for Provenance 6 (Mporokoso) but this will depend on future results from research.

3.3 *Schinziophyton rautanenii* Associated Problems

Microfungi associated with *Schinziophyton* (mungongo) in Zambia include some common moulds, such as *Fusarium sp.* and *Rhizopus stolonifera (Ehrenb. Fr.)* Lind isolated from seeds, and these sometimes adversely affect germination. Poor performance of cultivated seedlings is attributed to improper watering. *Special care in seed treatment and nursery management regimes is necessary to raise the plants successful.*

A *Hexagonia sp.* was reported by Parker (1978) on dead branchwood in Southern Province of Zambia; he relates Fanshawe's note to the effect that multiple-stemmed trees often contain an underground, water filled chamber of decay, which one imagines is due more to the activity of bacteria than fungi.

The plantation silviculture of *Schinziophyton* for a variety of end-uses has been proposed on several occasions. However, results from preliminary trials established from truncheons (cuttings) in the Kalahari Desert parts of Zambia, suggest that the greatest problem regarding their protection would be depredation by elephants in natural forests.

4. CONCLUSIONS AND RECOMMENDATIONS

The performance of each provenance shows to be better for Provenances 1 Kazembe; 2 Kabompo; 4 Mbala; 9 Nchelenge; 10 Namwala; and 11 Zambezi whose mean heights are or above the provenance mean at 4m spacing. For diameter growth, the same provenances except for Provenance 11 show vigorous growth. However, growth of Provenances 1 (Kazembe) and 9 (Nchelenge) are showing greater adaptability in the trials. *Assessments must continue at least four times per year to study provenance performance in terms of height, diameter, survival, flowering, fruiting and branching patterns.*

REFERENCES

1. Savory, A. (1964). *Variation of root systems in Zambian indigenous tree species.*
2. Fanshaw, D.B. (1959). *Vegetation of pattern of the western province of Zambia.*
3. Raussen, Th. (1997). *Field crop production guidelines for small-scale farmers.*
4. Maro, P.S. (1992). *Integrated land use planning in semi-arid areas.*
5. Longman, K.A., Jenik, J. (1987). *Tropical forest and its environment.*
6. Tiffen, M., Mortimore, M., Gichuki, F. (1995). *More people, less erosion.*

Part II

INTRODUCTION, DOMESTICATION AND DISSEMINATION OF ARID LAND PLANTS

INTRODUCTION AND SELECTION OF DROUGHT AND SALT TOLERANT PLANTS FOR AFFORESTATION AND LANDSCAPING IN ARID LANDS

Y. Ben Dov[1], M. Forti[1], R. Pauker[1], J.A. Aronson[2] and D. Pasternak[1]
[1]*The Institutes for Applied Research, Ben-Gurion University of the Negev, P.O.B 653, Beer-Sheva, 84305, Israel,* Tel: *972-7-6461972,* Fax: *972-7-647 2984, email: dovp@bgumail.bgu.ac.il*
[2]*Centre d'Ecologie Fonctionelle et Evolutive (CNRS-U.P.R.), 1919 route de Mende, 34293 Montpellier, France, Tel: 33 4 67613311, Fax: 33 4 67412138, email: aronson@cefe.cnrs-mop.fr*

1. BACKGROUND

Introduction and selection of plant species for afforestation and landscaping in arid regions was carried out in five locations in the Negev desert.

Three locations (Omer, Sede Teiman and Nir Oz) are located in the northern Negev where average rainfall ranges from about 250 mm/yr at Nir-Oz to 200 mm/yr at Omer and 180 mm/year at Sde Teiman. (*Tables 1 and 2*). The fourth location, Sapir, is situated in the hot and dry Arava Valley (summer average maximum ~38.5°C) where average annual rainfall is about 30 mm *(Table 3)*. No irrigation was applied at all in the first three locations. In Sapir the plot was irrigated with recycled saline water (TDS ~2000 ppm). Selection for salt tolerance of halophytes was carried out in Ashkelon by the Mediterranean Sea *(Table 4)*. The test plot was irrigated with full strength seawater (3.7% salts) and the control with 15% seawater.

The soil in Omer is a silty loam, in Sde Teiman silty, in Nir Oz, sandy loam, and in Sapir a fine-textured sodic soil (Solonetz). In Ashkelon the plants were established on a levelled sand dune 50m from the seashore.

Combating Desertification with Plants, edited by Dov Pasternak and Arnold Schlissel.
Kluwer Academic/Plenum Publishers, New York, 2001.

Table 1. Climatical Data for the Omer and Sdeh Teiman Introduction Sites[1]

Month	Temperature (°C)				Relative Humidity at 14:00	Precipitation (mm)	Evaporation (mm)
	Minimum		Maximum				
	Absolute	Daily average	Absolute	Daily average	Daily average	Monthly average	Daily average
January	-0.5	6.0	31.5	16.4	47	47	2.4
February	0.6	6.9	32.0	18.4	42	39	3.3
March	0.8	8.9	36.9	21.3	37	32	4.6
April	2.0	11.5	40.4	25.6	32	8	6.6
May	6.6	13.9	44.8	29.3	29	3	8.2
June	10.5	16.7	43.8	31.9	30	0	9.4
July	13.2	18.5	41.5	32.9	32	0	9.3
August	14.4	18.6	42.2	32.8	34	0	8.5
September	10.6	17.1	40.5	31.1	37	0	7.2
October	7.0	14.7	38.7	28.6	34	4	5.5
November	2.0	11.0	35.3	23.4	39	27	3.7
December	0.8	7.3	32.5	17.9	47	43	2.4
Total						203	

[1]Data from the near-by Beer Sheva Meteorological Station

Table 2 Climatical Data for the Nir Oz Introduction Site

Month	Temperature (°C)				Relative Humidity at 14:00	Precipitation (mm)	Evapora-t ion (mm)
	Minimum		Maximum				
	Absolute	Daily average	Absolute	Daily average	Daily average	Monthly average	Daily average
January	1.8	7.7	30.4	17.1	54	64	2.6
February	0.9	7.8	29.2	17.9	51	45	3.7
March	3.0	9.1	33.6	20.3	48	29	4.9
April	4.8	11.9	42.0	25.6	41	13	6.9
May	7.5	14.8	44.0	28.1	42	0.8	8.0
June	12.7	17.7	42.5	30.7	43	0	8.8
July	16.0	20.1	38.8	31.9	47	0	8.8
August	17.5	20.5	37.5	32.0	50	0	8.0
September	13.0	19.1	43.0	30.7	50	0	6.7
October	10.8	16.9	38.0	289	48	9	5.3
November	4.0	12.7	34.0	24.0	48	38	3.9
December	3.0	9.3	32.0	19.2		58	2.6
Total						256.8	

Table 3. Climatical Data for the Sapir Introduction Site

Month	Temperature (°C)				Relative Humidity at 14:00
	Minimum		Maximum		
	Absolute	Daily average	Absolute	Daily average	Daily Average
January	1.0	8.3	27.7	19.3	41
February	2.0	9.9	30.5	21.0	40
March	5.8	12.8	38.2	24.8	35
April	7.9	15.8	41.0	29.4	29
May	11.0	19.3	44.5	33.5	25
June	12.3	22.3	45.7	37.1	23
July	19.5	23.8	44.1	38.3	24
August	20.5	24.5	45.6	38.4	29
September	17.0	22.6	40.0	34.9	33
October	12.0	19.3	38.0	31.1	37
November	5.1	14.2	36.2	25.7	40
December	2.6	9.8	31.7	20.3	48
Total					

Table 4. Climatical Data for the Ashkelon Introduction Site

Month	Temperature (°C)				Relative Humidity at 14:00	Precipi-tation (mm)
	Minimum		Maximum			
	Absolute	Daily average	Absolute	Daily average	Daily average	Monthly average
January	-1.8	8.0	32.6	17.3	53	78
February	1.1	8.2	32.7	18.8	52	52
March	0.8	9.7	38.0	21.3	48	43
April	4.0	11.9	41.3	25.2	42	9
May	8.1	14.3	45.3	28.0	40	3
June	11.9	17.8	44.1	30.5		0
July	15.5	19.8	39.4	31.4	45	0
August	15.0	20.2	40.2	31.5	47	0
September	12.6	19.1	39.5	30.4	48	1
October	7.0	16.5	37.6	28.3	47	11
November	3.5	13.1	36.5	23.9	48	56
December	-0.2	9.4	33.2	18.8	52	75
Total						328

2. MATERIAL AND METHODS

The plant material was selected from previously established plant introduction plots, purchased from seed merchants, acquired from botanical gardens and research organizations, and collected through numerous seed collection missions in many arid regions.

Over the years several thousands of species were collected and tested. In all cases, they were planted in randomized duplicate plots, each plot containing 6-12 (for trees) and 12-24 (for shrubs) individual plants. The greater part of the Omer introduction plot was planted in 1966 and 1967 and hence data obtained there are more substantiated than those of the other locations.

The results are summarized in *Table 5* showing the most drought tolerant species, the most successful heat and drought tolerant species for hot deserts, and the most successful salt tolerant species. The table contains useful information on the species, such as special features, degree of tolerance to drought and salinity, and potential uses.

3. RESULTS

Table 5 summarizes results on the field trials in four sections. In the first, plants identified by their botanical name are organized in sections according to life form. In each of the sections, they are listed alphabetically, according to family, genus and species. In several cases synonyms are also mentioned.

Sector 3 (Uses) is self-explanatory, while sector 4 (Remarks) offers some complementary, relevant information. Sector 2 ("Tolerance to") requires a few comments. Here too, as in the previous sectors, + indicates that a species may fit the conditions mentioned at the head of the secondary columns. This potential has been tested in most instances in one or more of the experimental sites described above, in other experimental areas or in various landscaping projects.

It would be, however, irresponsible to claim that these evaluations, even if acquired experimentally, are absolute and applicable always and everywhere. Drought tolerance, for instance, cannot be related only to rainfall. Concurrent factors like soil characteristics, temperature, relative humidity, etc. may play, under isoclimatic conditions, an important role in facilitating or hindering survival and successful growth.

Moderate auxiliary irrigation, as applied by us in certain instances, does not exclude a species from the category of drought tolerance, if we compare its low water requirements with those of most other species.

In sum, the information presented in this paper can serve as a starting point to facilitate the search for suitable plant material in hot deserts around the world. It must be substantiated by experimentation *in situ* under the particular conditions of any given area. Further, we stress the importance of long-term follow-up to guarantee accurate and appropriate selection of trees and shrubs.

In some of our experiments, species have survived aridity conditions without any irrigation for over 35 years, thereby overcoming a broad range of weather fluctuations, reaching excellent dimensions and keeping in good shape. This seems an indication of reliable characteristics on the long range. In other cases only some individuals from the group have survived, exhibiting however good development. These surviving individuals represent a promising source of propagation material, to be exploited for further developments.

Many species which appear in the plant lists were used in various afforestation and landscaping projects in the Negev desert, along the Mediterranean coast and elsewhere in Israel. Others can be used as ornamentals and in the floriculture industry (potted plants, decorative branches and flowers). Their ornamental properties are therefore also presented here.

The following abbreviations are used in Table 5:

Tolerance to:
D = Drought
S = Salinity
Hs = Heat and salinity
Ss = Sea spray

Uses:
A = Afforestation
Dbf = Decorative branches and flowers
Ff = Food or fodder
Irm = Industrial raw materials
L = Landscaping
Pb = Pollen for bees
Sc = Soil conservation
Sls = Slope stabilization
Sas = Sand stabilization
Wb = Wind breaks

Table 5. List of trees, shrubs and prostrate plants for landscaping, afforestation and environmental improvement in arid lands.

TREES

Family and Species	Tolerance to:				Uses										Remarks
	D	S	Hs	Ss	A	L	Sc	Sas	Sb	Wb	Pb	Irm	Ff	Dbf	
Anacardiaceae															
Pachycormus discolor	+	+	+			+									Deciduous, used as potted plant(bonzai).
Pistacia atlantica	+	+	+		+	+									Deciduous, long lived.
P. palaestina	+	+	+		+	+									Deciduous, long lived.
P. vera	+	+	+		+	+							+		Deciduous, fruit tree.
Rhus lancea	+	+	+			+									Small tree.
Schinopsis lorentzii	+					+									Small tree, long lived.
Araliaceae															
Cussonia spicata	+					+									Used as potted plant (bonzai).
Bignoniaceae															
Chilopsis linearis	+					+									Deciduous willow-like tree.
Crescentia cujete		+	+			+									Medium-sized tree with inflated fruits used as gourds.
Kigelia pinnata (=K.africana)		+	+			+									Decorative fruit.
Tabebuia palmeri	+	+	+			+									
Bombacaceae															
Adansonia digitata (Baobab)		+	+			+									Very big tree. Edible fruit. Requires warm winters.
Burseraceae															
Bursera felicifolia	+	+	+			+									Low tree. Deciduous. Used as potted plant.
B. hindsiana	+	+	+			+									Low tree. Deciduous. Used as potted plant.
B. microphylla	+	+	+			+									Low tree. Deciduous. Used as potted plant.
Caesalpiniaceae															
Bauhinia caronii (=Lysiphyllum caronii)	+					+									
Caesalpinia spinosa	+					+									Small tree, reddish fruit, spiny.
Ceratonia siliqua	+				+	+						+	+		Long lived. Pods consumed by man and animals.
Cercidium floridum	+	+	+			+					+				Deciduous, beautiful green trunk with peeling bark, spiny.

Family and Species	Tolerance to:				Uses										Remarks
	D	S	Hs	Ss	A	L	Sc	Sas	Sls	Wb	Pb	Irm	Ff	Dbf	
Caesalpiniaceae (continued)															
Colvillea racemosa	+	+	+			+									Low tree.
Delonix (=Poinciana) regia		+	+			+									Suffers in cold regions. Impressive flowering. Gives heavy shade.
Geoffroea decorticans	+	+	+		+	+									Deciduous. Many invasive suckers. Possibly suitable for extreme erosion situations.
Haematoxylon brasiletto		+	+			+					+	+			Spiny. Source of furniture timber and coloring matters.
Parkinsonia aculeata	+	+	+			+					+				Deciduous, spiny, seeds itself.
Tamarindus indica		+	+			+						+	+		Fruit contains a pulp used for drinks and laxatives.
Caricaceae															
Carica papaya		+	+			+							+		Edible fruit. Suffers from frosts.
Casuarinaceae															
Allocasuarina decaisneana	+	+	+	+		+				+					Suitable for seaside locations. Used for wind breaks.
Casuarina equisetifolia	+	+	+	+		+				+					As above.
C. cristata	+	+	+	+		+				+					As above.
C. glauca	+	+	+	+		+				+					As above.
C. luehmannii	+	+	+	+		+				+					As above.
C. stricta (=C. quadrivalvis)	+	+	+	+		+		+	+	+					As above.
Celastraceae															
Maytenus emarginatus		+	+			+					+				Spiny.
Cupressaceae															
Callitris verrucosa	+				+	+		+	+		+				
Juniperus phoenicea	+				+	+									Slow growth, long lived.
Tetraclinis articulata	+				+	+		+	+						
Elaeagnaceae															
Elaeagnus angustifolia		+	+			+									
Flacourtiaceae															
Dovyalis (=Aberia) caffra		+	+			+							+		Edible fruit. Spiny. Suitable for fencing.

Table 5. *(Continued)*

TREES

Family and Species	Tolerance to:				Uses										Remarks
	D	S	Hs	Ss	A	L	Sc	Sas	Sls	Wb	Pb	Irm	Ff	Dbf	
Malvaceae															
Lagunaria patersonii		+	+	+		+						+			Urticating hairs on the seed pods. Host of cotton parasites
Thespesia populnea		+	+			+									
Meliaceae															
Melia azedarach		+	+			+					+				Shade tree. Umbrella canopy.
Mimosaceae															
Acacia gerrardii subsp. *negevensis*	+	+	+		+	+					+				Spiny.
A. homalophylla	+	+	+			+	+	+	+	+					Suckers. Volunteers.
A. horrida (=A. karroo)	+	+	+			+	+	+	+	+	+				Spiny, suitable for hedges.
A. loderi	+					+									
A. pendula	+	+	+			+	+	+	+						Weeping habit, shade tree. Produces suckers.
A. raddiana	+	+	+			+					+				Spiny. Long lived.
A. salicina	+	+	+	+	+	+	+	+	+	+					Morphologically variable. Fast growing. Produces suckers. Volunteers. May become a pest.
A. saligna(=A. cyanophylla)	+	+	+	+		+	+	+	+				+		Pioneer tree. Fast growing. Volunteers. Short-lived. Browsed.
A. seyal	+	+	+		+	+									Spiny .
A. stenophylla	+					+									Weeping habit. Thin canopy.
A. tortilis	+	+	+		+	+					+				Several stems. Spiny.
Albizzia lebbeck		+	+			+									Fruit dirties the ground.
Lysiloma candida	+	+	+			+									Deciduous.
Prosopis alba	+	+	+			+	+	+	+				+		Deciduous, thin tree. Pods eaten by animals.
Prosopis alpataco	+	+	+			+	+	+	+				+		Deciduous, small tree. Pods eaten by animals.
Mimosaceae															
P. caldenii	+					+	+	+	+				+		Deciduous, small tree. Pods eaten by animals.
P. chilensis	+					+	+	+	+				+		Deciduous, small tree . Pods eaten by animals.
P. glandulosa	+					+	+	+	+				+		Deciduous, small tree. Pods eaten by animals.

Family and Species	Tolerance to:				Uses										Remarks
	D	S	Hs	Ss	A	L	Sc	Sas	Sls	Wb	Pb	Irm	Ff	Dbf	
Mimosaceae (continued)															
P. juliflora	+	+	+			+	+	+	+				+		Deciduous. Wide canopy. Shade tree. Spiny or unarmed. Volunteers. Pods eaten by animals.
P. juliflora "28 A"	+					+	+	+	+				+		Deciduous, wide canopy, spiny.
P. nigra	+	+	+			+	+	+	+				+		Deciduous, spiny. Pods eaten by animals.
P. pallida	+	+	+			+	+	+	+				+		Deciduous, spiny.
P. pubescens	+					+			+						Deciduous.
P. sp. *" 146 "*	+					+	+	+	+						Deciduous, spiny.
P. tamarugo	+	+	+			+	+	+	+				+		Deciduous, slow growth shade tree. Pods eaten by animals.
Moraceae															
Ficus bengalensis		+	+			+									Shade tree.
F. palmeri	+	+	+			+									Used as potted plant (Bonzai) .
F. petiolaris	+	+	+			+									Used as potted plant (Bonzai) .
F. religiosa		+	+			+									Shade tree.
F. retusa (= F. nitida, F. microcarpa)		+	+			+									Shade tree.
F. rubiginosa (=F. australis)		+	+			+									Shade tree.
F. sycomorus	+	+	+			+	+	+					+		Edible fruit. Dirties the ground.
Moringaceae															
Moringa oleifera (=M. pterygosperma)		+	+			+									Deciduous tree. Fruit edible and used for water purification.
M. peregrina (=M. aptera)		+	+			+									Deciduous tree, fruit used for water purification. Requires mild winters.
Myrtaceae															
Eucalyptus angulosa	+				+	+	+	+	+		+				
E. annulata	+				+	+	+	+	+		+			+	Small, low branching tree.
E. astringens	+				+	+	+	+	+		+				
E. brachyphylla	+				+	+	+	+	+		+			+	
E. brockwayi	+				+	+	+	+	+		+				Tall and straight tree. Shiny foliage.

Table 5. *(Continued)*

TREES

Family and Species	**Tolerance to:**				**Uses**										**Remarks**
	D	S	Hs	Ss	A	L	Sc	Sas	Sls	Wb	Pb	Irm	Ff	Dbf	
Myrtaceae (Continued)															
E. burdettiana	+				+	+					+			+	Low branching tree.
E. calcicultrix	+				+	+									
E. camaldulensis (=E. rostrata)	+	+	+		+	+				+	+	+			Large tree. Variable. Used for afforestation. Provenances Broken Hill, No.6980 and No.6788 satisfactory in the N.Negev.
E. campaspe	+				+	+					+			+	Small, low branching tree. Decorative flowers buds. Smooth, reddish brown trunk.
E. cladocalyx (=E. corynocalyx)	+				+	+				+	+			+	Competes with other plants due to spreading roots. Suitable as avenue plant if planted sparsely.
E. clelandii	+				+	+					+				Low tree.
E. dundasii	+				+	+					+				Large erect tree. Shiny foliage.
E. eremophila	+				+	+					+				Low tree.
E. erythrocorys	+				+	+					+			+	Large yellow stamen flower. Brilliant scarlet bud cap. Large and most decorative fruit. Sensitive to soil type.
E. erythronema	+				+	+					+			+	Low tree. Red, pink or white flowers.
E. forrestiana	+				+	+					+			+	Low tree. Red calyx tube. Yellow stamens. Red decorative fruit. Suitable for planting under electricity lines.
E. gardneri	+				+	+					+			+	
E. gillii	+	+	+		+	+					+			+	Small, low branching tree.
E. globulus	+				+	+					+	+		+	Large tree. Grey elliptical leaves. Used for afforestation.
E. gomphocephala	+				+	+					+				Large tree, used for afforestation.
E. gracilis	+				+	+					+			+	Low tree with open wide canopy.
E. hemiphloia	+														
E.kondininensis	+				+	+									
E. kruseana	+	+			+	+					+			+	Low tree, with silver-grey leaves. Used for floral decoration.

Family and Species	Tolerance to:				Uses										Remarks
	D	S	Hs	Ss	A	L	Sc	Sas	Sls	Wb	Pb	Irm	Ff	Dbf	
Myrtaceae (continued)															
E. lansdowneana	+				+	+					+				Small tree, low branching tree. Pink, crimson or mauve flowers.
E. lesouefii	+				+	+									Large tree. Decorative fruit.
E. leucoxylon	+				+	+					+			+	Commercial use of flowering branches of the pink varieties, due to the delicate densely clustered flowers.
E. loxophleba	+				+	+									Large umbrella shaped tree.
E. megacornuta	+				+	+					+				Large canopy. Large flower. Attractive fruit.
E. microcarpa	+				+	+									Low tree, grows in deep soils in arid regions.
E. nutans	+				+	+					+			+	Small tree, low branching. Suitable for floral decoration.
E. occidentalis	+				+	+					+				Large tree, suitable for afforestation. Trunk covered with rough fibruous peeling bark. Grayish-black branches.
E. oleosa	+				+	+					+				Low branching. Umbrella-like canopy. Green shiny leaves.
E. platypus	+				+	+								+	Low tree.Potential for landscaping and decorative branches. Pinkish-brown trunk under the deciduous bark.
E. populnea	+		+		+	+								+	Erect tree with shiny leaves. Commercial use for decorative
E. pruinosa	+	+			+	+					+			+	Low tree. Bluish gray to bluish white leaves.
E. redunca	+				+	+									Suitable for sandy soils.
E. salmonophloia	+				+	+								+	Large tree.Smooth pinky-brown trunk.
E. salubris	+				+	+					+				Large tree.Shiny dark green leaves.Greenish-brown or redish brown trunk.
E. sargentii	+	+	+	+	+	+									Large canopy and short trunk.Tested under seawater irrigation.
E. spathulata	+	+	+		+	+								+	Erect tree. Silver-gray leaves. Commercial use for decorative branches.
E. stoatei	+				+	+								+	Low tree.
E. stricklandii	+				+	+									Large tree. Large yellow flowers.
E. torquata	+	+	+		+	+					+			+	Low wide canopy tree.Impressive white to red flowers. Successful in a broad range of environments. Commercial use of flowering branches.
E. torquata x woodwardii ("Torwood")	+	+	+		+	+					+			+	Large variability in plant shape, foliage and flower colours. Used for decorative branches bearing flowers or flower buds.

Table 5. *(Continued)*

TREES

Family and Species	Tolerance to:				Uses										Remarks
	D	S	Hs	Ss	A	L	Sc	Sas	Sls	Wb	Pb	Irm	Ff	Dbf	
Myrtaceae (continued)															
E. viminalis											+				Grows in deep soils.
E. woodwardii	+				+	+					+			+	Narrow weeping habit. Large impressive flower. Silver grey foliage. Commercial use for decorative branches.
Melaleuca bracteata			+			+					+			+	Commercial use for decorative branches.
M. lanceolata (=M. pubescens)	+	+	+			+	+	+	+		+			+	Wide small tree. Dense dark green foliage.
M. styphelioides						+									Single trunk. Peeling spongy bark.
Oleaceae															
Olea europaea			+			+						+	+		Silvery foliage. Agricultural crop (olive oil and pickled olives). Shaped also as bonzai like potted plant.
Papilionaceae															
Dalbergia sissoo		+	+			+									Deciduous.
Sesbania grandiflora		+	+			+									Sensitive to low temperatures.
Palmae															
Phoenix dactylifera			+										+		Very long feathery leaves. Single non branching. Trunk produces side shoots. Agricultural crop (dates).
Washingtonia filifera			+												Fan shaped leaf. Single trunk.
Pinaceae															
Pinus canariensis	+														Moderately drought tolerant.
P. halepensis	+														As above.
P. pinea	+														As above.
Punicaceae															
Punica granatum		+	+			+							+		Small deciduous tree. Yellow, brown or red edible fruit (pomegranates).
Rhamnaceae															
Zizyphus mauritiana	+	+	+			+					+		+		Loses its leaves after frosts. Edible fruit.
Z. spina-christi	+	+	+			+					+				Loses its leaves after frosts. Spiny. Fleshy, yellow edible fruit.

Family and Species	Tolerance to:				Uses										Remarks
	D	S	Hs	Ss	A	L	Sc	Sas	Sls	Wb	Pb	Irm	Ff	Dbf	
Rosaceae															
Crataegus sinaica	+	+	+			+									Orange-yellow, edible fruit.
Salicaceae															
Populus euphratica		+	+			+									Deciduous. Produces suckers.
Sapindaceae															
Atalaya hemiglauca		+	+			+					+				Small spiny,deciduous tree.
Salvadoraceae															
Salvadora persica	+	+	+			+		+							Edible small fleshy fruit. For warm regions only.
Sterculiaceae															
Brachychiton australe	+					+									
B. discolor	+					+									Deciduous. Flowers appear before leaves.
B. gregorii	+					+									Sold also as potted, bonzai-like plant.
B. populneum (=Sterculia diversifolia)	+					+									Evergreen.
B. rupestris	+					+									Thick stem tree. Sold also as potted, bonzai-like plant.
Tamaricaceae															
Tamarix aphylla (=T. articulata)	+	+	+	+	+	+	+	+	+	+	+	+			Afforestation tree.
T. aphylla (=T.articulata), var.erecta	+	+	+	+	+	+	+	+	+			+			As above
T. gallica	+	+	+	+	+	+	+	+	+			+			As above
T. jordanis	+	+	+	+	+	+	+	+	+						As above. Grows also as a tall shrub.
T sp. "Dead Sea"	+	+	+	+	+	+	+	+	+						An unidentified species from the Dead Sea area.
Zygophyllaceae															
Balanites aegyptiaca	+	+	+			+						+			Deciduous. Spiny. Edible bluish fleshy fruit. Suitable for warm winter regions.
Guaiacum coulteri		+	+			+									
TALL SHRUBS															
Agavaceae															
Yucca filamentosa	+	+	+	+		+									Large yellowish-white inflorescence.
Yucca variegata	+	+	+	+		+									

Table 5. *(Continued)*

TALL SHRUBS

Family and Species	Tolerance to:				Uses										Remarks
	D	S	Hs	Ss	A	L	Sc	Sas	Sls	Wb	Pb	Irm	Rf	Dbf	
Anacardiaceae															
Rhus integrifolia	+	+	+			+		+	+						Odorous. Reddish young growth.
R. crenata	+					+									
Schinus terebinthifolius		+	+			+		+	+						Light red fruit.
Apocynaceae															
Carissa grandiflora		+	+			+							+		Wine red, edible fleshy fruit.
Nerium oleander		+	+	+		+									Cultivars with simple or double flowers of various colors. Also used for flowering pots.
Thevetia peruviana (=T. neriifolia)		+	+	+		+									
Asclepiadaceae															
Calotropis gigantea	+	+	+	+		+									Purple corolla lobes. Fruit diameter up to 20 cm. Large, woolly leaves. Sensitive to frost. Volunteers.
C. procera	+	+	+	+		+									Violet to purple corrolla lobes. Fruit diameter up to 12 cm. Volunteers.
Caesalpiniaceae															
Caesalpinia (=Poinciana) gilliesii		+	+			+									Atractive yellow flowers with long red stamens. Volunteers.
Cassia auriculata		+	+	+		+			+						Large yellow flowers clustered in corymbs.
C. didymobotrya	+	+	+			+									Flowers clustered in long racemes.
C. eremophila	+	+	+			+		+	+						Profuse flowering over a long period in the cold season.
C. grandiflora (=C. laevigata)		+	+			+			+						
Casuarinaceae															
Allocasuarina humilis	+	+	+	+		+	+	+	+					+	Cone-like decorative fruit.
Casuarina lehmaniana	+	+	+	+		+	+	+	+					+	Cone-like decorative fruit.
Chenopodiaceae															
Suaeda monoica	+	+	+	+		+		+	+						
Combretaceae															
Conocarpus erectus		+	+	+	+	+		+	+						Can be developed into a small tree.
Compositae															
Felicia filiformis	+	+	+			+									Yellow flower. Profuse flowering.

Family and Species	Tolerance to:				Uses										Remarks
	D	S	Hs	Ss	A	L	Sc	Sas	Sls	Wb	Pb	Irm	Ff	Dbf	
Elaeagnaceae															
Elaeagnus pungens		+	+	+		+									Under side of leaves silver colored. Red fruit.
Euphorbiaceae															
Euphorbia cotinifolia		+	+			+								+	Deciduous. Bronze-red leaves. Sensitive to cold weather. Requires protection.
E. tirucalli	+	+	+	+		+		+	+						Potential industrial plant (fuel, medicinal products). The sap causes skin burns.
Ricinus communis	+	+	+			+		+	+			+		+	Industrial plant (castor oil). Red leaved types suitable for landscaping. Dark brown capsule. Volunteers.
Loganiaceae															
Buddleia madagascariensis		+	+			+									
Mimosaceae															
Acacia acuminata	+	+	+			+									Yellow spicate inflorescences.
A. aneura	+		+			+		+					+		Slow growth. Silver-grey foliage. Spicate inflorescence. Browsed.
A. bivenosa	+		+			+	+	+	+						Fast growing. Volunteers.
A. caven	+		+			+	+	+	+						
A. cyclops	+		+			+		+	+						Reaches a large diameter. Light green phyllods. Can develop into a small tree.
A. iteaphylla	+				+	+	+	+	+					+	
A. ligulata	+	+	+			+		+	+				+		Reaches a large diameter. Spicate inflorescence. Dark-green linear phyllods.
Acacia pycnantha	+					+									
A. sclerosperma	+	+	+			+		+	+						Reaches a large diameter. Rounded canopy.
A. sp. 106	+					+	+	+	+						Prostrate. Large diameter.
A. victoriae	+	+	+			+		+	+		+		+		Broad variability. Smooth and spiny types. Silver grey leaves. Develops suckers. Volunteers. Browsed.
Dichrostachys glomerata	+	+	+			+									Lower part of inflorescence yellow, top part pink to purple turning white with time. Deciduous. Spiny.
Leucaena leucocephala (=L. glauca)		+	+		+	+							+		Large variability (there are also tree shaped types). Fast growth. Cold sensitive. Deciduous in cold places. Fodder plant. Volunteers. May turn in a pest.

Table 5. *(Continued)*

TALL SHRUBS

Family and Species	Tolerance to:				Uses										Remarks
	D	S	Hs	Ss	A	L	Sc	Sas	Sls	Wb	Pb	Irm	Ff	Dbf	
Myoporaceae															
Eremophila bignoniiflora	+					+									
Myoporum acuminatum		+	+	+		+		+	+						
Myrtaceae															
Callistemon rigidus (C. linearifolius)		+	+			+									Suitable for flowering potted plants. Large spicate inflorescence. Some botanists distinguish between C. rigidus and C. linearifolius.
C. citrinus		+	+			+									Suitable for flowering potted plants. Large spicate inflorescence. Several cultivars as "Harkness" and "Kings Park" (="Kings Park Special".
Eucalyptus grossa	+				+	+								+	Large flower. Reddish operculum. Large shiny leaves with yellow or red central vein.
Leptospernum laevigatum		+	+	+		+		+	+						
Melaleuca acuminata	+		+			+	+	+	+					+	
M. cuticularis		+	+			+									
M. elliptica		+	+			+								+	Suitable for bonzai like potted plants.
M. sp. 29	+	+	+			+	+	+	+						Previously recorded erroneously as M. glaberrima.
M. halmaturorum	+	+	+	+		+		+	+						
M. huegelii	+					+		+	+					+	Dark mauve flower buds. White flowers. Fast growing. Fragile
M. nesophila	+	+	+	+		+		+	+					+	Pink-purple flowers in terminal heads. Yellow-green obovate leaves. Reddish young growth.
M. aff. microphylla				+		+		+	+					+	
M. viminea		+	+			+		+	+					+	Fragrant flowers. Short lived.
Papilionaceae															
Colutea istria	+	+	+			+		+	+				+		Light brown decorative siliqua. Browsed.
Retama raetam	+	+	+	+		+		+	+					+	Flowers in racemes. Mauve calyx. White corolla. Delicate fragrance. Volunteers.
Sophora secundiflora	+	+	+	+		+					+				Slow growth. Impressive fragrant blue flowers in large inflorescences. Large red poisonous seeds.

Family and Species	Tolerance to:				Uses										Remarks
	D	S	Hs	Ss	A	L	Sc	Sas	Sls	Wb	Pb	Irm	Ff	Dbf	
Pittosporaceae															
Pittosporum phillyreoides	+	+	+	+		+					+				Orange capsules revealing red decorative seeds when fruit splits.
Resedaceae															
Ochradenus baccatus	+	+	+			+	+	+	+						Suitable for regions with warm winters.
Sapindaceae															
Dodonaea lobulata	+					+	+	+	+					+	Red to pink capsules.
D. viscosa		+	+			+		+	+					+	Brown capsules. Dirties ground. A red leaved type is suitable for decorative branches.
Tamaricaceae															
Tamarix chinensis		+	+	+		+	+	+	+						Pink flowers. Escaped from cultivation in Israel.
T. nilotica		+	+	+		+	+	+	+						As above.
T. "Mapu" (=ramosissima?)		+	+	+		+	+	+	+						Doubtful indentification. Largely used and commonly known in Israel as "Mapu". Impressive profuse bloom.
Verbenaceae															
Vitex agnus-castus	+	+	+	+		+		+	+						Deciduous.
V. trifolia (=V. trifoliata)		+	+	+		+	+	+	+						Deciduous. Suffers in cold places.
Zygpophyllaceae															
Bulnesia retama	+					+	+	+							Well adapted to sand, rocky ground and river-beds.
SHRUBS															
Agavaceae															
Agave spp.	+	+	+	+		+									Growing under extremely harsh conditions in arid areas. Several species of economic importance e.g. A.Sisalana (fibers).
Anacardiaceae															
Pistacia lentiscus	+	+	+			+	+	+	+						Reddish fruit, turning to black. Yields mastic.
Rhus deptata	+					+									
R. leptodictya	+					+									
Apocynaceae															
Acokanthera spectabilis (=A. oblongifolia)		+	+	+		+	+		+					+	Variegated leaves. Ruby-black fruit. Poisonous.
Carissa grandiflora		+	+			+	+	+	+						Fleshy red wine colored edible fruit.

Table 5. *(Continued)*

SHRUBS

Family and Species	Tolerance to:				Uses										Remarks
	D	S	Hs	Ss	A	L	Sc	Sas	Sls	Wb	Pb	Irm	Ff	Dbf	
Caesalpiniaceae															
Cassia artemisioides	+					+	+	+	+						Pinnate silver-grey leaves.
C. mexicana		+	+	+		+		+	+						Volunteers.
C. nemophila var. coriacea	+					+		+	+				+		Browsed.
Cassia phyllodinea	+					+		+							Yellow fragrant flowers.
C. sturtii	+	+	+	+		+		+	+				+		Well browsed by animals. Volunteers.
Capparaceae															
Capparis sinaica (=C. cartilaginea)		+	+			+									Fleshy, large fruit, red outside and orange inside when ripe.
C. spinosa		+	+			+							+		Purple stamens. Flower buds pickled (capers).
Chenopodiaceae															
Atriplex barclayana		+	+			+							+		Short-lived.
A. canescens	+	+	+	+		+		+	+				+		Golden decorative fruit. Browsed.
A. cinerea	+	+	+	+		+		+	+				+		Many types and shapes. Browsed.
A. halimus	+	+	+	+		+		+	+				+		Browsed.
A. nummularia	+	+	+	+		+		+	+				+		Well browsed. Suitable for fences.
A. polycarpa	+	+	+	+		+		+	+				+		Used for fences. Browsed.
Maireana astrotricha	+	+	+			+	+	+	+				+		Compact grey plant.
M. pyramidata	+	+	+	+		+	+	+	+				+		Bluish-green leaves.
M. (=Kochia) sedifolia	+	+	+	+		+	+	+	+				+	+	Fleshy silver-grey leaves. Excellent for decorative branches.
Suaeda fruticosa		+	+	+		+									Salt marsh plant.
Compositae															
Artemisia monosperma	+	+	+	+		+	+	+	+						Sand dunes plant. Volunteers.
Baccharis aff. sarothroides	+	+	+	+		+		+	+		+			+	
Chrysanthemoides monilifera	+	+	+	+		+	+	+	+						Flowers almost all year round.
Olearia axillaris	+	+	+	+		+	+	+	+						
Cupressaceae															
Juniperus chinensis	+					+	+	+	+						Evergreen conifer. Cold and drought resistant. Grows in various well drained soils. Several varieties and hybrids as "Hetzii", "Gold Coast", "Mint-Julip", pfitzeriana"Aurea", pfitzeriana Glauca, pfitzeriana"Green".

Family and Species	Tolerance to:				Uses										Remarks
	D	S	Hs	Ss	A	L	Sc	Sas	Sls	Wb	Pb	Irm	Ff	Dbf	
Cupressaceae (continued)															
Juniperus virginiana 'Grey-Owl'	+	+	+			+	+	+	+						Evergreen conifer. Cold and drought resistant. Grows in various well drained soils.
Ephedraceae															
Ephedra foemina (=E. campylopoda)	+	+	+	+		+	+	+	+						Red fruit. Volunteers.
Euphorbiaceae															
Acalypha wilkesiana (=A. tricolor)		+	+			+									Reddish, brown or copper colored leaves.
Euphorbia balsamifera	+		+			+									
Jatropha cinerea			+			+									Grows in hot regions.
Fouquieriaceae															
Fouquieria columnaris			+			+									Grows in hot regions.
F. splendens			+			+									Grows in hot regions.
Labiatae															
Rosmarinus officinalis	+					+	+	+	+		+			+	Aromatic. Used for spices, perfumery and medical preparations. The "Blue-Lagoon" variety has dark blue flowers.
Westringia fruticosa(=W.rosmariniformis)	+	+	+	+		+		+	+		+				The type with white is lower than the type with blue flowers.
Liliaceae															
Aloe spp.	+	+	+			+									Several species used in the pharmeceutical and cosmetic industries (e.g. A. vera).
Malvaceae															
Hibiscus rosa-sinensis		+	+			+									
Mimosaceae															
Acacia ashbae	+					+	+	+	+						Grey-green shrub with densely packed linear leaves. Yellow spherical flower heads.
A. maginii	+		+			+									
Myoporaceae															
Eremophila divaricata	+														
E. maculata	+	+	+			+		+			+				Pink and yellow flowering varieties. Tends to crouch. Several c.v. as "Aurea" and "Goondiwindi".
E. racemosa B.G	+					+									
E. subfloccosa	+	+				+		+	+						
Myrtaceae															
Melaleuca armillaris (=M. alba)	+					+	+	+	+					+	Commercial use for decorative branches.

Table 5. *(Continued)*

SHRUBS

Family and Species	Tolerance to:				Uses										Remarks
	D	S	Hs	Ss	A	L	Sc	Sas	Sls	Wb	Pb	Irm	Ff	Dbf	
Myrtaceae (continued)															
Melaleuca coccinea	+	+	+			+	+	+	+					+	Commercial use for decorative branches.
M. decora	+					+								+	As above.
M. decussata	+					+								+	As above.
M. erubescens	+	+	+			+	+	+	+						
M. lanceolata (=M. pubescens) "dwarf"	+					+	+	+	+						A prostrate c.v. with dense foliage suitable for ground cover.
M. wilsonii	+					+	+	+	+						
Myrtus communis						+						+		+	Very fragrant and aromatic. Yields a resinous exudate (mastic). Used for medicinal and perfumery preparations.
Papilionaceae															
Castela sp.		+	+			+									
Polygonaceae															
Calligonum comosum	+	+	+			+	+	+							Flower covered by brown bristles. Deep rooted. Suitable for ground cover.
Proteaceae															
Grevillea oleoides	+					+	+	+	+		+				Reaches a wide diameter.
Hakea rubra	+					+									
H. oleifolia	+					+									
Scrophulariaceae															
Leucophyllum candidum	+	+	+			+	+	+	+						
L. frutescens	+	+	+	+		+	+	+	+						Silver-grey foliage. Drought and salt resistant. Grows in a variety of soils.
Simmondsiaceae															
Simmondsia chinensis (jojoba)	+	+	+	+		+	+	+	+			+			Dioecious. Industrial crop (liquid wax). There are types with ornamental potential.
Solanaceae															
Lycium shawii (=L. arabicum)	+	+	+			+	+	+	+						Red berries. Native in the Middle East. Growing in hot and saline regions.

Family and Species	Tolerance to:				Uses										Remarks
	D	S	Hs	Ss	A	L	Sc	Sas	Sls	Wb	Pb	Irm	Ff	Dbf	
Thymelaeaceae															
Thymelaea hirsuta	+					+	+	+	+						Native in the Middle East. Grows in sandy, rocky and stony soils.Deep roots.
Zygophyllaceae															
Nitraria retusa	+	+	+	+		+	+	+	+						Edible fleshy red fruit. Salt marsh plant. Grows mainly in
Zygophyllum dumosum	+	+	+	+		+	+	+	+						stony slopes.
LOW AND DWARF SHRUBS, HERBACEOUS PLANTS															
Acanthaceae															
Ruellia ciliosa		+	+	+		+		+	+	+					Volunteers.
Amaranthaceae															
Alternanthera bettzickiana		+	+	+		+									Yellow-cream to red leaves.
Apocynaceae															
Catharanthus roseus (=Vinca rosea)		+	+	+		+									There are a var."Alba" and prostrate cultivars.
Nerium oleander "nanum"		+	+	+		+	+	+	+						Used for flowering potted plants.
Asclepiadaceae															
Pergularia (=Daemia) tomentosa	+	+	+			+		+							Grows in warm regions. Felted white shrubs. Follicles covered by prickles.
Caryophyllaceae															
Gymnocarpos decander (=Gymnocarpus fruticosum)	+	+	+			+			+						Grows on stony slopes of basalt, granite and flint rocks.
Chenopodiaceae															
Arthrocnemum macrostachyum	+	+	+	+		+	+	+	+					+	Succulent salt marsh plant.
Atriplex glauca		+	+	+		+	+	+	+						
Enchylaena tomentosa	+	+	+	+		+	+	+	+				+		May turn into an infestant plant. Browsed.
Halimione portulacoides	+	+	+	+		+	+	+	+						
Halosarcia(=Arthrocnemum) halocnemoides		+	+	+		+									Tested as a potential oil source.
H. pergranulata(=Arthrocnemum halocnemoides,var. pergranulata)	+	+	+	+		+	+	+	+						Blue-green salt marsh plant. Succulent. Turns red in winter.

Table 5. *(Continued)*

LOW AND DWARF SHRUBS

Family and Species	Tolerance to:				Uses										Remarks
	D	S	Hs	Ss	A	L	Sc	Sas	Sls	Wb	Pb	Irm	Ff	Dbf	
Chenopodiaceae															
Salicornia europaea		+	+	+		+	+	+	+						Annual. Succulent salt marsh plant. Wide-spread from Mediterranean to Atlantic Ocean.
Sarcocornia fruticosa (=Arthrocnemum fruticosum)	+	+	+	+		+	+	+	+						Succulent erect salt marsh plant.
S. perennis (=Arthrocnemum perenne)	+	+	+	+		+								+	Prostrate with erect branches. Fleshy. Grows in salt marshes.
Compositae															
Achillea fragrantissima	+			+		+									Odorous.
Anthemis maritima	+			+		+		+							
Arctotis stoechadifolia (=A. grandis)		+	+	+		+									
Artemisia sieberi (=A. herba-alba)	+	+	+	+		+							+		Spreads from mountains of Central Asia to North West Africa. Grows in stony habitats. Grey. Aromatic. Used in popular medicine. Browsed.
Aster alpinus		+	+	+		+									Originates from hilly habitats.
Asteriscus graveolens	+	+	+			+									Fleshy.
Borrichia frutescens	+	+	+	+		+	+	+	+						Volunteers.
Calocephalus brownii		+		+		+									Silvery foliage. Short-lived.
Centaurea cineraria		+	+	+		+									
C. cineraria "candidissima"		+	+	+		+									
Encelia farinosa	+	+	+	+		+									Grey foliage.Short-lived.
Eriocephalus africanus	+	+	+	+		+									
Gazania longiscapa		+	+	+		+									Black dot at the base of the corolla.
G. nivea		+	+	+		+									Volunteers.
G. rigens (=G. splendens)		+	+	+		+									Black or brown dot at the base of the flower. Volunteers.
Inula crithmoides	+	+	+	+		+	+	+	+						
Osteospermum pachypteris	+	+	+	+		+							+		Volunteers.Browsed.
O. sinuatum	+	+	+	+		+							+		Volunteers.Browsed.
Otanthus maritimus	+	+		+		+									Seaside plant (Mediterranean and Atlantic Ocean).
Santolina chamaecyparissus		+	+	+		+		+							
Senecio cineraria (=Cineraria maritima)		+	+	+		+									Grey leaves.
Viguiera tomentosa				+											

Family and Species	Tolerance to:				Uses										Remarks
	D	S	Hs	Ss	A	L`	Sc	Sas	Sls	Wb	Pb	Irm	Ff	Dbf	
Crassulaceae															
Sedum sp.		+	+	+		+									An unidentified fleshy sedum species.
Cruciferae															
Moricandia nitens	+	+	+			+									Browsed.
Cyperaceae															
Cyperus alternifolius var. gracilis		+	+	+		+								+	A low cultivar different from the common tall type.Commercial use for decorative branches.
Euphorbiaceae															
Euphorbia antisyphilitica (Candelilla)	+	+	+			+						+			Utilized for extraction of wax.
E. milii (=E. splendens var. prostrata)		+	+	+		+									Various color shades. Used as flowering potted plant. Spiny.
Poinsettia pulcherrima "nana"		+	+	+		+									Top leaves are red, pink or white. Used as potted plant.
Geraniaceae															
Pelargonium capitatum (=P.drummondii, Geranium capitatum)		+	+	+		+									Short lived.
Pelargonium fragrans		+	+	+		+						+			Used in the perfume industry.
Goodeniaceae															
Scaevola crassifolia		+	+	+		+		+	+		+				
S. taccada	+	+	+	+		+		+							Dark-blue fruit.
Gramineae															
Cortaderia selloana (=Gynerium argenteum)		+	+			+								+	Used for dry decorative branches.
Festuca elatior	+	+	+	+		+									
F. glauca (=F. ovina var. glauca)	+	+	+	+		+									Bluish leaves.
Pennisetum asperifolium (=P. setaceum)		+	+	+		+		+	+						Reddish-creamy spikes. Volunteers.
Juncaceae															
Juncus acutus (=J. littoralis)		+	+	+		+								+	Grows often in salty marshes. Potential source of pulp for paper.
J. maritimus		+	+	+		+								+	Coastal marshes. Salt tolerant. Potential source of pulp for paper.
Labiatae															
Coleus spicatus		+	+	+		+		+							Aromatic. Suffers from cold conditions.
Lavandula officinalis (=L. spica, L. vera, L. angustifolia)	+					+						+	+		Aromatic. Used in the perfume and spice indusrtry. Cover plant.
Plectranthus sp.	+					+									
	+					+									Used as a medicinal plant.Aromatic.

Table 5. *(Continued)*

LOW AND DWARF SHRUBS

Family and Species	Tolerance to:				Uses										Remarks
	D	S	Hs	Ss	A	L	Sc	Sas	Sls	Wb	Pb	Irm	Ff	Dbf	
Liliaceae															
Asparagus falcatus		+	+			+		+	+					+	Commercial use for decorative branches.
Papaveraceae															
Glaucium flavum (=G. luteum)		+	+	+		+		+							Yellow flowers without a dark dot at the base of petals.
Papilionaceae															
Crotalaria aegyptiaca	+					+									Yellow corolla with red-striped standard.
Plumbaginaceae															
Armeria pseudo-armeria (=Statice pseudo-armeria)		+	+	+		+									Types with either white or red flowers.
Limonium angustifolium	+	+	+	+		+									Blue or purple-blue corolla. White or bluish calyx. Grows in marshes or near salty springs.
L. axillare	+	+	+	+		+									
L. latifolium	+	+	+	+		+									Purple corolla.White calyx.
L. oleifolium	+	+	+	+		+									Seaside plant.
L. perezii		+	+	+		+									White or yellow corolla.Blue or ruby calyx.
L. pruinosum	+	+	+	+		+									Desert saline soils.Delicate pink flowers.
L. sinuatum	+	+	+	+		+									Yellow-white corolla. Blue, ruby or white calyx.
L. tataricum	+	+	+	+		+									Red corolla.White calyx.
L. vulgare	+	+	+	+		+									Purple-blue corolla. Blue or white calyx.
Portulacaceae															
Portulacaria afra "Malekar"	+	+	+	+		+									Bushy c.v with wide leaves. Potted as bonzai like plant.
Scrophulariaceae															
Kickxia (=Linaria) floribunda	+					+									Volunteers.
Umbelliferae															
Crithmum maritimum		+	+	+		+									Succulent seaside plant, tolerant to seawater spray.
Pituranthos tortuosus	+					+									Aromatic plant used as a spice.
P. triradiatus	+					+									
Zygophyllaceae															
Fagonia arabica	+					+									Covered by glandular hairs.

PROSTRATE PLANTS

Family and Species	Tolerance to:				Uses										Remarks
	D	S	Hs	Ss	A	L	Sc	Sas	Sls	Wb	Pb	Irm	Ff	Dbf	
Aizoaceae															
Aptenia cordifolia		+	+	+		+			+						Succulent.Suitable for ground cover.
A. lancifolia		+	+	+		+			+						As above
Carpobrotus acinaciformis (=Mesembryanthemum acinaciforme)		+	+	+		+		+	+						Ground cover. Succulent. Tolerant to seawater spray.
C. edulis (=Mesembryanthemum edule)		+	+	+		+		+	+						As above.
Drosanthemum hispidum		+	+	+		+		+	+						Ground cover succulent. Tolerant to seawater spray.
Galenia secunda		+	+	+		+		+	+						Ground cover succulent. Tolerant to seawater spray. Volunteers.
Glottiphyllum aff. depressum		+	+	+		+									Flat, fleshy, round plant. Suitable for rockeries.
Lampranthus aurantiacus		+	+	+		+		+	+						Succulent. Ground cover.
L. blandus (=L. albus)		+	+	+		+		+	+						Succulent. Ground cover. Tolerant to seawater spray.
Lampranthus multiradiatus (=Mesembryanthemum multiradiatum)		+		+		+		+							Succulent.Ground cover.
L. roseus (=Mesembryanthemum roseum)		+	+	+		+		+							As above.
Sesuvium portulacastrum		+	+	+		+		+							Succulent ground cover. Tolerant to seawater spray.
S. verrucosum		+	+	+		+		+	+						Succulent ground cover.
S. sp. *"Green"*		+	+	+		+		+	+						Unidentified species. Succulent green leaves. Ground cover. Tolerant to seawater spray.
S. sp. *"Red"*		+	+	+		+									Unidentified species. Reddish succulent leaves. Suffers from cold weather. Rockery plant.
Amaranthaceae															
Iresine rhizomatosa		+	+	+		+		+	+						Ground cover. Tolerant to seawater spray. Colorful reddish leaves.
Batidaceae															
Batis maritima		+	+	+		+		+	+						Grows in seaside salt marshes. Ground cover. Tolerant to seawater spray.
Chenopodiaceae															
Atriplex semibaccata	+	+	+	+		+		+	+				+		Browsed.Volunteers.May turn in a weed.
Suaeda vera		+	+	+		+		+	+						Ground cover. Tolerant to seawater spray. Bluish-green leaves.
Compositae															
Kleinia repens (=Senecio serpens)		+	+	+		+									Succulent.Cylindrical grey-bluish leaves.
Wedelia paludosa		+	+	+		+									Ground cover. Tolerant to seawater spray. Prefers places with warm winters.

Table 5. *(Continued)*

PROSTRATE PLANTS

Family and Species	Tolerance to:				Uses										Remarks
	D	S	Hs	Ss	A	L	Sc	Sas	Sls	Wb	Pb	Irm	Ff	Dbf	
Convolvulaceae															
Ipomoea pes-caprae (=I. biloba)		+	+	+		+		+	+						Cold sensitive. Grows in tropical coasts.
I. stolonifera		+	+	+		+		+	+						Succulent seaside plant. Tolerant to seawater spray.
Frankeniaceae															
Frankenia pauciflora		+	+	+		+									Grows on sandy salty soils.
Graminaeae															
Cynodon dactylon (=Capriola dactylon)		+	+	+		+	+	+					+		Bermuda. Several varieties tolerant to high salinity and salt spray (Suwanee, Coastal, etc.). Cultivated for fodder, pasture and lawns.
Distichlis spicata		+	+	+		+							+		
Paspalum distichum(=P. vaginatum)		+	+	+		+	+	+					+		Salt and salty spray tolerant. Cultivated for fodder and lawns.
Labiatae															
Rosmarinus officinalis var.prostratus	+					+	+	+	+		+				
Mimosaceae															
Acacia redolens	+					+	+	+	+						The "Ein Vered" c.v excels for the decorative foliage.
Myoporaceae															
Eremophila glabra		+	+			+		+	+						Several c. vars of different flower color. The orange c.v is more drought tolerant.
E. serpens		+				+		+	+						Ruby colored new growth. Short lived.
E. subteretifolia	+					+		+	+						
Myoporum debile		+	+	+		+									Short lived.
Myoporaceae															
M. ellipticum	+	+	+	+		+		+	+		+				Tolerant to seawater spray.
M. floribundum		+	+	+		+		+	+		+				As above.
M. parvifolium		+	+	+		+		+	+		+				Tolerant to seawater spray. Several c.vars ("broad", "fine", "purple", "mid leaf") differing in shape or color of leaves.
Nyctaginaceae															
Commicarpus plumbagineus (= C. africanus)	+	+	+	+		+		+	+						Tolerant to seawater spray.
Onagraceae															
Oenothera drummondii	+	+	+	+		+	+	+							Suitable as seaside plant. Tolerant to seawater spray.

Family and Species	Tolerance to:				Uses										Remarks
	D	S	Hs	Ss	A	L	Sc	Sas	Sls	Wb	Pb	Irm	Ff	Dbf	
Papilionaceae															
Lotus berthelotii (=*L. peliorhynchus, L. pelyorensis*)		+		+		+		+							Short leaved. Red or orange flowers.
L. creticus	+	+	+	+		+		+	+						Suitable for seaside landscaping.
L. cytisoides	+	+	+	+		+		+							Seaside sand dune plant.
L. lanuginosus	+	+	+	+		+		+							Grows in hot deserts. Red corolla.
Portulacaceae															
Portulacaria afra	+	+	+	+		+		+	+						Prostrate and erect varieties. Green or coloured leaves. Late blooming stage.
Verbenaceae															
Lippia sp. *lanceolata*		+	+	+		+	+	+	+		+				Tolerant to seawater spray. Large leaves. 20-40 cm high.
Phyla (=Lippia) nodiflora		+	+	+		+	+	+	+		+				Resists tramping. Tolerant to seawater spray.
Vitex ovata		+	+	+		+		+	+						Declines after a short time.

REFERENCES

1. Aronson, J. (1989). *Haloph, a data base of salt tolerant plants of the World*, Office of Arid Lands Studies, The University of Arizona, Tucson, Arizona, U.S.A. 75 pp.
2. Ben-Dov, Y., Forti, M. and Pasternak, D. (1993). *Drought-and salt-tolerant plants for water-saving landscap*ing (In Hebrew). The Institutes for Applied Research, Ben-Gurion University of the Negev, Beer-Sheva, Israel. 110 pp.
3. Benjamin, R.W., Lavie, Y., Forti, M., Barkai, D., Yonatan, R., and Hefetz, Y. (1995). Annual regrowth and edible biomass of two species of *Atriplex* and *Cassia sturtii* after browsing, *J. Arid Environ.*, 29: 63-84.
4. Benzioni, A. 1995. *Jojoba domestication and commercialization in Israel.* In: Janick, J. (Ed.), Horticultural Reviews, Vol. 17, Wiley, New York, pp. 231-264.
5. Bitan, A. and Rubin, S. (1991). *Climatic Atlas of Israel for physical and environmental planning and design* (In Hebrew). Israel Meteorological Service, Ministry of Energy and Infrastructure, Tel Aviv University Publishing House, Tel-Aviv 244 pp.
6. Forti, M. (1967). *Plant introduction, a biennial report for the years 1965 and 1966.* Negev Institute for Arid Zone Research, Beer-Sheva, Israel, 59 pp.
7. Pasternak, D., Aronson, J., Ben-Dov, Y., Forti, M., Mendlinger, S., Nerd, A. and Sitton, D. (1986). Development of new arid zone crops for the Negev Desert of Israel. *J. Arid Environ.* 11: 37-59.
8. Pauker, R. (1996). *Water-wise landscaping,* Proceedings of the 7th International Conference on water and irrigation, Tel-Aviv, Israel. pp 308-315.

THE SCREENING OF MULTIPURPOSE TREE SPECIES FOR AGROFORESTRY PRACTICES IN THE SEMIARID ECOZONE OF GHANA

Joseph Cobbina
Forestry Research Institute of Ghana, Agroforestry Section, University P.O.Box 63, Kumasi, Ghana, Tel: 233-51-60123/2, Fax: 233-51-60121, email: forig@africaonline.com.gh

1. INTRODUCTION

The semiarid ecozone of Ghana forms part of the extensive West African savanna which covers an area of 3.5 x 10^6 km^2 and stretches for about 4,000 km from the Guinea-Senegal cost in the west to Chad and Central Africa Republic in the east (Jones and Wild, 1975). In general, the amount of rainfall received decreases as one moves from south to north and also trees become smaller and more widely separated from one another.

Most soils in the zone are sandy and thus have poor moisture holding capacity. Soil structure is weak. Consequently clods are easily broken down to their constituent particles which fills soil pores thus forming a crust which restricts infiltration, aeration and seed emergence, and increases runoff. Because of the climate and the sparce vegetative cover, soil organic matter contents are low. Most soils have low cation exchange capacity. Total nitrogen contents of soils are low and there is also acute phosphorus deficiency (Adu, 1972). Due to the widespread depletion of trees, many of the rural people have resorted to burning sorghum and millet straw and also cowdung to provide energy for domestic use and thereby exacerbating the soil fertility problem.

Although a wide range of tree species are reported to be adapted to the environmental conditions of semiarid ecological zones of the world (NAS,

Combating Desertification with Plants, edited by Dov Pasternak and Arnold Schlissel.
Kluwer Academic/Plenum Publishers, New York, 2001.

1980), no systematic study has ever been undertaken in Ghana to determine which species possess all of the following characteristics of an ideotype agroforestry multipurpose tree species (Owino, 1992) :(i) ability to grow rapidly at early stages, (ii) ability to regenerate readily after pruning, (iii) ability to produce large quantity of leafy and woody biomass. But in view of the widespread degradation of soils in the ecozone and the acute shortage of fuelwood, there is the urgent need to screen to find best-bet tree species to be used for diverse agroforestry practices.

2. MATERIAL AND METHODS

2.1 Study Site

The study was carried out in the Upper Tamne Forest Reserve near Bawku, Upper East Region of Ghana (11° 10'N; 0° 15' W; 260m above sea level). The vegetation consists of short deciduous trees often widely spaced and a ground flora comprising different grass species of varying heights. The commonest tree species include *Acacia spp., Adansonia digitata, Vitellaria paradoxa (syn. Butyrospermum parkii)* and *Parkia biglobosa (syn. Parkia clapertoniana)*(Adu, 1972).

The topography is gently undulating to gently rolling. Soils are derived from parent rocks known as Voltaian, Birrimian and also very old granite. The soils are red and brown sandy loam classified as Luvisol (UNESCO/FAO Soil Legend). Surface soils have slightly acidic pH values and total nitrogen as well as extractable phosphorus contents are very low. The area has a single rainy season and the average total annual precipitation ranges from 900mm to 1000mm. The mean monthly temperatures vary from 25°C to 34°C.

2.2 Experimental Design

Sixteen weeks old seedlings of eight multipurpose tree species (*Table 1)* were transplanted in field plots arranged as randomized complete block with three replications in July, 1990. Two tree planting arrangements namely, dispersed and alley cropping or hedgerow intercropping planting arrangements, were used.

In the dispersed tree planting system on a unit plot (5m x 6m) there were two rows of trees 6m long spaced 2.5m apart with a within row spacing of 2m. On the other hand, in the alley cropping planting system, each unit plot (4.8m x 6m) contained two rows of trees also 6m long at spacing of 2.4m apart within row spacing of 0.5m.

At four months and again 12 months after transplanting the seedlings a complete assessment of tree survival was made on all the tree species. The height of the leader shoots of all trees was also measured at monthly intervals starting from the 4th month after transplanting to the 12th month. Then at 24 months after transplanting, the primary shoot of all the trees in the alley cropping system was cut back to a height of 0.6m above ground level and biomass (leaf, stem) yield was assessed from oven-dried, weighed samples. The coppice shoots were cut back to 0.6m height above ground level after 10 months of continuous, uninterrupted growth and biomass yield was estimated. The number of coppice shoots and length of leader shoot were measured at 3 and 9 months after cutting the trees.

Data accumulated from all measurements were analysed as randomized block design. Following significant F test, the protected Least Significant Difference (LSD) procedure (Steel and Torrie, 1980) was used to compare the treatment means.

3. RESULTS AND DISCUSSION

3.1 Tree Survival and Height

In general survival of the trees at 4 and 12 MAP was good, ranging from 72% to 100% (*Table 1*).

Table 1: Survival rate (%) of tree species at 4 and 12 months after transplanting (MAP) seedlings

Species	Dispersed planting		Alley Cropping	
	4 MAP	12 MAP	4 MAP	12 MAP
Acacia auriculformis	91.7	88.3	88.4	87.2
Acacia senegal	91.7	87.5		
Albizia lebbeck	96.7	95.8	93.6	92.3
Faidherbia albida	92.5	80.8		
Gliricidia sepium	100.0	100.0	80.8	80.8
Leucaena leucocephala	76.7	72.5	91.0	85.9
Prosopis juliflora	95.8	95.8		
Senna siamea	100.0	100.0	96.2	94.9
Mean	93.1	90.1	90.0	80.2
LSD (0.05)	17.1	18.2	11.3	11.2

By and large trees in dispersed planting arrangement had higher survival rate than those in the alley cropping planting arrangement with *A. lebbeck*,

G. sepium and *S. siamea* clearly showing that trend. It is possible that intra-specific competition in the narrowly spaced (0.5m) alley cropping planting arrangement was responsible for the relatively lower survival rates.

The trend of height growth from the 4th month to the 12th month after transplanting is shown in *Figure 1*.

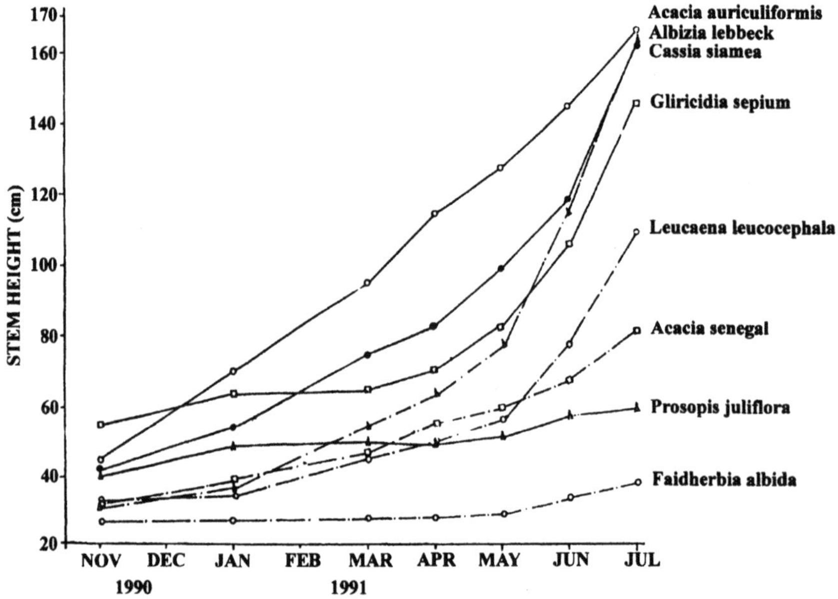

Figure 1. Seedling growth rates between the 4th andl2th month after transplanting.

All the species showed slow growth rate initially. However, by the 9th month after transplanting, the height growth rates of some of the species had increased dramatically and continued to show sharp increases thereafter. At 12 months after transplanting, *A auriculiformis, A. lebbeck,. G. sepium* and *S. siamea* had the most vigorous shoot growth rates. The outstanding performance of *A.auriculiformis* in the semiarid ecological zone is of great significance since the species is referred to as a humid ecozone tree by NAS (1980). In a study in a humid zone of India (Mathew et al., 1992), *A. auriculiformis* exhibited greater vigorous growth than *L. leucocephala* underscoring the species superior growth vigour in diverse ecological zones.

3.2 Biomass Production

Leaf and stem dry matter yields of the primary growth were significantly greater (P<0.05) for *A. auriculiformis* and *S. siamea* than *for A. lebbeck, G. sepium* and *L. leucocephala* (*Table 2*).

Table 2. Biomass yield (kg DM/tree) of primary and 10 months old coppice growth on five multipurpose trees planted in alley cropping system

Species	Primary growth (24 MAP)		10 months coppice growth	
	Leaf	Stem	Leaf	Stem
Acacia auriculiformis	0.412	0.858	0.468	1.238
Albizia lebbeck	0.132	0.376	0.106	0.655
Gliricidia sepium	0.188	0.364	0.242	0.868
Leucaena leucocephala	0.055	0.161	0.022	0.206
Senna siamea	0.508	0.915	0.422	1.404
Mean	0.259	0.535	0.252	0.874
LSD (0.05)	0.105	0.376	0.159	0.548

The biomass (leaf and stem) production for the 10 months old coppice regrowth also followed similar trend as for the primary growth (*Table 2*). *A auriculiformis and S. siamea* were very productive in coppice regrowth and significantly ($P < 0.05$) out-yielded *A. lebbeck, G. sepium* and *L. leucocephala* in leaf and stem dry matter. *A. lebbeck* and *S. sepium*, in turn, significantly out-performed *L. leucocephala* in both stem and leaf dry matter yields.

The poor performance of *L. leucocephala* is an indication that it is not adapted to the droughty conditions of the semiarid confirming the report (NAS, 1980) that it is a humid zone species. On the other hand, the outstanding performance of *A. auriculiformis* might suggest that it is reasonably adapted to the semiarid climatic conditions since it is also referred to as humid zone species (NAS, 1980). However, the slow litter mineralization of pure stands of *A. auriculiformis* which may probably be due to the thick leathery consistence and high tanin content of the leaves(Drechsel at al., 1991), preclude its use in alley cropping with the aim to improve soil fertility and organic matter content. Instead *A. auriculiformis* may probably be more suitable for woodlots and protective cover for topsoil of badly degraded lands and as mulching material to conserve moisture.

3.3 Coppice Shoot Growth

The coppice shoot length measured at 3 and 9 months after cutting back to 0.6m above ground level in June showed significant ($P<0.05$) differences among the species evaluated (*Table 3*).

Table 3: Length of coppice shoot (cm) of five multipurposes trees planted in alley cropping system

Species	Age of coppice growth (months after cutting)		
	3	9	Mean
Acacia auriculiformis	164.7	278.8	221.6
Albizia lebbeck	224.3	268.6	246.5
Gliricidia sepium	210.0	250.2	230.1
Leucaena leucocephala	205.7	218.9	212.3
Senna siamea	274.3	364.5	319.4
Mean	215.7	276.2	
LSD (0.05)	31.0	43.8	

On the whole over a 6-month period *A. auriculiformis* and *S. siamea* grew faster (70% and 33% respectively) than *A. lebbeck* and *G., sepium* (20%) and *L. leucocephala* (6%). Generally the growth rates obtained during the first 3 months after pruning which coincided with rainy season were seven times higher than those obtained during a period of 6 months in the dry season (data not presented). It is clear that even during the dry season *A. auriculiformis* and *S. siamea* had the highest growth rates whilst *L. leucocephala* had the least. *A. lebbeck* and *G. sepium* attained comparable growth rates and were intermediate. It can be inferred from these results that *A. auriculiformis* and *S. siamea* are species well adapted to the long dry season of the semiarid ecozone.

The number of coppice shoots counted on the stumps at 3 and 9 months after pruning also differed significantly ($P<0.05$) among the species (*Table 4*). Generally, the number of coppice shoots decreased from large numbers at 3 months after cutting to relatively smaller numbers 9 months after cutting. At 3 months after cutting, *S. siamea, G. sepium* and *A. auriculiformis* had, respectively, 13, 17 and 21 coppice shoots.

These results indicate that these species would cast intense shade on associated food crop if left uncut in arboreal forms. *A. lebbeck* is an exception since the species always maintained only few (2-3) coppice shoots. Besides the fewer number of shoots, *A. lebbeck* coppice shoots often grew almost vertical at a wide angle (>45°) from ground level. In contrast *A. auriculiformis, G. sepium* and *S. siamea* coppice shoots grew at an angle of 45° or less. It appears *A. lebbeck* will be suitable for a tree/food crop intercropped system whereby the coppice shoot will be left in arboreal form to be harvested after a longer period of growth (more than 9 months) for firewood or used as support for climbing crops.

Table 4: Number of coppice shoots on stumps of five multipurpose trees planted in alley cropping system

Species	Age of coppice growth (months after cutting)		
	3	9	Mean
Acacia auriculiformis	21.3	7.8	14.6
Albizia lebbeck	3.0	2.1	2.6
Gliricidia sepium	16.7	11.2	13.9
Leucaena leucocephala	8.7	3.9	6.3
Senna siamea	12.7	4.9	8.8
Mean	12.5	5.9	
LSD (0.05)	3.2	1.6	

4. CONCLUSION

In conclusion, the results of the study has shown that *S. siamea* is a promising species for alley cropping in semiarid zone of Ghana in view of the species rapid initial growth and large quantity of leaf biomass production. In addition, *S. siamea* has immense potential to be used as mulching material and to recycle nutrients. *A auriculformis* also has some potential; however, the slow decomposition rate of the leaf litter makes the species suited to only woodlot agroforestry and as protective cover for soil. *A auriculformis* can also be used for live fencing since the foliage is not palatable. *A. lebbeck* is yet another promising species which would be especially suitable for a tree and food crop intercrop system whereby the coppice shoots are left in arboreal form to be cut after several months of growth for fuelwood.

REFERENCES

1. Adu, S.V. (1972) *Eroded savanna soils of the Navrongo-Bawku area, northern Ghana.* Ghana J. Agric. Sci. 5: 3-12.

2. Drechsel, P, Glaser, B. and Zech, W. (1991) *Effect of four multipurpose tree species on soil amelioration during tree fallow in Central Togo*. Agroforestry Systems 16: 193-202.
3. Drechsel, P. and Zech, W. (1991) *Foliar nutrient levels of broad-leaved tropical trees: A tabular review*. Plant and Soil 131: 29-46.
4. Jones, M.J. and Wild, A. (1975) *Soils of West African savanna*, Technical Communication No. 55. Commonwealth Agricultural Bureau, England, pp. 246.
5. Mathew, T., Mohan Kumar, B., Swresh Babu, K.V. and Umamaheswaran, K. (1992) *Comparative performance of four multipurpose trees associated with four grass species in the humid regions of Southern India*. Agroforestry Systems 17: 205-218.
6. NAS (National Academy of Sciences) (1980) *Firewood crops: Shrub and tree species for energy production*. National Academy of Sciences, Washington, D.C., U.S.A. pp. 237.
7. Owino, F. (1992) *Multipurpose tree screening and evaluation*. In: Tripathi, B.R. and Psychas, P.J. eds., The AFNETA Alley Farming Training Manual, Volume 1: Core Course in Alley Farming. Alley Farming Network for Tropical Africca, IITA, Ibadan, Nigeria.
8. Steel, R.G.D. and Torrie, J.H. (1980) *Principles and Procedures of Statistics: A Biometrical Approach*. 2nd edition, McGraw-Hill Book Co., New York. pp. 633.

RESULTS OF 50-YEAR STUDIES ON AFFORESTATION AS RELATED TO THE PROBLEM OF DESERTIFICATION CONTROL IN CISCASPIAN LOWLAND IN RUSSIA

M.L. Sizemskaya, A.V. Bykov, N.U. Kulakova, G.V. Lindeman, I.N. Olovjannikova, M.K. Sapanov, N.G. Senkevich, T.A. Sokolova, T.S. Vsevolodova-Perel
Institute of Forest Research of Russian Academy of Sciences, Uspenskoe, Moscow region, 143030 Russia, Tel/Fax: 7-095-5616590, email: root@ilan.msk.su

This work contains results of 50-years research on Dzhanybek Forestry Station (the part of the Institute of Forest Research of RAS) located in the Caspian lowland within the Trans-Volga clay semidesert, 30 km to the north of the Elton lake, on the border of Russia and Kazakhstan. Natural conditions are characteristic for the region in the western part of Volga-Ural interfluve. The local climate is arid and continental. The territory represents a closed basin plain. Solonetz soil complex and vegetation are predominant. Underground waters lie on the depth from 4 to 6 m. The capacity of capillary fringe is 3 m. Two presented ecotops differ in main parameters limiting the ability of plants to live: mineralization of underground waters, salinity and fertility of soils. The solonchakous solonetzes and ligth-chestnut soils determine one of them, while non-saline meadow chestnut soils with the fresh water lenses in depressions are common for the second one.

The technology of afforestation without irrigation was developed on the Dzhanybek station. Afforestation is possible only after reclamation of solonetz: its dealcalization and desalinisation by means of deep (50 cm) ploughing which destroys the solonetz horizon and moves gypsum from the

Combating Desertification with Plants, edited by Dov Pasternak and Arnold Schlissel.
Kluwer Academic/Plenum Publishers, New York, 2001.

lower strata into the upper soil layer. Desalinisation is caused by additional humidification due to snow accumulation by shelter belts formed from high-stem crops in the first years and by elm tree strips in succeeding years. These belts and strips provide the flushing of soils and the removing of salts from the top soil layer.

The necessary conditions for life duration of afforestation are the large feeding area of trees and the possibility of snow accumulation. Therefore we have to plant one or two strips only and to cultivate them over the course of their life. Moreover there is a need to maintain space between rows in the state of black fallow. The fulfilment of these conditions allows elm strips to be able to work up to their age of 45 years at least.

In afforestation on the solonetz soil complex the nonpercolative water regime is replaced by periodical flushing regime owing additional snow accumulation, in doing so salts are washed and the specific soil profile consisting of homogeneous desalinised ploughing horizon and alluvial-salt horizons begins to be formed. In the initial stage of reclamation the redistribution of soluble salts proceeds within the soil profile only.

As cereals, so perennial herbs can be grown on ameliorated lands. Their productivity is 2-2,5 times higher, in average, comparing with non-forested fields.

Much more favorable opportunities for afforestation are there in the local depressions where it is possible to use the wider list of tree species. If the choice of species with high reproductive and competitive ability is successful, there is no need for soil maintenance providing the sufficient vitality and life duration of afforestation up to 100-years age.

From 1954 more than 200 species of trees and shrubs from different continents and geographical zones are being tested in the mesodepressions on the meadow-chestnut soils, where underground water is available to the plants. In spite of hot and dry summer and severe winter conditions (thin snow covering) many tested species could adapt to the new environment. Some of them even grow well, bring fruit and viable seeds and regenerate.

The mesodepressional environment is favourable for the species of the broad-leafed forest zone and of catchments and depressions of forest-steppe zone origin (for example, *Quercus robur L., Betula pendula Roth., Acer platanoides L., A. saccharinum L.,Tilia cordata Mill., Populus alba L., Malus sylvestris Mill., Pyrus communis L., Sorbus aucuparia L.)*,. The best ecotypes are shade- and semi-shade-proof species with mesoxerophitic, mesophitic, and microtermic properties. The species of the steppe zone developed best of all in the reclaimed solonetz complex (*Ulmus pumila L., Ribes aureum Pursh., Berberis vulgaris L., Lonicera tatarica L., Syringa vulgaris L., Elaeagnus oxycarpa Schlecht., Amelanchier spicata (Lam.) C. Koch., Rosa sp., Cotinus coggygria Scop., Tamarix ramosissima Ledeb.*).

Mainly they are light-demanding shrubs with xerophitic, oligotrophic and mezotermic properties, originating from steppe depressions and their slopes.

As the afforestation in depressions grow, conditions favorable for forest species are induced. Soil acidification takes place, forest fauna and mycoflora emerge, forest-like phytocenoses with forest litter and shrub undergrowth are being formed.

Created afforestations have a significant beneficial effect on environment during their long functioning.

Creation of small tree-shrub groups in shallow depressions and modest-sized stands in larger ones disturbs the course of natural processes inherent in these regions to the least degree. It complies with requirements of adaptive afforestation based on the principle of rational nature management and allows to use the potential fertility of arid regions to the most extent.

We suggest the conceptual model of such landscape which meets these conditions. The next step is the realization of this model and its experimental verification.

The unique landscape objects situated in the Dzhanybek Forest Station alongside with the obtained scientific results are instructive for a number of arid and semiarid regions which are plagued with the same ecological problems as those of the Caspian Planes. The experience of afforestation at this station is especially important considering the menace of global desertification due to the greenhouse effect.

CITRULLUS COLOCYNTHIS, AN ISRAELI DESERT PLANT, AS A POTENTIAL SOURCE OF EDIBLE OIL

Dan Schafferman, Ella Shabelsky, and Zohara Yaniv
Dept. of Genetic Resources and Seed Research, A.R.O., The Volcani Center, Bet Dagan 50250, Israel. Tel: 972-3-9683201/2, Fax: 972-3-968 966 5327, email: vayaniv@volcani.agri.gov.il

1. INTRODUCTION

Citrullus colocynthis (L.) Schrad. (wild gourd) is a desert plant of the Cucurbitaceae, naturally adapted to arid environments. It was known in biblical times as a source of seed oil and its fruits were used as an efficient laxative.

Seed samples have been collected throughout the Negev, Arava and Sinai Deserts in Israel and kept in the Israeli Gene Bank. The aim of the present study was to evaluate this gene pool for chemical and agronomical characters, and to test the seeds as candidates for a potential new oilseed crop in Israel - a crop adapted to arid zones.

Botanical and ethnobotanical features (1,2,4) are listed below.

1.1 Botany

- Common names: colocynth, wild gourd, bitter-apple
- Non hardy, herbaceous vine, branched from base.
- Origin: Tropical Asia and Africa, widely distributed in the Saharo-Arabian region
- Solitary pale yellow blooms.

Combating Desertification with Plants, edited by Dov Pasternak and Arnold Schlissel.
Kluwer Academic/Plenum Publishers, New York, 2001.

- 15-30 round fruits on plant, 3-4 inch in diameter, green all with undulate yellow stripes yellow all over when dry.
- Small seeds (1/4 inch or less in length) smooth and brownish when ripe.
- In Israel, it occurs in sandy soils and wadiis and flowers between May and August.

1.2 Ethnobotany and Medicinal Uses

- During biblical times, fruits were gathered and considered as a deadly poison.
- Fruits are widely used medicinally, especially for stomach pains.
- The pulp is a drastic hydragogue, cathartic and laxative (in the early 20th century fruits were exported as a laxative from the Gaza Strip to Europe).
- Seeds are edible when ground and provide a rude bread for the desert Bedouins.
- Seeds have a high oil content (17-19%); in ancient times it was among the oils permited to be used for candle light.

2. MATERIALS AND METHODS

2.1 Seed Collection

Seeds of wild *Citrullus colocynthis* were collected during 1983/84 throughout the Negev, Arava and Sinai Deserts in Israel.

2.2 Field Observation

Twenty-eight accessions of *Citrullus colocynthis* were studied in 1995 at the Bet Dagan Experimental Farm. Fruits were harvested according to maturation of each accession, and the seeds were analyzed for their fatty acid composition. (3)

2.3 Field Evaluation

Based on the field-observation trial conducted in the 1995 season, four accessions were selected and tested at the Bet Dagan Experimental Farm during the 1996 season..

Data regarding the agronomical and chemical parameters were collected and evaluated during the growing season.

2.4 Statistical Analysis

All data were analysed by means of procedures of the SAS package (SAS Institute 1985). Statistical differences were calculated as Least Significance Difference (LSD) and evaluated with Duncan Multiple Range Test (DMRT) at p=0.05.

3. RESULTS AND DISCUSSION

3.1 Evaluation of C. colocynthis Wild Accessions

The results of the chemical analyses are summarized in *Table 1*. The predominant fatty acid of the seed oil was linoleic acid (C18:2) (67-73%).

The fatty acid composition of the seed oil of *C. colocynthis* is very similar to that of safflower oil - thus, rendering its use as valuable edible oil.

Table 1. Chemical variability in seed oil of 28 accessions of *Citrullus colocynthis* collected in Israel and grown at Bet Dagan experimental farm during 1995.

Fatty acids	% of total oil	
	Mean	Range
Palmitic (C16:0)	10.1	8.6-12.0
Stearic (C18:0)	6.7	5.2-8.2
Oleic (C18:1)	13.1	10.1-16.0
Linoleic (C18:2)	70.1	67.0-73.0

3.2 Field Evaluation of Four Selected Lines

Table 2 presents agronomical data collected during the cultivation of four selected lines of C. colocynthis in the 1996 season.

The number of fruits harvested per plot (10 m^2) varied within lines from 112 to 129, with Line 4 being the highest. Fruit size varied from 5.7 cm (Line 1) to 6.3 cm (Line 2) in diameter.

Fruits of C. colocynthis contain a large number of seeds. Numbers varied, in lines, between 291 and 404 seeds per fruit; 1000-seed weight was the same in Lines 2, 3 and 4 (about 44 g) while Line 1 had significantly smaller seeds (40 g). Lines 2 and 4 excelled in seed yield (2.1 kg/10 m^2) while Line 1 yielded only (1.5 kg/10 m^2).

Table 2. Agronomic data of *Citrullus colocynthis* lines evaluated in Bet Dagan during the 1996 season.

Line	Fruit no. on 10 m^2	Fruit size (diameter) (cm)	Seed no. per fruit	1000-seed weight (gr)	Seed yield (gr 10 m^{-2})
1. IN 34250	127.3 a	5.70 b	291 b	40.0 b	1,479 c
2. IN 34256	120.6 b	6.32 a	396 a	43.4 a	2,107 a
3. IN 34262	112.5 c	6.18 a	404 a	44.0 a	1,893 b
4. IN 34267	128.8 a	6.12 a	370 a	43.6 a	2,143 a
LSD ($p<.05$)	7.1	0.28	68	1.44	208

Statistical differences are given by Least Significance Difference (LSD) and presented by Duncan's Multiple Range Test (DMRT) at P<.05.

Table 3 presents the results of chemical analyses of oil content and fatty acids composition of four selected lines of C. colocynthis in the 1996 season. The two major fatty acids in seedoil of C. colocynthis are oleic acid (C18:1) and linoleic acid (C18:2). This oil contains only traces of (C18:3) linolenic acid. Thus could provide a good source of edible oil.

Table 3. Chemical data of *Citrullus colocynthis* lines evaluated in Bet Dagan during the 1996 season.

Line	Oil content *	Oil yield (calculated)	Oleic FA content	Linoleic FA content	Oleic + Linoleic FA yield
	(%)	(l ha^{-1})	(%)	(%)	(l ha^{-1})
1. IN 34250	17.1 c	253 b	11.7 c	70.9 a	209 b
2. IN 34256	19.0 ab	400 a	15.0 a	66.7 c	327 a
3. IN 34262	19.5 a	369 a	14.2 b	68.2 b	304 a
4. IN 34267	18.5 b	396 a	14.3 b	67.9 b	325 a
LSD ($p<.05$)	0.68	103	0.66	0.75	89

* Oil content is based on dry weight of the seeds.
Statistical differences are given by Least Significance Difference (LSD) and presented by Duncan's Multiple Range Test (DMRT) at P<.05.

3.3 The Effect of The Ripening Period on Yield Parameters

Seed yield - Fig. 1. Seed yield was maximal in the first month of ripening and declined drastically until the end of the ripening period. In addition, differences among lines were observed, indicating biodiversity.

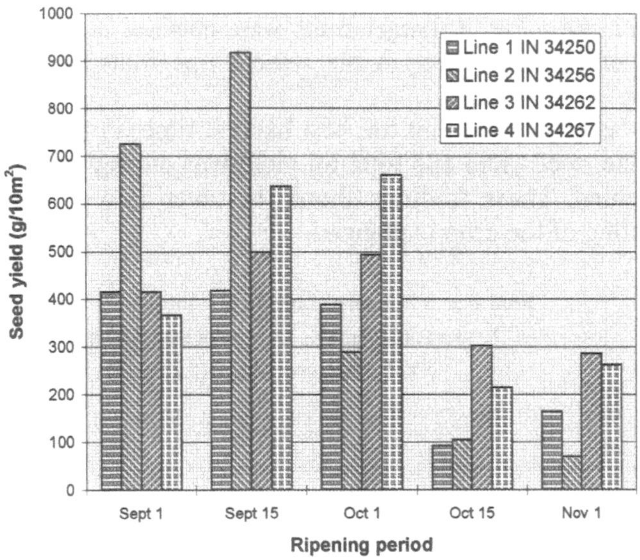

Figure 1. The relation of seed yield to ripening period

Seed weight - Fig. 2. Early-ripening fruits (first month) contained larger seeds than later-ripening fruits, as measured by 1000-seed weight. It could reflect the fact that late ripening seeds contain less water then early ripening seeds.

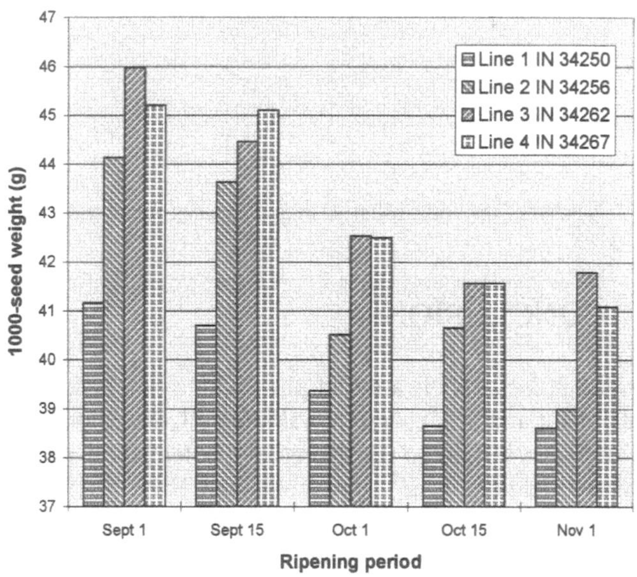

Figure 2. The relation of 1000-seed weight to ripening period

Fruit size - Fig. 3. Larger fruits were obtained mostly during the first month of ripening. Later in the season, ripe fruits became smaller and smaller.

It was concluded that the best harvest time, in terms of fruit yield, fruit size, seed yield and total oil yield was during the first four weeks of ripening. These findings should be taken into consideration when harvesting of the crop is planned.

Fruit size: S = small (0.2cm Ø); M = medium (5.2 cm Ø);
L = large (6.2 cm Ø); XL = extra large (7.2 cm Ø)
XXL = extra-extra-large (8.2 cm Ø)

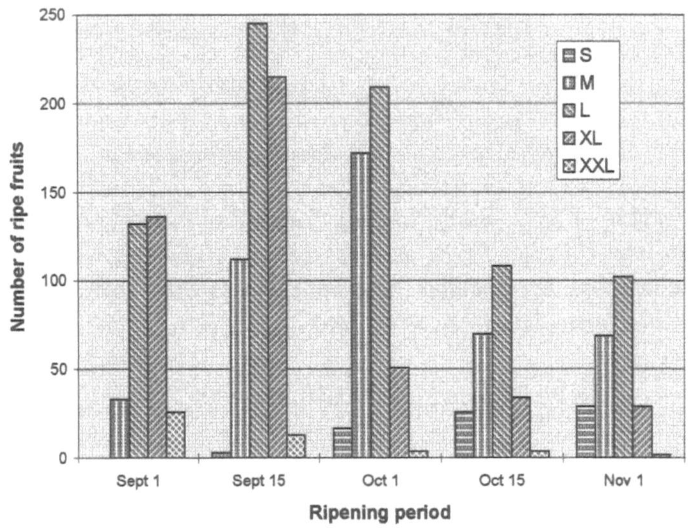

Figure 3. Number of ripe fruits collected in each ripening period and arranged by fruit size

4. CONCLUSIONS

Citrullus colocynthis is a desert plant with a rich history as an important medicinal plant and as a source of valuable oil. Based on our field observations, *C. colocynthis* can be considered as a potential oil crop for edible uses. A calculated yield of 2 t ha^{-1} (!) was obtained. With 17-19% oil in the seeds, this yield is equivalent to 400 l ha^{-1} oil. This oil contains a total of 80-85% unsaturated (oleic and linoleic) fatty acids, thus making it a high quality oil for human nutrition.

C. colocynthis is a plant of dry climate and its cultivation should now be evaluated under arid conditions. A special attention should be given to evaluate its medicinal and diuretic potential as a source of high value by-products.

REFERENCES

1. Dafni, A., Yaniv, Z. & Palevitch, D. (1984). *Ethnobotanical survey of medicinal plants in Northern Israel.* Journal of Ethnopharmacology, 10:295-310.
2. Palevitch, D. & Yaniv, Z. (1991). *Medicinal Plants of the Holyland.* Tamus Modan Press, Tel Aviv. 440 pp. 56-58. (in Hebrew).
3. Yaniv, Z., Elber, Y., Zur, M. & Schafferman, D. (1991). *Differences in fatty acids composition of oils of wild Cruciferae seeds.* Phytochemistry, 30 (3): 841-843.
4. Zohary, M. (1982). *Plants of the Bible.* Cambridge: 185 pp. Cambridge University Press.

EUPHORBIA TIRUCALLI FOR HIGH BIOMASS PRODUCTION

Patrick L. J. Van Damme
Faculty of Agricultural and Applied Biological Sciences, Coupure Links 653, B 9000 Gent, Belgium, Tel: 32 9 264 6089, Fax: 32 9 264 6241, email: Patrick.VanDamme@rug.ac.be

1. INTRODUCTION

Euphorbia tirucalli L. is a somewhat bizarre looking plant. It is a succulent plant growing to a height of about 10 m occurring in the tropics. The main trunk and branches are woody and brown but the younger branches are green and cylindrical, looking like so many pencils, hence its vernacular name of pencil tree.

Especially in resource-poor conditions, farmers have not enough capital to purchase barbed wire; as an alternative, farmers have always used the plant as a live fence. In most of its area, it is also used as fish poison, or in several toxic and/or medicinal preparations.

In the western world, it was until lately probably best (and only) known to succulent or xerophytes fanatics and collectors. This changed when Melvin Calvin developed the hypothesis that its latex might well be a source of diesel. According to him, the hydrocarbons it contains might be capable of producing the equivalent of 10-50 barrels of oil per acre (Calvin, 1978). Following this, research was started on how to grow *E. tirucalli*.

In 1980, the Belgian administration for development cooperation together with the Kenyan government, started a huge project in Baringo district, Kenya, with the idea to eventually grow 10,000 ha of *E. tirucalli* and produce oil for local consumption. Even though the project was to become a failure, mainly due to political manoeuvring and mismanagement on the Belgian side (De Coninck, 1996, and personal observations), the research

Combating Desertification with Plants, edited by Dov Pasternak and Arnold Schlissel.
Kluwer Academic/Plenum Publishers, New York, 2001.

that was conducted in its wake both in Belgium and Kenya showed the plant's potential as a biomass crop for marginal land in arid and semiarid environments using low external inputs.

This article wants to highlight the major findings on *E. tirucalli* collected by the author over the past decade and briefly present its potential as an energy/biomass crop.

2. CENTRE OF ORIGIN AND SUBSEQUENT DISTRIBUTION

There is still some discussion as to where *E. tirucalli* originated. Its centre of origin most probably lies in East Africa (Leach, 1973), more specifically Tanganyika (Tanzania) from which it was collected in 1862 by John Hanning Speke and J.A. Grant when they were looking for the sources of the Nile. An alternative theory, mentioned by Leach (1973), puts its centre more to the west, in Angola. It seems almost certain that the suggestion formulated by Croizat (1965) that Madagascar was the centre of origin can be discarded.

Even though Ponsinet and Ourisson (1968a and b), Mahlberg *et al.* (1983) and Mahlberg and Pleszczynska (1984) support Croizat's thesis, Leach (1973) seems to be right in stating that *E. tirucalli* cannot have come from Madagascar as it is the only *Euphorbia* having two minute stipules at the basis of each leaf, making it somewhat of an exception on an island which has been documented to present numerous examples of very close relationships between species in a number of genera.

E. tirucalli has spread from its centre of origin all over the African continent including Madagascar, basically because it is easily propagated vegetatively through stem cuttings. Bantus probably carried it south some 2000 years ago (Moll, 1978). Natural stands occur in such diverse areas as South Africa (where it covers whole areas with dense tirucalli forests), Kenya and Senegal. It is used for fencing in Rwanda, Burundi, Congo, Namibia, and most of southern and eastern Africa.

Van Damme (1989a) has compiled a list of 28 African countries in which literature references have mentioned *E. tirucalli* to occur. Personal observations in some 10 other countries would suggest that it indeed occurs in all African countries.

From Africa, the plant spread into the other continents which have subtropical to tropical climates, although it is still unclear as to how, when and what ways it followed. The species epithet is derived from *tiru-calli*, a local name given to the plant on the Malabar Coast of India (White *et al.*, 1941; Berhaut, 1975). It was from this area that Hendrik Adriaan van Rheede in his

Hortus Malabaricus first described the plant in 1679. At that time, the indigenous people already used *E. tirucalli* as a hedge plant. There is a hypothesis that the plant came to that part of the continent with Portuguese travellers who had travelled from Portuguese East Africa. J. Commelin produced the first drawings of the plant in 1697. He used plant material from Ceylon.

The American continent is somewhat underrepresented: only Brazil, Haiti and the USA (Florida) seem to harbour any significant populations. In a number of other countries, the plant probably escaped from (botanical) gardens into the wild, to form relatively small stands.

Our personal experience, however, shows that records of *E. tirucalli* might be rather scarce due to the fact that outside the African continent it is not or hardly known from the wild, and has thus not been taken up in local/national floras. As a consequence, it is often presented as *Euphorbia* spp. In a number of other cases, authors have mistakenly determined it as another species also on the African continent. In West Africa, *E. tirucalli* is often referred to as *E. balsamifera,* which bears little resemblance to the plant, however.

Its present area of distribution only seems to be limited by the minimum temperature in any zone: temperatures should not drop below 0 C, although it has been reported from areas where frost sometimes occur (Japan: Goo *et al.*, 1984; Cochabamba, Bolivia - personal observation). Leach (1973) indicates that *E. tirucalli* does not occur in deserts, although he does not further define the desert environment.

3. DESCRIPTION

There is no descriptors' list available to uniformly describe *E. tirucalli* germplasm. We will, however, present the results of our own work.

3.1 Taxonomy

Order	Euphorbiales (Cronquist, 1981)
Family	Euphorbiaceae
Subfamily	Euphorbioideae (Webster 1967 and 1987)
Tribe	Euphorbieae (Wheeler, 1943)
Genus	Euphorbia L.
Subgenus	Euphorbia (Webster *et al.*, 1975)
Section	aphyllis (Webster *et al.*, 1975)
Subsection	tirucalli (Webster, personal communication)
Species	tirucalli L.

3.2 Vegetative Characteristics

3.2.1 Stems and Leaves

This succulent, for the layman cactus-like, tree is devoid of spines. It produces white latex, which earned it its common English name of milkbush. It can grow into 5 - 10 m high trees. It is thus not a woody vine as stated by Baslas and Gupta (1983; one can even wonder whether they did indeed perform their chemical analysis work on *E. tirucalli*...).

The trunk's diameter can reach maximum values of 75 cm. Its branches are often arranged in pseudowhorls. Their typical arrangement along stems and trunk gives it the impression of brush-like masses (White *et al.*, 1941) or to consist of a collection of brooms, which is referred to by yet another vernacular name: pencil tree, and puts them in a group of so-called coralliformous euphorbias (Leandri (1952-53) *fide* Cremers, 1983).

Biesboer and Mahlberg (1981) add to this that *E. tirucalli* belongs to a group of corraliformous and succulent euphorbias, which they claim are highly specialised. Branches are shed in extreme dry conditions as a kind of survival strategy.

E. tirucalli wood is soft and very unlike the typical non-succulent dicotyledonous wood. Carlquist (1962) stated that the secondary development of *E. tirucalli* follows a pathway he called pseudomorphosis (following in this de Beer (1930) which he cited). As a result, the wood is rather useless as source of timber or even firewood.

Notwithstanding the numerous botanical descriptions stating that the plant is leafless, it has indeed minute and often short-lived leaves on freshly developed branches. These leaves are small, early deciduous and alternate, only 1 - 2.5 cm long, and ca. 3 - 4 mm wide, oblanceolate, acute at the tip and tapering to a sessile base (Van Damme, 1985). Their photosynthetic function is taken over by the green branches when the leaves fall.

On the stem, stomata are situated in grooves where they are protected from extreme and desiccating conditions. Stem stomatal frequency is 40/mm^2, whereas in the grooves there are some 12 stomata per mm.

Normally, the plant grows in seasonal flushes during the moist period of the year. At that moment, young branches are produced on stems of the previous year, or on older plant material. This is called articulated growth by Deghan and Webster (1979).

The young branches carry short-lived leaves that are shed when weather conditions get drier. When leaves are present, they represent only some 2 – 3 % of the plant's physiologically active surface.

3.2.2 Latex

As mentioned before, an important feature of *E. tirucalli* is the milky-white latex that can be found in all plant parts. The latex oozes from the plant when it is damaged. Kopaczewski (1947) analysed the latex and found it to have a density of 1.054, a dry matter content (in %, at 65 °C) of 27.6, a surface tension (10^{-5} N/cm) of 32.8 – 33.6, specific viscosity of 1.75 – 1.9, and pH (colorimetric) of 4.4. Lynn & Clevette-Radford (1987) found protein content (mg/ml) is 4.3, carbohydrates (mg/ml) amount to 0.8 whereas the amount of solid particles (mg/ml) is 100. Their pH value was slightly different from the previous value: 4.9.

By comparison with other Euphorbiaceae, *E. tirucalli* is relatively poor in proteins and carbohydrates, but rich in solids. The pH is below the optimum pH for the proteolytic enzymes occurring in the latex to be active. These enzymes are thus probably not active in the latex but elsewhere; the latex would only be a vehicle for them. White *et al.* (1941) mention the latex contains 15.7 % of rubber, 82.1 % of resin, 1.3 % proteins, and 0.9 % insoluble substances. The latex has been used as a source of rubber: it was marketed in the beginning of the century (from South Africa into the UK) but never became a commercial success because of its rather limited properties. It has, however, a number of properties that make it an interesting source of biochemicals. It has high enzymatic activity and contains a number of secondary metabolites.

The latex is caustic and irritant (Henke, 1986; Berhaut, 1975). Even honey derived from bees having visited *E. tirucalli* flowers can be irritant (Wauters, personal communication). The active principles are derived from esters belonging to a group of tetracyclic diterpenes (Evans & Kinghorn, 1974). When the latex hits the eye, it causes intense pain and temporary blindness for several days (Raymond, 1936; Watt & Breyer-Brandwijk, 1962) and keratoconjunctivitis (Henke, 1986). Cornea and iris are hurt in the process. More in general, the latex causes irritation of and blisters on skin, which can initiate secondary infections (Evans & Kinghorn, 1974) and dermatitis.

Apart from the enzymes and tetracyclic diterpenes already mentioned, the latex also contains a number of triterpenes which are rich in energy and can be the source of a fuel with a higher energy content than diesel oil. Triterpenes basically occur as alcohols. The best researched amongst them are tirucallol (euphorbol) and taxaxosterol (Haines & Warren, 1949), euphorol, taraxerol and cycloeuphorol (Afza *et al.*, 1979) and euphol (McDonald *et al.*, 1949). Euphol and tirucallol have been produced using *in vitro* techniques (Ulens, 1986; Dillen, 1986). The latex also contains a number of polycyclic diterpenes which are cocarcinogenic: they promote tumor growth in mice

(Baslas & Gupta, 1983). According to these latter authors, the active principles here are diterpenic phorbol and resiniferonol derivates. Ingenol and phorbol are, according to Fürstenberger & Hecker (1977, 1985 and 1986) the precursors of a number of a number of irritant 12,13-diterpene esters belonging to the ingenane and tigliane group. For a thorough review of the chemicals present in the latex, consult Beckstrom-Sternberg & Duke (1994).

Even though the latex is not without negative points, there seems to exist no major danger in manipulating it if frequent contacts and oral intake can be avoided. Chemically speaking the irritant principles and cocarcinogenic substances are different from one another.

3.3 Generative Characteristics

The colourless to yellow flowers are borne in typical Euphorbiaceae (and within the family the Euphorbieae tribe) inflorescences at the end of twigs of the previous year. Cremers (1977) describes them as subterminal cyathia. Plants are dioecious. The flowers consist of an involucre that bears five independent nectaries. Nectar production is profuse and attracts insect pollinators. Flowers have been studied in detail (electronic microscopy) by Van Damme (1989a). Seeds are borne in capsules. Seed set is rather limited.

3.4 Germplasm

The Gent University (Belgium) *E. tirucalli* germplasm collection contains some 18 accessions from Senegal, Morocco, Rwanda, South Africa, Kenya, and Sri Lanka. Accessions differ as to vegetative characteristics, e.g. relative dimensions of stems and leaves, general plant habit, and growth rate. Preliminary tests as to differences in biochemistry, including secondary metabolites, and stress tolerance were, however, not conclusive (Van Damme, unpublished data).

3.5 Ecology

Literature data on ecological characteristics of *E. tirucalli* are rather scarce. As stated before, *E. tirucalli*'s distribution seems to be limited by the minimum temperatures which may not descend under 0°C, as frost is causing death when plants are unprotected. According to Goo *et al.* (1984), *E. tirucalli* grows best at high temperatures (*sic*); in their experiments, temperatures ranged 25 - 30°C. The same authors were able to protect *E. tirucalli* plants from frost damage (-1°C) by covering them by polyvinyl film. Pruning and covering with straw, as done with other plants/crops, was unable to protect plants from dying at 3°C.

Leach (1973) tells us that *E. tiruçalli* does not thrive in desert conditions, whereas Spoon (1927) had indicated that *E. tirucalli* does not occur in the "real" tropics, but prefers more temperate environments.

From the above, it is clear that more research is needed to be able to define the ecological range of *E. tirucalli*.

3.6 Soils

According to Kerharo (1974) and Viaud and Teisseire (1979), *E. tirucalli* grows best on sandy and salty soils. Its occurrence on the coasts of Senegal would seem to prove their point: they grow in the dunes that receive a lot of salt sprays from the sea.

Our own research results (Verloo & Van Damme, unpublished) indicate that *E. tirucalli* grows better in heavy, clay soils with high salt, i.e. sodium and potassium, concentrations than in soils with the same granulometry but lower salt concentrations. This is probably due to the fact that clay soils loose their structure when exposed to salt. This loss of structure would seem to improve aeration and thus plant development (see Poljakoff-Mayber & Gale, 1975). However, by adding salt to a soil, the soil's water potential lowers which would normally hamper normal plant growth, through a physiological drought stress phenomenon. Because in the case of *E. tirucalli* this was apparently not the case, we started a research into the species' salt tolerance mechanisms and limits.

3.7 Ecophysiology

From the above, it is clear that *E. tirucalli* is well-equipped to tolerate dry climatic conditions, and can often be found growing in saline soil conditions. Morphologically speaking, the plant looks succulent (stems), but clearly has mesophytic, i.e. non-drought tolerant, leaves. The physiological basis for these tolerance mechanisms, however, is not so clear. Bender (1971), Bender *et al.* (1973) and Webster *et al.* (1975) suspected *E. tirucalli* to follow the Crassulacean Acid Metabolic pathway (CAM). They based their assumption on C-isotope measurements. These $^{13}C/^{12}C$ measurements showed *E. tirucalli* to have $\delta^{13}C\%_0$ values of -15.3, which is a mean value for CAM plants, and a borderline value for C_4 plants.

Ting *et al.* (1985) gave a general range of values of -15.4 to -16.2 for CAM plants grown in greenhouses and in the wild (Costa Rica), whereas C_4 plants had values between -12.6 and -11.3, resp. Smith *et al.* (1985) showed CAM bromeliads to have values between -12.3 and -15.4. Lower values for CAM plants would indicate a relatively bigger share of daytime CO_2 fixation (and thus indicate plants follow C_3 metabolism), whereas higher values would indicate the carbon dioxide is taken up during the night. Somewhat lower values in CAM plants would then be an indication of somewhat better growing conditions as the latter promote C_3 uptake (at the end of the light phase, Kluge *et al.*, 1982).

As shown above, the values obtained for *E. tirucalli* are far below the values for C_3 stems that are in the -20 to -30 range, so that it is clear that the stems do not follow the C_3 metabolism. Smith (1982) using $^{13}C/^{12}C$ data put *E. tirucalli* with the C_4 plants, hereby 'forgetting' Trumpke's (1914) finding that Kranz syndrome (indicative of C_4 metabolism) does not occur in *E. tirucalli* stems... Confronted with these data and a number of other arguments, Smith acknowledged in a personal communication having put *E. tirucalli* with the group of C_4 plants was a mistake.

To find out about the 'real' metabolism of the plant, we devised a phytotron experiment measuring gas exchanges (CO_2 and water vapour) from complete stems (i.e. stems + leaves), and stems from which the leaves' gas exchanges had been blocked by applying nail polish ('leafless treatment'). Relative humidity during the experiments was 50 %. Further details about the experimental set-up can be found in Van Damme (1989a).

The results show that the succulent stems are indeed following the CAMetabolic pathway, whereas the short-lived, non-succulent leaves are following C_3. The data also show that CAM-efficiency increases with the stem material's age up to the point where the outer layers start to lignify. This ontological strengthening of CAM has also been documented for other CAM plants: *Cissus quadrangularis* (Ting *et al.*, 1983), *Mesembryanthemum crystallinum* (von Willert *et al.*, 1976), and a number of *Kalanchoë* spp. (Brulfert *et al.*, 1982).

It is important to point out that this combination of CAM stems and C_3 leaves is almost unique in the plant world. Ting *et al.* (1983) and Lange & Zuber (1977) described analogous examples for *Cissus quadrangularis* and *Frerea indica*. It can be expected that a number of other *Euphorbia* spp. and succulents that have succulent stems and 'normal' mesophytic leaves might have the same combination.

E. tirucalli maximum CO_2 uptake rates compare favourably with values measured in other CAM plants: 0.1 – 0.12 mg $CO_2/m^2.s$ against 0.0278 – 0.1111, resp. (the latter values coming from Kluge and Ting, 1978). *E. tirucalli* thus fixes CO_2 rather efficiently.

The relative importance of the leaves in metabolism and more particularly in CO_2 uptake is quite high: they only represented 4 % of the total assimilating surface, but were able to have the CO_2 uptake increase by 24 % when compared with the 'leafless' treatment. When considering daytime CO_2 uptake, the increase was much higher still, amounting to 93 %. Higher CO_2 uptake usually also results in higher water loss through transpiration.

Figures showed water loss was 14 % higher on a 24-hours basis (and 33 % higher when taking daytime light phase figures). The combination of both yields an increased Water Use Efficiency (WUE) of 9 % for the 24-hours cycle (and 45 % for the light phase). In other words, the (small) leaves have a very high affinity for CO_2 but tend to spend water moderately.

This combination of CAM stems and C_3 leaves is quite beneficial for the plant, and might explain its high production potential. In natural and normal, i.e. moist, conditions, the C_3 leaves increase the plant's ability to maximise growth: CO_2 is taken up during the day and the night, whereas water is basically only spent during the day. Overall WUE is thus quite high (and in natural conditions during the rainy season presumably higher than in our set-up with its relative humidity of 50 %; the lower the value, the higher the evaporative demand and thus the lower the WUE). During the dry season, the plant looses its leaves and will then only depend on CAM for its CO_2 uptake.

4. DROUGHT AND SALT STRESS RESISTANCE (Van Damme, 1989, 1990 a and b)

E. tirucalli grows naturally in arid to semiarid areas. In a number of cases, the plants can be seen to grow in coastal areas which have higher salt loads than average, both at soil and air levels (salt spray). *E. tirucalli* can thus be considered as (potentially) highly tolerant to even resistant for drought and salt conditions.

The ecophysiological basis for this tolerance was presented by Van Damme (1989a). His laboratory research findings were integrated in and confronted with data obtained from the field (Kenya, unpublished) and general literature.

As pointed out above, *E. tirucalli* has CAM stems and C_3 leaves. CAMetabolism implies nocturnal synthesis of C_4-acids from the CO_2 taken up during the dark phase. These acids are then again decomposed into carbon dioxide and the initial C_3 compounds. These organic acids also occur in *E. tirucalli*. They probably act as osmotic agents thus promoting higher salinity tolerance than would be the case if the plant were a "normal" C_3 or C_4 plant. At higher soil salt concentrations, however, osmotic effects at root level seem to hamper correct water uptake explaining why growth is slowed down. The succulent stems also have a buffering function: when Na and/or Cl enter the stems' tissues, the ions are diluted by the water present in parenchyma and vacuoles.

Slowing down of growth processes occurs at conductivity levels of 10 mS/cm. The lethal level is reached at conductivity levels of 22 – 25 mS/cm.

It would seem that the exact value at which *E. tirucalli* stops growing or even dies, would depend upon other environmental parameters (climate, soil type) in combination with germplasm characteristics. As seen in other plants/crops fertilisation increases production at any conductivity level, but does not increase salt tolerance.

Increased NaCl levels in the soil do not result in increased Na and/or Cl concentrations in the above ground biomass. Moreover, *E. tirucalli* excludes Na and Cl from entering young, growing stem material: the ions are sequestered in older stem material, and do thus not interfere with CO_2 metabolism. Only growth rates are depressed, hence plants growing in soils with increasing NaCl concentrations will tend to remain smaller. NaCl in the soil induces stems to loose turgor pressure as can be seen from wrinkling stems.

Increased salt stress lowers CO_2 uptake and nightly acid accumulation only at soil conductivity levels > 21 mS/cm. WUE is lowered at lower conductivity values, because of increased transpiration.

From these findings, it can be concluded that *E. tirucalli* could be a good candidate for arid and semiarid zone cropping, and as a complement crop in saline agricultural systems. It could increase the productivity of farming systems situated on marginal lands, without decreasing the area of non-saline land sown to crops. If these lands are waterlogged, then they could help lower the water tables beneath the salt-affected land and the marginally affected land that is still cropped. Thus the land resource would become more productive because of decreased water logging and salinity. As *E. tirucalli* is a source of biomass and thus green energy, it may ensure that farmyard manure, which is increasingly used as fuel and thus not used as fertiliser, can be returned to the fields as fertiliser. This would further improve crop yields.

5. FORMAL AND INFORMAL USES

Information on the web (http://www.rain-tree.com/aveloz.htm) gives some (incomplete) info on the plant's ethnobotanical uses:

Country	Ethnobotany Worldwide Uses
Brazil	Abscess, Wart
India	Abscess, Ache(Ear), Asthma, Colic, Cough, Emetic, Gastralgia, Neuralgia, Purgative, Rheumatism, Rubefacient, Toothhache, Vesicant, Wart
Java	Leprosy, Dermatosis, Fracture, Ostealgia, Wound
Malaya	Cancer, Osteosis, Piles, Piscicide, Rhinosis, Scurf, Swelling, Thorn
Tanzania	Piscicide, Poison

The uses of *E. tirucalli* have been thoroughly reviewed by Van Damme (1989b and 1990c). *Grosso modo*, its uses can be summarised as follows:

- it is a source of rubber, or more technically polyisoprene, and varnish, for local use and export;
- through its irritants (see above), plants are appreciated throughout the world for fences around houses and fields: thieves and roaming animals will get irritated when the latex hits their eyes and skins, and thus be 'demotivated' from entering the premises;
- plant material can be burnt and the so-called pearl ash used to make soap (Watt & Breyer-Brandwijk, 1962);
- the (light) wood can be used to make floaters (Watt & Breyer-Brandwijk, 1962);
- at one moment, the latex was used to make paints, varnish and resins (Spoon, 1928);
- the latex contains secondary metabolites that can be used as fish poison; in a number of cases they are used against:
    - bacteria (*Staphylococcus aureus*; Karel & Roach, 1951);
    - amoeba (*Entamoeba histolytica*; Dhar *et al.*, 1968);
    - molluscs (*Lymnaea natalensis*, which is a host for *Fasciola gigantica*, is killed for a 100 % by fresh or dried plant material; Vassiliades, 1984) (*Biomphalaria glabrata*, host for *Schistosoma mansoni*, is killed (LD90) at concentrations of 85 ppm; Jurberg *et al.*, 1985);
    - nematodes (*Hoplolaimus indicus*, *Helicotylenchus indicus* and *Tylenchus filiformis* are killed for 100 % after 72 h with a 1:1 solution of latex and water; Siddiqui *et al.*, 1984);
    - insects; and
    - mammals;
- the latex also contains active principles that are used as medicine in treating a variety of ailments: ascites, oedemes, and constipation (Senegal; Berhaut, 1975), as a purgative (Van Giffen, 1924; Raymond, 1939), against stomach pain and to induce vomiting (Van Giffen, 1924), against rheumatism, toothache and neuralgia (Khan *et al.* (1987) cited by Van Damme, 1989a and b). Skin diseases (caused by fungi) may be treated by them, whereas open wounds may heal quicker by applying latex to them.

Even though the latex is a cocarcinogenic, it is used in India against tumors and cancer in general, and in cases of asthma, leucorrhoea and digestive disorders (Baslas & Gupta, 1983). The anti-cancer properties have also been demonstrated by Abbott *et al.* (1966). As so many other plants, it is an abortifaciens (Saha *et al.*, 1961) and, in the same sector, a remedy against sexual impotence and even sterility. Syphilis is also treated by it (Watt & Breyer-Brandwijk, 1962). Snake bites are treated with *E. tirucalli* roots in Tanzania (Raymond, 1936). Excessive internal use may, however, cause problems such as gastro-intestinitis (followed by death) and stomach perforation.

> Spotlight Magazine published an article about cancer and natural products in March 1999. In this article, it recommended the sap of *E. tirucalli* as an anti-tumor agent for cancer patients. The present state of the art does not permit to sustain this recommendation. Actual documented independent clinical research has indicated that the sap of *E. tirucalli* has actual weak tumor-promoting properties. There is, however, absolutely no evidence or research supporting that taking the sap internally will impact a cancerous tumor inside the body. The only thing that could be possibly misconstrued through documented indigenous or folklore use of the sap is possibly for skin cancers.
>
> The sap is very caustic and irritating and has been used to "burn" off warts and thus possibly skin cancers from the skin, but there is no evidence that taking it internally for a tumor elsewhere in the body is beneficial, indicated or prudent. The sap and the entire plant has been shown clinically to be immunosuppressive, and as such can't be recommended in treating cancer patients. It has also been demonstrated in several clinical studies (Ito, *et al.*, 1981; Osato, 1987; Mizuno, 1986; Aya, *et al.*, 1991; Imai, *et al.*, 1994; Van den Bosch, *et al.*, 1993) to activate Epstein Barr Virus and has been linked to possibly causing cancer - Burkitt's lymphoma.
>
> (http://www.rain-tree.com/aveloz.htm)

As mentioned before, the latex contains energy-rich chemicals that can be processed into diesel oil-like products. The technology to transform latex into diesel fuel is rather expensive though, and it is commonly accepted crude petroleum barrels should fetch at least 30 US $/barrel before it becomes economically interesting to start processing *E. tirucalli*. As the plant will mostly be grown in marginal conditions in Third World conditions the high investments might not be offset by equally high benefits due to difficulties in commercialisation. However, our own experiments with *biogas* production look rather promising and might offer a good alternative for diesel oil production from latex.

E. tirucalli has an energy content of 15.9 MJ/kg dry matter (De Wilde, 1983) which can easily be used to produce biogas. The low N content and easily decomposed organic components results in high output rates of methane. A 2-phased methane fermentation yields better results than a 1-phased system. A field producing 500 t/ha/y biomass, would result in 137 kg/dry matter/day, and 31 m^3 methane/day. Its yield (53.8 %) can be increased by an alkali pre-treatment of *E. tirucalli*. Ollevier (personal communication) was able to obtain 323 l biogas from 1 kg of dry *E. tirucalli* material; De Wilde (1983) using a similar set up obtained 226 l methane. Pilot installations have been tested in field conditions in Senegal by Farinet *et al.* (1988).

6. EUPHORBIA TIRUCALLI AS A CROP

There is only limited information available on how to grow *E. tirucalli* as a crop. As has been discussed before, the plant is an ideal candidate for marginal soils and climates. It can be grown without additional fertiliser gift, and irrigation is not required if and when rainfall does not fall below 200 mm/year (with acceptable water retention capacity of the soil).

6.1 Propagation and Plant Establishment

Propagation can be from cuttings that can be planted rather densely. Plant density will depend on climatic and edaphic characteristics of the area but can be in the 10,000 – 20,000 plants/ha range (Wauters, 1981a and b). Cuttings can best be left to dry for at least a day before planting. Benincasa *et al.* (1980) claim *E. tirucalli* grows better in lighter soils than in heavy soils. In Kenya, people protect freshly planted cuttings by covering them with thorny stems from *Acacia* trees. This provides protection against browse and shadow. The time of planting is during, and more specifically at the beginning of the rainy period. Once established, the plants will quickly form a closed canopy. This combined with a litter of stems, will dominate and eliminate weeds. Weeding is thus only needed in the beginning, after planting.

6.2 Pests and Diseases

6.2.1 Plant Parasites

E. tirucalli is sometimes attacked by *Cuscuta* spp. (witch weed).

6.2.2 Pests

6.2.2.1 Nematodes

E. tirucalli has been shown it can be heavily infested by *Meloidogyne* spp. (Prot, 1981, and personal communication). In this respect, one should be careful not to promote *E. tirucalli* as a hedge plant in *Meloidogyne*-infested soils, as the plant seems to literally attract the nematodes, which may then spread into crop fields.

Volckaert (1984) mentions *E. tirucalli* is also infested by *Helicotylenchus*, *Scutellonema*, *Pratylenchidae* and *Criconematidae* as these were found in soil samples collected from *tirucalli* plots in Kenya.

6.2.2.2 Insects

Aphids, mealy bugs and grass hoppers (species not identified) have been seen to live on and off *E. tirucalli* plants in a number of countries.

6.2.2.3 Mites (Van Damme, 1989a)

Mites have been found to occur on young leaf and stem material in both greenhouse and field conditions.

6.2.3 Diseases

Nectria euphorbiana has been determined on dead stem material in India (Tilak and Talde, 1979). Dutrecq and Mbuthia (1986) found a number of fungi on *E. tirucalli* material in Kenya.

Granulodiplodia spp. was found on stems, together with *Aspergillus* spp., *Sclerotium* spp., *Bispora* spp., *Rhizoctonia* spp., *Blastomyces* spp., *Tubercularia* spp. *Trichaegum* spp. and *Dotichiza* spp., and the *Botryodiplodia*-complex (including *Lasiodiplodia*, *Granulodiplodia* and *Macrophoma*) which cause necrotic spots on the stems, dry root rot and die back of young stem material, and may result in massive death.

During and after a moist period, necrotic spots caused by *Alternaria* may appear on the stems. Ghaffar (1968) mentioned that *E. tirucalli* is a host plant for *Botryodiplodia theobromae* in Pakistan.

7. CONCLUSION

Even though *Euphorbia tirucalli* has only a limited crop record, the plant seems well-suited to be grown in marginal areas in semiarid to arid areas using limited inputs.

Further research is needed into possible phytosanitary problems whereas biogas production efficiency should be improved and low-cost technology tested.

REFERENCES

1. Abbott, B.J., Hartwell, J.L. Iltis, H.H., Kupchan, S.M., Leiter, J. Perdue, R.E. & Schepartz, S.A. (1966). *Screening data from the cancer chemotherapy national service center screening laboratories*. XXXVI. Plant extracts. Cancer Research, 26 (9), 1131 - 1277.
2. Afza, N., Malik, A. & Siddiqui, S. (1979). *A new triterpenoid from Euphorbia tirucalli*. Pakistan Journal of Scientific and Industrial Research, 22 (3), 124 - 127.
3. Aya, T., Kinoshita, T., Imai, S., Koizumi, S., Mizuno, F., Osato, T., Satoh, C., Oikawa, T., Kuzumaki, N., Ohigashi, H. (1991). *Chromosome translocation and c-MYC activation by Epstein-Barr virus and Euphorbia tirucalli in B lymphocytes*. Lancet 337 (8751), 1190.
4. Baslas, R.K. & Gupta N.C. (1983). *Chemical investigation of Indian medicinal plants possessing anticancer activity: roots of Euphorbia tirucalli Linn.* Indian Journal of the Chemical Society, 60 (5), 506 - 507.
5. Beckstrom-Sternberg, S.M. & Duke J. (1994). The Phytochemical Database. http://probe.nalusda.gov:8300/cgi-bin/browse/phytochemdb; ACEDB version 4.3 - data version.
6. Bender, M.M. (1971).*Variations in the* $^{13}C/^{12}C$ *ratios of plants in relation to the pathway of photosynthetic carbon dioxide fixation.* Phytochemistry, 10, 1239 - 1244.
7. Bender, M.M. & Smith, D. (1973). *Classification of starch- and fructosan-accumulating grasses as C-3 or C-4 species by carbon isotope analysis.* Journal of the British Grassland Society, 28, 97 - 100.
8. Benincasa, M.M.P., Rodrigues, T.J.D., Leite, I.C., Damasceno, M.M. & Banzatto, D.A. (1980). *Estudos biologicos de Euphorbia tirucalli (Avelos). Efeitos do tipo de solo e adubao sobre indices de crescimento de plantas.* Revista Brasileira de Botanica, 3 (1/2), 49 - 53.
9. Berhaut, J. (1975). *Flore illustrée du Sénégal.* In: Ministère du Développement Rural. Dicotylédones. Tome III. Connaracées - Euphorbiacées. Direction des Eaux et Forêts, Dakar, 634 pp.
10. Biesboer, D.D. & Mahlberg, P.G. (1981). *A comparison of alpha-amylases from the latex of three selected species of Euphorbia (Euphorbiaceae).* American Journal of Botany, 68 (4), 498 - 506.
11. Brulfert, J., Mueller, D., Kluge, M. & Queiroz, O. (1982). *Photoperiodism and Crassulacean acid metabolism. I. Immunological and kinetic evidences for different patterns of PEPC isoforms in photoperiodically inducible and noninducible CAM plants.* Planta, 154, 332 - 338.
12. Calvin, M. (1978). *Green factories.* Chemical Engineering News, 56 (12). 30 - 36.
13. Carlquist, S. (1962). *A theory of paedomorphosis in dicotyledonous woods.* Phytomorphology, 12 (1), 30 - 45.
14. Cremers, G. (1977). *Architecture végétative de quelques espèces malgaches du genre Euphorbia L.* Bulletin van de Nationale Plantentuin van Belgie, 47 (1/2), 55 - 81.
15. Cremers, G. (1983). *Etude des grains d'amidon du latex chez quelques Euphorbes malgaches.* Bulletin van de Nationale Plantentuin van België, 53, 405 - 415.
16. Croizat, L. (1965). *An introduction to the subgeneric classification of Euphorbia L., with stress on the South African and Malagasy species.* Webbia, 20 (2), 573 - 706.
17. Cronquist, A. (1981). *An integrated system of classification of flowering plants.* Columbia University Press, New York.

18. Deghan, B. and Webster, G.L. (1979). *Morphology and infrageneric relationships of the genus Jatropha (Euphorbiaceae).* University of California Press in Botany, 94, 73 pp.
19. De Coninck, D. (1996). *Witte Olifanten. De miljardenschandalen van de Belgische ontwikkelingssamenwerking.* Van Halewyck, Leuven, 198pp.
20. De Wilde, B. (1983). *Tweefasige methaanfermentatie van plantenmateriaal, toegepast op Eichhornia crassipes (Mart.) Solms en Euphorbia tirucalli L.* Thesis, RUG, 99 pp.
21. Dhar, M.L., Dhar, M.M., Dhawan, B.N., Mehrotra, B.N. & Ray, C. (1968). *Screening of Indian plants for biological activity: part I.* Indian Journal of Experimental Biology, 6, 232 – 247.
22. Dillen, W. (1986). *Produktie van eufol and tirucallol in kallus- en suspensiekulturen van Euphorbia tirucalli. L.* Thesis, RUGent, 109 pp.
23. Dutrecq, A. & Mbuthia, G. (1986). *Quelques aspects de souches de Granulodiplodia spp., de Lasiodiplodia spp., de Macrophoma spp. et de Phyllosticta spp. isolées d'Euphorbia tirucalli au Kénya.* Mededelingen van de Faculteit van de Landbouwwetenschappen, 51 (2b), 557 – 562.
24. Evans, F.J. & Kinghorn, A.D. (1974). *The succulent Euphorbias of Nigeria.* Lloydia, 38 (4), 363 – 365.
25. Farinet, J.L., Forest, F. & Boquien, C.Y. (1988). *Valorisation énergétiques des résidus lignocellulosiques hétérogènes par le procédé en continu transpaille.* CIRAD/IRAT, Montperllier (France), 16 pp.
26. Fürstenberger, G. & Hecker, E. (1977). *New highly irritant euphorbia factors from latex of Euphorbia tirucalli L.* Experientia, 33, 986 – 988.
27. Fürstenberger, G. & Hecker, E. (1985). *On the active principles of the Euphorbiaceae, XI. The skin irritant and tumor promoting diterpene esters of Euphorbia tirucalli L. originating from South Africa.* Zeitschrift NaturForschung (C), 40 (9 - 10), 631 - 646.
28. Fürstenberger, G. & Hecker, E. (1986). *On the active principles of the Euphorbiaceae, XII. Highly unsaturated irritant diterpene esters from Euphorbia tirucalli originating from Madagascar.* Journal of Natural Products, 49 (3), 368 – 397.
29. Ghaffar, A. (1968). *Some new hosts of Botryodiplodia theobromae Pat.* The Pakistan Journal of Forestry, 18 (2), 233 – 234.
30. Goo, M., Kato, S., Ikeda, H. and Yagi, K. (1948). *Arboriculture of Euphorbia tirucalli.* Bulletin of the Tokyo University Forests, 73, 129 - 138.
31. Haines, D.W. & Warren, F.L. (1949). *The Euphorbia resins. II. The isolation of taraxasterol and a new triterpene, tirucallol, from E. tirucalli.* Pakistan Journal of Scientific and Industrial Research, 2554 – 2556.
32. Henke, G. (1986). *Ueber den Milchsaft einiger Euphorbiaceen.* Archiv der Pharmacie, 24 (17), 729 – 759.
33. Imai, S., , Sugiura, M., Mizuno, F., Ohigashi, H., Koshimizu, K., Chiba, S. & Osato, T. (1994). *African Burkitt's lymphoma: a plant, Euphorbia tirucalli, reduces Epstein-Barr virus-specific cellular immunity.* Anticancer Research, 14(3A), 933-936.
34. Ito, Y., Kawanishi, M., Harayama, T. & Takabayashi, S. (1981). *Combined effect of the extracts from Croton tiglium, Euphorbia lathyris or Euphorbia tirucalli and n-butyrate on Epstein-Barr virus expression in human lymphoblastoid P3HR-1 and Raji cells.* Cancer Letters, 12 (3), 175-180.
35. Jurberg, P., Cabral Neto, J.B. & Schall, V.T. (1985). *Molluscicide activity of the avelos plant (Euphorbia tirucalli L.) on Biomphalaria glabrata, the mollusc vector of schistosomiasis.* Mem. Inst. Oswaldo Cruz, 80 (4), 423 – 427.

36. Karel, L. & Roach, E.S. (1951). *A dictionary of antibiosis*. Columbia University Press, New York, 119.
37. Kerharo, K. (1974). *La pharmacopée sénégalaise traditionnelle. Euphorbia tirucalli* L. Vigot, Paris, 416 - 417.
38. Kluge, M., Fischer, A. & Buchanan-Bollig, I.C. (1982). Metabolic control of CAM. In: Ting, I.P. & Gibbs, M. Crassulacean Acid Metabolism. Proceedings of the fifth annual symposium in botany. American Society of Plant Physiologists, Rockville (Maryland), 31 – 50.
39. Kluge, M. & Ting, I.P. (1978). *Crassulacean Acid Metabolism. Analysis of an ecolgical adaptation*. Ecological Studies, 30. Springer-Verlag, Berlin, 209 pp.
40. Kopaczewski, W. (1947). *Etude physico-chimique du latex. V. Le latex d'Euphorbia tirucalli*. Bulletin de la Société de Chimie Biologique, 10 – 12, 924 – 926.
41. Lange, O.L. & Zuber, M. (1977). Frerea indica, a stem succulent CAM plant with deciduous C_3 leaves. Oecologia, 31, 67 – 72.
42. Leach, L.C. (1973). *Euphorbia tirucalli L.: its typification, synonymy and relationships, with notes on "Almeidina" and "Cassoneira"*. Kirkia, 9, 69 - 87.
43. Lynn, K.R. & Clevette-Radford, N.A. (1987). *Biochemical properties of latices from the Euphorbiaceae*. Phytochemistry, 26 (4), 939 – 944.
44. Mahlberg, P. & Pleszczynska, W. J. (1984).). *Evolution of succulent Euphorbia as interpreted from latex composition*. Tropische und Subtropische Pflanzenwelt, 45, 94 – 108.
45. Mahlberg, P., Pleszczynska, J. Rauh, W. & Schnepf, E. (1983). *Evolution of succulent Euphorbia as interpreted from latex composition*. Bothalia, 14 (3/4), 857 – 863.
46. McDonald, A.D., Warren, F.L. & Wiliams, J.M. (1949). *The Euphorbia resins*. I. Euphol. Pakistan Journal of Scientific and Industrial Research, 155 – 157.
47. Mizuno, F., Osato, T, Imai, S., Koizumi, S., Aya, T., Kinoshita, T., Hirai, N., Hirota, M., Ohigashi, H. & Koshimizu, K. (1986). *Epstein-Barr virus-enhancing plant promoters in east Africa*. AIDS Research, 2, S151-S155.
48. Moll, E.J. (1978). *A new method of collecting Euphorbia latex*. Veld en Flora, 64 (1), 2 – 1.
49. Osato, T. (1987). *African Burkitt's lymphoma and an Epstein-Barr virus-enhancing plant Euphorbia tirucalli*. Lancet 1 (8544), 1257-1258.
50. Poljakoff-Mayber, A. and Gale, J. (1975). *Intermediary remarks on response of plants to salinity*. In: Poljakoff-Mayber, A. and Gale, J. Plants in saline environments. Springer, Berlin, 91 - 94.
51. Ponsinet, D. & Ourisson, G. (1968a). *Etudes chimio-taxonomiques dans la famille des Euphorbiacées. III. Répartition des triterpènes dans les latex d'Euphorbia*. Phytochemistry, 7, 89 98.
52. Ponsinet, D. & Ourisson, G. (1968b). *Les triterpènes des latex d'Euphorbia: contibution à une étude chimio-systématique du genre Euphorbia*. Adansonia, 8 (2), 227 – 239.
53. Prot, J.-C. (1981). *Introduction aux nématodes phytoparasites. Les nématodes parasites des cultures maraîchères*. ORSTOM, BP 1386, Dakar, Senegal, 28 pp.
54. Raymond, W.D. (1936). *The poisonous effects of some local species of Euphorbia*. East African Medicinal Journal, 12, 369 – 374.
55. Raymond, W.D. (1939). *Native poisons and native medicines of Tanganyika*. The Journal of Tropical Medicine and Hygiene, 42, 295 – 303.
56. Saha, J.C., Savini, E.C. & Kasinathan, S. (1961). *Ecbolic properties of Indian medicinal plants*. Indian Journal of Medical Research, 49 (1), 130 – 151.

57. Siddiqui, M.A., Haseeb, A. & Alam, M.M. (1984). *Toxicity of plant latex to some plant parasitic nematodes*. National Academy of Science Letters, 7 (1), 1 – 2.
58. Smith, B.N. (1982). *General characteristics of terrestrial plants (agronomic and forests) -- C_3, C_4, and Crassulacean Acid Metabolism plants.* CRC Handbook of Biosolar Resources, 1 (2), 99 – 113.
59. Smith, J.A.C., Griffiths, H., Bassett, M. & Griffiths, N.M. (1985). *Day-night changes in the leaf water relations of epiphytic bromeliads in the rain forests of Trinidad.* Oecologia, 67, 475 – 485.
60. Spoon, W. (1927). *Eenige opmerkingen over exploitatie van euphorbia's in Zuid-Afrika; winning van hars en caoutchouc.* Berichten van de Afdeeling Handelsmuseum van de Koninklijke Vereeniging Koloniaal Instituut, 32, 3 - 28.
61. Spoon, W. (1928). *Berichten van het handelsmuseum van het Koloniaal Instituut*, Amsterdam, 34, 3 - 14.
62. Tilak, S.T. & Talde, V.K. (1979). *Contribution to our knowledge of ascomycetes of India.* Indian Journal of Mycological Plant Pathology, 9 (1), 17 – 21.
63. Ting, I.P., Sternberg, L.O. & De Niro, M.K. (1983). *Variable photosynthetic metabolism in leaves and stems of Cissus quadrangularis* L. Plant Physiology, 71, 677-679.
64. Ting, I.P., Bates, L., Sternberg, L.O. & Deniro, M.J. (1985). *Physiological and isotopic aspects of photosynthesis in Peperomia.* Plant Physiology, 78, 246 – 249.
65. Trumpke, H. (1914). *Beiträge zur Anatomie der Sukkulenten Euphorbien.* Doctoral thesis, University of Breslau, 92 pp.
66. Ulens, H. (1986). In vitro van *Euphorbia tirucalli* L. Thesis CTL, Gent, 56 pp.
67. Van Damme, P.L.J. (1985). *Destructive and non-destructive methods for determining stem and leaf areas on Euphorbia tirucalli L.* Mededelingen van de faculteit van de landbouwwetenschappen, 50 (4), 1369 - 1381.
68. Van Damme, P.L.J. (1989a). *Studie van Euphorbia tirucalli L., morfologie, fysiologie, teeltvoorwaarden.* PhD thesis, University Gent, 375 pp.
69. Van Damme, P. (1989b). *Het traditioneel gebruik van Euphorbia tirucalli.* Afrika Focus, 5 (3/4), 176 - 193.
70. Van Damme, P. (1990a). *Reasons for the Extreme Salt Tolerance of Euphorbia tirucalli.* Proceedings of the First Congress of the European Society of Agronomy, Paris, 5 - 7 December 1990, 1 P 3.
71. Van Damme, P. (1990b). *Reasons for the Extreme Drought Tolerance of Euphorbia tirucalli.* Proceedings of the First Congress of the European Society of Agronomy, Paris, 5 - 7 December 1990, 3 O 5.
72. Van Damme, P. (1990c). *Gebruik van Euphorbia tirucalli als rubberleverancier en energiegewas.* Afrika Focus 6 (1), 19 - 44.
73. van den Bosch, C., Griffin, B.E., Kazmebe, P., Dziweni, C. & Kadzamira, L. (1993). *Are plant factors a missing link in the evolution of endemic Burkitt's lymphoma ?* British Journal of Cancer, 68 (6), 1232 - 1235.
74. Van Giffen, (1924). *Vergiftiging met het melksap van Euphorbia tirucalli L.* Pharmacologisch Tijdschrift van Nederlandsch-Indië, 1 (12), 376 - 387.
75. Vassiliades, G. (1984). *Note sur les propriétés molluscicides de deux Euphorbiacées: Euphorbia tirucalli et Jatropha curcas.* Essais en laboratoire. Revue de Médecine Vétérinaire des Pays Tropicaux, 37 (1), 32 – 34.
76. Viaud, P. and Teisseire, D. (1979). *Latex d'Euphorbes, une source possible d'hydrocarbures et de caoutchouc.* Caoutchouc et plastiques, 593, 181 - 185.
77. Volckaert, V. (1984). *Studie van enkele aspecten van de teelt van Euphorbia tirucalli L.* Afstudeerwerk, Faculteit van de Landbouwwetenschappen, RUGent, 107 pp.

78. von Willert, D.J., Kriest, G.O., Treichel, S. & von Willert, K. (1976). *The effect of leaf age and salt stress on malate accumulation and phosphoenolpyruvate carboxylase in Mesembryanthemum crystallinum.* Plant Science Letters, 7, 341 – 346.
79. Wauters, N.M. (1981a). *Euphorbia projekt.* Rapport 2. Belgische ontwikkelingssamenwerking (ABOS) – MENR Kenya, 35 pp.
80. Wauters, N.M. (1981b). *Euphorbia projekt.* Rapport 3. Belgische ontwikkelingssamenwerking (ABOS) – MENR Kenya, 28 pp.
81. Webster, G.L. (1967). *The genera of Euphorbiaceae in the southeastern United States.* Journal of the Arnold Arboretum, 48, 303 – 430.
82. Webster, G.L. (1975). *Conspectus of a new classification of the Euphorbiaceae.* Taxon, 24 (5/6). 593 – 601.
83. Webster, G.L., Brown, W.V. & Smith, B.N. (1975). *Systematics of photosynthetic carbon fixation pathways in Euphorbia.* Taxon, 24 (1), 26 – 33.
84. Wheeler, L.C. (1943). *The genera of the living Euphorbieae.* The American Midland Naturalist, 30, 456 – 503.
85. White, A., Dyer, R.A. and Sloane, B.L. (1941). *Euphorbia tirucalli. The succulent euphorbieae of southern Africa,* 101 - 106.

BORAGINACEAE AS POTENTIAL SOURCES OF GAMMA-LINOLENIC ACID

Hava F. Rapoport, Mercedes del Rio and Antonio de Haro
Instituto de Agricultura Sostenible (CSIC), Apartado 4084; 14080 Córdoba, Spain, Tel: 34-957-499217, Fax: 34-957-499252, email: ag2rapop@uco.es

1. INTRODUCTION

Gamma linolenic acid (GLA, 18:3 Δ6, 9, 12) is an essential fatty acid in demand for its clinical and pharmaceutical applications. The most common commercial source of GLA for pharmaceutical use has been the oil of evening primrose seeds, *Oenothera biennis* (Galwey and Shirlin, 1990; del Rio et al., 1993). A possible alternative GLA source is the seed of borage, *Borago officinalis*, which is characterized by high oil content (24-34% as compared to 14-25% in *Oenothera*), with high GLA (23% as compared to 10.5%) and low erucic acid composition (Janick et al., 1989; del Rio et al., 1993). Additional advantages of borage are its annual life cycle and larger seeds as compared to *Oenothera* (Galwey and Shirlin, 1990). In addition to borage, other genera of the Boraginaceae, or borage family, also represent potential sources for GLA (Galwey and Shirlin, 1990; del Rio et al., 1993; de Haro and del Rio, 1998).

Borage and other Boraginaceae represent potential new crops with characteristics desirable for growth in Mediterranean and arid environments. The family Boraginaceae contains a large number of native Mediterranean species with both annual and perennial growth habits. Many of the annuals grow during the winter and are thus able to take advantage of available rains. Borage has an initial rosette vegetative growth form, suitable for over-wintering, followed by the subsequent development of flowering shoots

Combating Desertification with Plants, edited by Dov Pasternak and Arnold Schlissel.
Kluwer Academic/Plenum Publishers, New York, 2001.

when climatic conditions are adequate. Wild borage is a weedy plant adapted to marginal conditions.

A major limitation to commercial borage production is seed shattering, or the lack of seed retention (Janick et al., 1989; de Haro and del Rio, 1998). The borage ovary is highly lobed and develops into a fruit composed of four hard-shelled, single-seeded nutlets or mericarps, which readily separate from the fruit receptacle when mature (Roth, 1977). Harvest is further complicated by indeterminate flowering and non-uniform nutlet ripening (de Haro and del Rio, 1998), which lead to an extended, sequential maturation period. Janick and co-workers have attempted to find solutions to the seed-shattering problem with the development of vacuum harvesters and *in vitro* production of borage seed (Janick et al., 1989; Quinn et al., 1989).

2. EVALUATION AND MUTAGENESIS IN SPANISH BORAGE POPULATIONS

Wild and cultivated Spanish populations of borage, *Borago officinalis*, have been identified and evaluated as potential sources of GLA (del Rio et al., 1993). The wild germplasm was collected from spontaneous blue-flowered roadside populations in different localities of Andalusia, southern Spain.

Table 1. Mean and range of 1000 seed weight, oil content and fatty acid composition (%) of white and blue flowered borage populations.

Borage Populations	Seed weight (g)	Oil content (%)	Fatty acid content GLA	Fatty acid content Erucic
WF I (23)	15.6	34.3	22.2	1.7
	12.9-20.1	32.1-36.3	14.1-26.6	1.3-2.3
WF II (26)	15.2	34.1	22.1	1.6
	12.7-19.2	31.7-35.8	8.7-27.1	1.0-2.0
WF III (28)	15.1	34.5	22.3	1.7
	10.8-20.0	32.4-37.8	19.6-25.7	1.4-2.1
WF IV (22)	14.7	34.2	23.6	1.6
	12.4-17.8	32.1-37.2	15.9-27.1	1.2-1.9
WF V (17)	15.1	34.4	25.1	2.5
	13.8-17.7	31.2-38.0	21.9-28.6	1.3-3.3
WF VI (14)	15.3	34.4	23.9	1.8
	12.8-19.0	32.4-37.2	21.3-26.8	1.2-2.7
BF (55)	14.0	32.5	18.9	1.8
	9.3-18.1	26.7-35.4	12.6-26.5	0.6-2.9

WF = white flowered, BF = blue flowered. Each population type is followed by the (number of entries) for that type. Adapted from del Rio et al., 1993.

White-flowered borage is cultivated as a vegetable in the north of Spain, but has not been evaluated as a source of GLA. Populations of vegetable borage differing in plant habit, maturity and seed production were obtained from the breeding program of the Center for Agricultural Research of Rioja, Spain.

Means and ranges for seed weight, oil content and GLA and erucic acid composition for Spanish cultivated (WF) and wild (BF) borage populations are presented in *Table 1*. Seed weight showed an overall range of variation from 9.3 to 20.1 g per 1000 seeds. The wild, blue-flowered (BF) material had lower values for this character than the cultivated, white-flowered populations, as well as somewhat lower values for GLA and total oil content.

Oil content and GLA composition were generally high and erucic acid low in the Spanish populations (*Table 1*). However a significant range of variability was found in these traits, which has been useful for breeding. Entries with oil content higher than 34%, GLA greater than 27% of oil, and erucic acid values lower than 2.5% have been selected for further use in breeding programs.

As part of the borage breeding program chemical mutagenesis with ethyl methane sulfonate (1% aqueous vol/vol for 16 h) was performed on white-flowered germplasm with high self-fertility (de Haro and del Rio, 1998).

Table 2 Variability in flowers and seeds in untreated and mutant borage plants[a]

		Untreated	Mutant plants			
		plants	Type B[b]	Type B1[c]	Type C1[d]	Type C2[e]
Petals	Mean[f]	5b	8a	5.1b	5b	5b
	Range	5	7-9	5-6	5	5
Sepals	Mean	5c	8a	5.2b	5c	5c
	Range	5	7-9	5-7	5	5
Ovules	Mean	4c	18.3a	7.9b	4c	4c
	Range	4	16-25	7-8	4	4
Seeds per	Mean	2.8b	2.2c	4.5a	1.2d	2.4b,c
flower	Range	0-4	0-8	0-7	0-4	0-4

[a] Number of flowers inspected in each mutant type and in untreated plants: 50.
[b] Type B : plants with flowers larger than normal and, often, with the style and stigma malformed.
[c] Type B1 : plants with normal shaped style and high fertility, obtained after crossing type B plants with normal plants.
[d] Type C1 : plants with flowers closed and seed retention.
[e] Type C2 : plants with flowers partially opened and seed retention.
[f] Means within the same row with the same letter are not significantly different. Significance was calculated by the Duncan's Multiple Range Test (α=0.05 level).

Adapted from de Haro and del Rio, 1998.

Galwey and Shirlin (1990) previously attempted this technique in borage, applied during shorter time periods, with limited results. In the second generation following mutagenesis several mutant types were clearly identified: a chlorotic mutant (type A); a mutant with an increased number of sepals, petals and ovules but reduced fertility (type B); and mutants with closed flowers (type C1) or partially open flowers (type C2), both with increased seed retention (de Haro and del Rio, 1998). In *Table 2* flower and seed characteristics of the type B and C mutants are contrasted with those of untreated plants. Mutant type B, in addition to the increased numbers of floral parts, frequently had stigma and style malformations and, probably related to this aberrant morphology, low fertility. Type B was crossed with normal plants, producing type B1, in which the number of floral parts, including ovules, is less than in type B but still higher than in untreated plants. Furthermore, in mutant type B1 stigma and style morphology are normal and fertility high (de Haro and del Rio, 1998).

3. ANATOMICAL STUDIES OF SEED RETENTION

Anatomical studies of borage fruits were performed in order to identify the factors involved in seed retention.

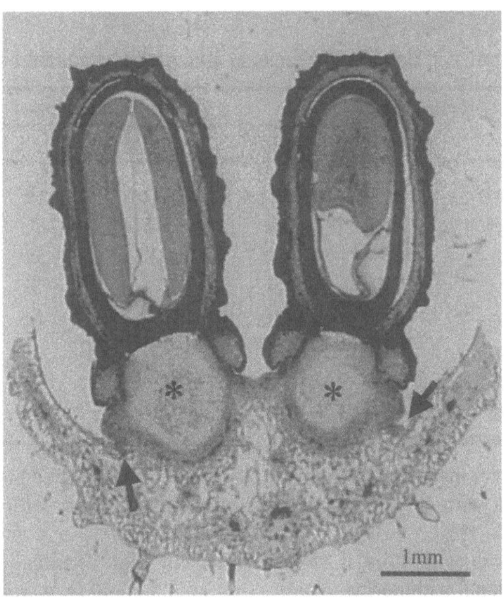

Figure 1. Longitudinal section of wild-type borage fruit showing two of the four nutlets or mericarps, each covered by the typical stony pericarp. The basal part of the nutlet consists of an elaiosome (*), which is attached to the fruit receptacle. At the exterior of the elaisome base is a small remnant of petal tissue (arrow). An embryo is present in the single seed of each nutlet.

Developing fruits of wild types, C1 mutants (closed flowers) and C2 mutants (partially open flowers) were fixed in formalin: acetic acid: 60% ethanol (2: 1: 17 by volume) under vacuum, embedded in paraffin, sectioned at 12 μm and stained with tannic acid, iron chloride, safranin and fast green (Jensen, 1962).

Figure 1 shows a longitudinal section of a fruit of wild-type borage. Two of the four nutlets or mericarps are present in the section. The basal part of each nutlet consists of a rounded, non-sclerified elaiosome, which is attached to the common receptacle or gynobase. Detachment of the mature nutlet occurs when an abscission or separation zone forms between the elaiosome and the receptacle (Roth, 1977). At the exterior of the elaiosome base is a small remnant of the petal base (*Fig.1*); in wild-type borages the petals senesce and abscise early in fruit development.

Longitudinal sections of closed flower (type C1) and partially-open flower (type C2) mutants are presented in *Figures 2* and *3*, respectively, and a transverse section of type C2 in *Figure 4*. In both of these mutants the petals and petal accessory structures remain present throughout fruit development and maturation, and appear to physically retain the nutlets on the fruit receptacle. In the closed flower (C1) mutant, petal structures completely surround and cover the developing nutlets (*Fig. 2)*.

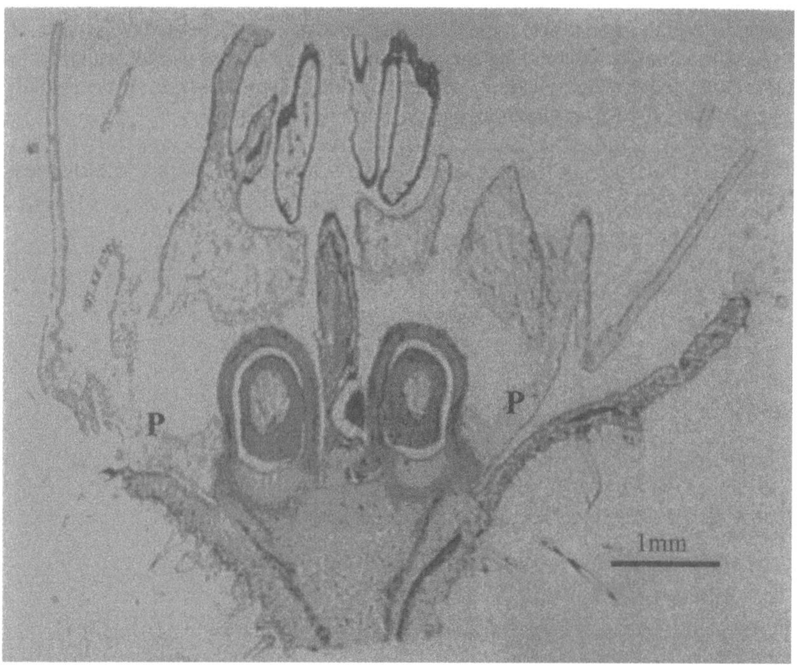

Figure 2. Closed flower (type C1) borage mutant fruit in longitudinal section. Two nutlets and the lower part of the style are present in this section. The maturing fruit is completely surrounded by petal and accessory tissues (P), and the petal bases are pressing against the nutlet bases.

In both mutants pressure of the petals against the nutlet base and elaiosome is evident from the close proximity of these structures and from external indentation of the nutlet base (*Fig. 2, 3*)

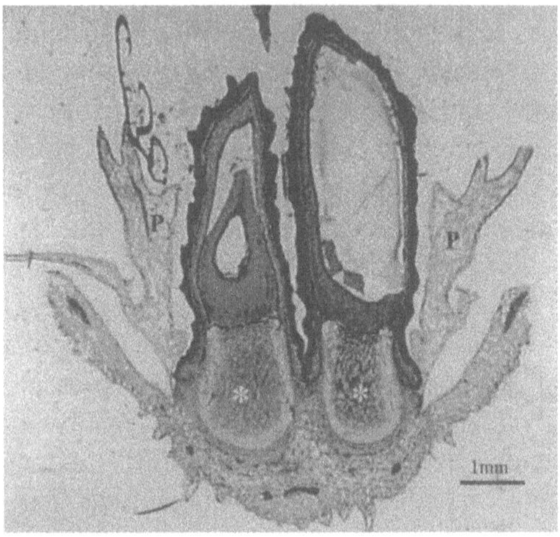

Figure 3. Partially open flower (type C2) borage mutant fruit in longitudinal section. The petals (P) are present and pressing against the nutlets, as indicated by indentations at the nutlet bases and the close proximity of the two nutlets. The elaiosomes (*) are partially embedded in the fruit receptacle.

Retention activity by the petals can also be deduced from the closeness of the C2 nutlets (*Fig. 3,4*) as compared to the wild-type (*Fig. 1*), and the tightly surrounding petal tube (*Fig. 4*)

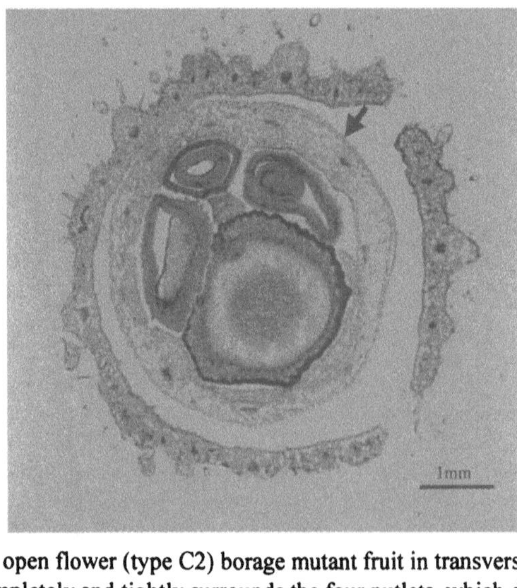

Figure 4. Partially open flower (type C2) borage mutant fruit in transverse section. The petal tube (arrow) completely and tightly surrounds the four nutlets, which differ in degree of development and sectioning angle

An additional component of nutlet retention in the mutants might be the greater degree to which the elaiosomes are embedded in the fruit receptacle (*Fig. 3*) as compared to the wild-type (*Fig.1*). This factor, as well as the development of the nutlet separation layer, is currently under study.

4. BORAGINACEAE GERMPLASM EXPLORATION

Table 3. Fatty acids (mean and range) in Mediterranean Boraginaceae

Species	**Fatty acid content (%)**				
	Oleic	**Linoleic**	**α-Linolenic**	**GLA**	**Erucic**
Anchusa	26.9	41.3	--	10.7	6.6
azurea "52.	16.9-38.1	31.5-51.2	--	4.4-23.1	2.2-8.5
Anchusa	21.8	29.6	13.2	15.8	3.3
officinalis "7.	15.9-29.2	23.7-33.6	8.6-16.7	12.6-20.5	2.0-5.6
Anchusa	23.2	28.3	12.8	13.1	3.5
undulata "4.	18.7-27.2	28.1	10.8-15.3	11.4-15.0	3.4-3.8
Cynoglossum	63.5	1.9	8.6	--	10.7
creticum "1.					
Echium	12.4	14.2	41.8	8.2	--
albicans "2.	11.6-13.3	14.1-14.4	39.8-43.9	7.7-8.7	
Echium "6.	14.1	10.0	47.7	5.1	0.2
boissieri	10.8-17.8	8.3-12.0	44.7-49.5	4.6-5.8	0-0.4
Echium	21	27.2	20.5	12.1	1.3
gaditanum "3.	15.7-30.9	23.7-33.0	12.1-28.0	11.5-12.9	0-4
E. plantagineum "1.	17.1	19.6	33.1	9.6	--
Heliotropium	22.6	66.5	--	--	--
europaeum "1.					
Lithospermum	34.8	37.2	0	9.9	5.3
sp. "2.	33.7-36.0	36.6-37.8	0	9.9-10.0	5.3
Nonea	27.6	26.4	9.9	12.5	1.8
vesicaria "1.					
Omphalodes	40.9	13.7	8.5	6.0	10.5
commutata "1.					
Onosma	20.1	19.9	35.5	5.6	0.4
tricerosperma "1.					
Symphytum	15.8	43.7	1.3	26.6	1.4
officinale "3.	14.3-16.8	42.8-44.5	1.2-1.4	25.5-27.6	1.3-1.5

Each species is followed by the number of entries analyzed.
Adapted from de Haro et al., 1998.

In addition to the breeding program involving borage, different genera and species of Mediterranean Boraginaceae are being collected and tested for seed fatty acid content (*Table 3*) and a germplasm collection is being established (de Haro et al., 1998). These species and genera represent a wide range of oil storage characteristics and growth habits, potentially available for the development of new crops. *Symphytum officinale* is notable for GLA content greater than 20% of oil. That species, however, requires a well-watered environment. *Anchusa azurea*, *Echium boissieri* and *Nonea vesicaria* are three species with promising oil composition and well-adapted to arid Mediterranean climates.

5. CONCLUSIONS

Both borage and other Boraginaceae have potential for development as new Mediterranean crops, which serve as sources of GLA. The wild weedy Spanish borage and cultivated vegetable types offer potential and variability for high seed oil content and desirable fatty acid composition of that oil, but like all *Borago officinalis* their potential for commercial harvest is limited by the problem of seed shattering.

The Spanish borage populations have formed the basis of a breeding and selection program. As part of this program, chemical mutagenesis with ethyl methane sulfonate has provided mutants with greater agronomic potential due to seed retention and increased seed production. Seed retention appears to result from altered floral structure and fruit morphogenesis: modified petals and petal accessory structures are maintained throughout fruit maturation and hold the nutlets onto the fruit receptacle. An additional factor, still to be examined, may be reduced development of the nutlet-receptacle separation layer. Increased seed production can be achieved from the mutants with multiple floral parts, including multiple ovules.

Current research is being directed, on one hand, toward additional exploration and testing of promising Mediterranean members of the Boraginaceae, and on the other hand toward the borage breeding and selection program, including further studies of the morphogenetic factors determining seed retention and the heritability of desirable characteristics from borage mutants and other germplasm.

ACKNOWLEDGEMENTS

Collection of Mediterranean Boraginaceae is under the supervision of Antonio Pujadas, University of Córdoba. Vicente Domínguez, Center for Agricultural Research, Rioja, provided the white-flowered vegetable borage germplasm. The authors wish to acknowledge the assistance of Ester Garcia-Cuevas in the preparation of the anatomical sections and of Gloria Fernández in the chemical analyses.

REFERENCES

1. De Haro, A., M. D. Rio, G. Fernández and A. Pujadas. "*Desarrollo y evaluación de un banco de germoplasma de borragináceas. En busca de nuevas fuentes de ácido gammalinolénico (Formation and evaluation of a boraginaceae germplasm collection. Looking for new sources of gamma linolenic acid).*" XI Jornadas de selección y mejora de plantas hortícolas, Córdoba, 320-326.
2. De Haro-Bailón, A. and M. D. Rio. (1998). *Isolation of chemically induced mutants in borage (Borago officinalis L.).* Journal of the American Oil Chemists Society. 75: 281-283.
3. Del Río, M., J. M. F. Martínez and A. D. Haro. (1993). *Wild and cultivated Borago officinalis L.: Sources of gamma-linolenic acid.* Grasas y Aceites. 44: 125-126.
4. Galway, N. W. and A. J. Sherlin. (1990). *Selection of borage (Borago officinalis) as a seed crop for pharmaceutical uses.* Heredity. 65: 249-257.
5. Janick, J., J. E. Simon, J. Quinn and N. Beaubaire. (1989). *Borage: A source of gamma linolenic acid.* In: Herbs, Spices, and Medicinal Plants; Recent Advances in Botany, Horticulture, and Pharmacology, Vol. 4, L. E. Craker and J. E. Simon, eds, Oryx Press, Phoenix, Arizona, 145-168.
6. Jensen, W.H. (1962). Botanical Histochemistry. W.H. Freeman, San Francisco.
7. Quinn, J., J. E. Simon and J. Janick. (1989). *Recovery of g-linolenic acid from somatic embryos of borage.* Journal of the American Society for Horticultural Science. 114: 511-515.
8. Roth, I. (1977). Fruits of Angiosperms. Gebruder Borntraeger, Berlin-Stuttgart.

ROOT INVOLVEMENT IN PLANT DEVELOPMENT AND PRODUCTIVITY UNDER SEMIARID CONDITIONS

Avner Carmi
Volcani Center, P.O.Box 6, Bet Dagan, 60250, Israel, Tel: 972 3 968 3226/7, Fax: 972 3 966 5327

1. INTRODUCTION

Sorghum genotypes with different root traits were selected to study the contribution of the roots to the build-up of plant tolerance to drought. Two lines with genetic similarity were chosen for a comparative study. The SCO-233 is characterized by a fast root growth, long tap and adventitious roots and deep root system, which was a consequence of the root length and their vertical growth. The SCO-322 line had a much shallower root system which resulted from its shorter roots and their tendency to a more horizontal growth direction, as compared to SCO-233. The overall length of the 322 roots was shorter than that of the 233. The biomass of the root system, and the average number of adventitious roots per plant were greater in the 322 type. The root to shoot ratio, in dry matter, was similar in both genotypes. The above-stated differences were also expressed under water stress conditions, which were imposed by adding PEG to the roots nutrient solution medium. The dry matter production of the two genotypes was reduced by the stress, but the effect on the 233 was less pronounced. The ability to reduce transpiration rate, in response to drought, which is vital for survival in drying soils, was strongly pronounced in 233. In 322 this capability for adjustment of the stomata was much weaker, and the high transpiration levels continued after imposing the water stress. The next stage of this research, studied under

Combating Desertification with Plants, edited by Dov Pasternak and Arnold Schlissel.
Kluwer Academic/Plenum Publishers, New York, 2001.

field conditions, showed the implication of the above-stated differences between the two genotypes on growth, productivity and survival as being under management of minimal irrigation, or under dry land conditions.

2. RESULTS

2.1 Experiments in nutrient solution

2.1.1 Vegetative development in early stages

The differences in elongation rates of the seminal roots between the SCO-322 and the SC0-233 genotypes are presented in *Table 1*.

Table 1. Elongation rate of seminal roots growing in nutrient solution (± SD).

	Seminal root length (cm) at age (days):		
Genotype	**18**	**25**	**32**
322	10.9 ± 1.6	17.7 ± 1.3	23.4 ± 2.6
233	16.3 ± 2.3	30.6 ± 4.6	38.9 ± 8.4

During a-32 d period following sowing, the 233 line exhibited an elongation rate, which was almost doubled than that of the 322 line. This phenomenon occurred under conditions in which the roots were grown in aerated nutrient solution. On the other hand, the initiation of adventitious roots, over a 20 d period, following sowing, was greater in the 322 line (*Table 2*).

Table 2. Number of adventitious roots and root depth of sorghum plants growing in nutrient solution. Measurements were carried out 20 days after sowing (± SD).

Genotype	**Adventitious root Number/plant**	**Root depth (cm)**
322	10.0 ±1.4	19.7 ± 4.0
233	6.7 ± 0.9	30.2 ± 2.6

Measurements of dry matter production, assimilate allocation to roots, stems and leaves, and the dry matter ratio of roots to leaves, are shown in *Table 3*. Over a 20 d period both genotypes accumulated similar amount to dry matter per plant, and the distribution of assimilates between the vegetative organs was similar (*Table 3*). The quantitative ratio in dry matter, as pronounced by root dry matter per leaf area unit was also similar.

Table 3. Dry matter production, assimilate allocation and quantitative root/leaf ratio of 20 day-old-plants (± SD).

Genotype	Total dry wt. (g/plant)	Percentage of dry matter Allocated to:			Root dry matter/leaf area unit (g/dm^2)
		Roots	stems	leaves	
322	0.447 ± 0.087	**28.1 ± 4.4**	**25.7 ± 2.7**	**46.2 ±2.0**	0.12 ± 0.026
233	0.393 ± 0.058	**28.3 ± 1.5**	**27.8 ± 0.8**	**43.8 ± 1.7**	0.12 ± 0.014

2.1.2 Effects of Polyethylene Glycol (PEG)

The responses of the two genotypes to water stress were studied, under conditions in which the stress was imposed by adding PEG to the root nutrient solution medium. The effects of the PEG addition given to 19-d old plants on initiation of adventitious roots and their total elongation value is summarized in *Table 4*. Both genotypes responded to the stress with similar percentage of decrease in the initiation of new roots.

Table 4. The effects of 8% polyethylene glycol (PEG) given to 19-day-old plants on initiation* of adventitious roots and their total elongation. With (+) or without (-) PEG (±SD).

Age	Number of adventitious roots/plant			
(days)	**322(+)**	**322(-)**	**233(+)**	**233(-)**
30	2.2 ± 1.3	4.3 ± 1.2	2.2 ± 0.4	2.0 ± 1.0
32	4.7 ± 1.3	5.2 ± 0.4	4.0 ± 1.4	4.0 ± 0.9
34	5.0 ± 1.2	7.2 ± 1.8	4.7 ± 1.2	5.1 ± 1.3
36	8.3 ± 2.2	9.5 ± 2.5	6.8 ± 1.4	5.8 ± 1.4
38	12.0 ± 3.1	13.7 ± 2.3	12.0 ± 3.7	10.8 ± 2.1

Age	Total length of adventitious roots (cm/plant)			
(days)	**322(+)**	**322(-)**	**233(+)**	**233(-)**
30	9.4 ± 2.8	12.1 ± 2.6	6.1 ± 2.7	6.4 ± 2.5
32	17.4 ± 5.3	30.5 ± 8.1	20.2 ± 8.1	17.6 ± 4.3
34	33.4 ± 8.6	50.7 ± 13.2	39.6 ± 10.0	35.3 ± 6.5
36	50.7 ± 13.3	102.6 ± 24.3	61.7 ± 12.9	69.4 ± 10.1
38	111.6 ± 26.4	181.1 ± 32.1	110.9 ± 27.8	123.6 ± 30.2

*Observations were done on roots that were less than 5 cm length at age of 30 days, and roots that were initiated later on.

A difference in the response of the two genotypes was demonstrated, concerning root elongation. Under the water stress conditions, the total elongation of the new adventitious roots, during a 38-d period was dramatically reduced in plants of the 322 line, and just mildly in plants of the 233 line (*Table 4*).

Both genotypes responded to the stress in reduction in dry matter accumulation (*Table 5*), but the relative reduction in the 322 line was considerably greater than in the 233 line was considerably greater than in the 233 line. The effect of the PEG on assimilate allocation to the vegetative organs was also similar in both lines.

Table 5. The effects of 8% PEG, given to 21 day-old plants on dry matter production and assimilate allocation. Plants were harvested 40 days after PEG addition. With (+) or without (-) PEG. (± SD).

Treatment	Total dry matter (g/plant)	Percentage of dry matter allocated to:			
		Roots	Stems	Leaves	Panicles*
322 (+)	16.1 ± 1.9	23.9 ± 4.4	31.7 ± 2.6	44.4 ± 4.3	-
322 (-)	25.2 ± 2.4	16.1 ± 2.2	51.6 ± 3.7	28.8 ± 4.6	3.4 ± 2.6
233 (+)	15.2 ± 0.7	22.3 ± 2.0	40.5 ± 5.2	27.1 ± 4.5	10.1 ± 1.2
233 (-)	18.4 ± 1.7	14.7 ± 2.7	65.9 ± 7.2	12.6 ± 3.9	6.6 ± 2.5

*Just a third of the plants had panicles

A dramatic change in the effect of the water stress on the transpiration rate of both genotypes was shown, as presented in *Table 6*. While the leaves of the SCO-322 line continued to transpire as normal in spite of the PEG presence in the root zone solution, the leaves of the SCO-233 demonstrated a considerable reduction in their transpiration rates, under the same conditions. Consequently, the leaves of the 322 plants wilted, while such a phenomenon was absent, as concerns the 233 genotype.

Table 6. The effects of 8% PEG given to 21 day-old plants during a 40-day period on total foliage area and transpiration rates. With (+) or without (-) PEG (± SD).

Treatment	Total foliage area (dm^2/plant)	Transpiration rate* (g H_2O/dm^2 x day)
322 (+)	8.1 ± 1.5	70.2 ± 13.6
322 (-)	12.8 ± 2.0	69.5 ± 5.0
233 (+)	14.7 ± 2.7	93.6 ± 11.9
233 (-)	38.5 ± 6.5	131.2 ± 18.9

* was measured over 24 hours on a sunny day.

2.2 Experiments under field conditions

The possibility that the better adaptation of the 233 line, to water stress, as shown in nutrient solutions, will repeat in soil and under field conditions, was studied during several years. The two lines were grown in semiarid regions in Israel, under practical commercial conditions and different regimes of irrigation.

The dryland treatments existed mainly on the water reservoir in the soil profile, which remained after the winter rains, while the other treatments were irrigated by sprinklers during the growth season, according to the commercial irrigation management.

Measurements of roots location in the soil profile are presented in *Table 7*. The lowest depth in which a significant fraction of thin roots (less than 0.5 mm diameter) was observed in soil excavations, which were done 105-d after emergence, were 51, 67, 95 and 105 cm, for the treatments of 322 irrigated and non irrigated and 233 irrigated and non-irrigated, accordingly. The 233 line was characterized by much deeper root systems under irrigated or non-irrigated conditions, as compared to the 322 line.

Table 7. The lowest depth in which a significant fraction of thin roots (less than 0.5 mm diameter) was observed in soil excavations which were carried out between adjacent plants, inside the row. Measurements were taken 105-d after emergence.

Genotype	Irrigation	Root depth (cm)
322	+	51
322	-	67
233	+	95
233	-	105

The reproductive development of the 233 line was much quicker as compared to the 322, under both irrigated and non irrigated conditions (*Table 8*). The suppressing effect of drought on flowering was shown very dramatically in the 322, but was almost absent in the 233.

The yield of the 233 line without irrigation ranged around 3780 kg/hectare, as compared to 4250 kg/hectare in the irrigated plots. The 322 line gave, under the same environmental conditions and under the same length of growth period just an amount of 130 kg/hectare of harvestable yield in the irrigated plots at no-yield in the non-irrigated treatments (*Table 8*). Under arid conditions, in the southern part of Israel, plots of the 233 line, which were raised following irrigated potatoes, gave grains yield of 4732 kg/hectare, in a short growth period of only 88-d, while being irrigated with total amount of 80 mm.

Table 8. Reproductive development and yield parameters in irrigated and non-irrigated plots which were harvested 120 d after emergence.

Genotype	Irrig.	Days to heading	Kernel weight (mg)	Grain yield (g/m^2)
322	+	115	11.5	13
322	-	-*	-	-
233	+	75	16.7	425
233	-	76	17.0	378

* Most of the plants did not flower

3. SUMMARY

A compact line of sorghum, the SCO-233, was studied under hydrophonics or under field conditions. This line demonstrated high adaptation to drought. The genotype is characterized by the ability to develop a deep root system and to absorb water from deep soil layers. The line is indifferent to photoperiod and it can flower and give fertile seeds under extreme drought. Its growth period is short and it might give harvestable yield of more than 5000 kg/hectare in a period of less that 90 days, following sowing.

The potential of adaptation and productivity of this line to growth conditions in expanded semiarid regions in Asia, Africa and South America needs to be studied.

ACACIA MEARNSII FOR INCOME GENERATION AND ENVIRONMENTAL PROTECTION IN KENYA

Naftali Manddy Onchere
Perimart International, Biashara Street, Lakhamshi House, 1st Floor, P.O.Box 33976, Nairobi, Kenya, Tel/Fax: +254-2-243989, email : onchere@iconnect.co.ke

1. INTRODUCTION

Black Wattle trees (*Acacia mearnsii*) were introduced by missionaries into the then Machakos district around 1928. The then Machakos district where the black wattle trees were introduced has only very recently been administratively subdivided into the current Machakos, Kitui, Mwingi and Makueni districts. By the 1950's a large number of families were growing wattle trees for fuel wood and for structural purposes. In the late 1950's, a cotton ginnery was converted into a wattle bark shredding, compressing and baling factory. The wattle factory bought dry wattle bark (called stick bark) from farmers in Machakos, processed it without extracting tannin, and exported it

In 1968 the wattle factory received more stick bark from the farmers than it could handle with its capacity. The factory management took the draconian decision of burning a large quantity of the stick bark belonging to the farmers of Machakos. The wattle bark farmers were incensed and sought legal redress. In the quest for redress, it came to light that the farmers were being economically exploited. There was a huge price disparity between the price that the wattle bark fetched in the export market and that which the factory paid the farmers.

Combating Desertification with Plants, edited by Dov Pasternak and Arnold Schlissel.
Kluwer Academic/Plenum Publishers, New York, 2001.

The government decided that the farmers should be assisted to purchase the plant through the Cooperative Bank and with technical assistance from the Ministry of Cooperative Development and the Ministry of Agriculture. This was envisaged to benefit the wattle bark farmers who will get better prices for their wattle bark and share the profits from the factory. This was also at the time when the recently independent Kenyan State was strongly enforcing its Africanization Policy and assisting her native citizens to purchase property and businesses from non-native Kenyans. This was being facilitated through the cooperative movement, indigenous land buying companies, and specifically set up or facilitating financial mechanisms and institutions.

The farmers, in 1973, formed the Machakos Wattle Bark Cooperative Society (MWB) in preparation for the intended purchase. In the meantime, they still continued to sell their wattle bark to the factory. The MWB was officially registered as a cooperative society in January 1974 as Cooperative Society Number 2199, under the Cooperative Societies Act (Cap 490 Sec 7). The members of the MWB were drawn from nine locations from the Central, Western and Eastern administrative divisions of Machakos district. By the end of the membership recruitment drive, in 1979, these locations were Iveti, Mitaboni, Muboti, Kilungu, Mukaa, Kikima, Kitundu, Wamunyu, and Kalama.

As the farmers organized themselves to purchase the factory, the factory management bought a posho (flour) mill and a sisal mill. Before the farmers finalized the purchase of the wattle bark factory, its management had not only purchased the posho and sisal mills but had in the meantime sold the factory to a third party. In 1977 the factory was valued at Kenya Shillings 1.9 million. The farmers agreed to purchase the factory from the third party, inclusive of the posho and sisal mills, in 1979, at a total value of Ksh 2.175 million. It was agreed that the members of the MWB should raise 25% of the total purchase cost of the factory and the Cooperative Bank raise the remaining 75%. The MWB factory was officially opened in September 1979.

Since then, the MWB offers members and non-members dry wattle bark processing and marketing facilities, until these operations seized. It has leased the sisal mill to commercial companies to clean and market sisal from their own sources. The MWB also offers maize milling services to the people living around the Konza MWB factory.

Over the formative years, the operations of the MWB grew from strength to strength and enabled the society to meet its principal service objectives. The executive management committees and the MWB staff achieved clearly visible and commendable progress during this period. The members of the MWB and other farmers who had sizeable acreage of black wattle recorded clearly visible and tangible economic gains. The effects of drought and desertification were mitigated and the farmers concerned lived dignified and self-supporting lives. They were not dependent on famine relief and handouts even during reasonably severe droughts.

The advent of economic liberalization introduced hitherto unheard of market forces and competition. The MWB was not prepared for the ensuing onslaught and could not cope. The operations of the wattle factory that generated the bulk of the income for the MWB closed down in 1989 and have not been revived by November 1999. The only economic activities of the MWB since 1989 have been leasing the sisal mills, providing maize milling services to local residents around the factory and the leasing of former staff houses to private individuals.

Despite formidable odds, the MWB has somehow continued operating, with executive management committee members being regularly elected and a clearly defined management structure being in place. The assets of the MWB have not been plundered. There have been no recorded incidents of corruption or misappropriation of funds. These positive achievements are a testimony of commitment to the activities and operations of the MWB by the local communities.

2. MANAGEMENT OF THE MACHAKOS WATTLE BARK COOPERATIVE SOCIETY

The management of the MWB is composed of an executive management committee, factory staff, and members' delegates. Initially each of the three administrative divisions from where the members of the MWB are drawn elected three members to the executive management committee. Office bearers were then elected, including the chairman, vice chairman, treasurer, vice treasurer and secretary. Later, a delegate system was started where members from each of the locations elected delegates. These delegates in turn elected local office bearers amongst themselves. The delegates from all the locations then met and elected the executive management committee members for the society from amongst themselves.

The delegates normally meet four times in a year, once per quarter. Delegates' meetings are usually held at the factory premises at Konza town, Machakos district. The executive management committee members are charged with overall management of the factory with the assistance of the factory staff. The executive management committee and delegates are reimbursed transport costs and are paid a sitting allowance.

An Annual General Meeting is held once a year based on the delegates' method. The executive management committee is elected every three years during which time, a third of the committee members step down and a new committee elected. Old office bearers are not barred from contesting and can thus be re-elected. This is supposed to provide for continuity of MWB operations and ensures that institutional memory is not lost.

The entire membership has met only once and the meeting was such that another one has never been called by November 1999. A delegate system has since been enforced as a means of getting members' opinions and views. At local level, general meetings are rare and far in between. Knowledge of the operations and running of the factory is not common amongst the ordinary members. There are currently three factory staff members: a factory manager, a posho mill operator and a watchman.

3. SILVICULTURAL PRACTICES OF BLACK WATTLE

Black wattle trees are known for the good quality of structural wood and fuel wood that they produce. The trees have a fast growth rate and are not prone to infestation by pests and diseases. The tree bark, when young, is used widely in Kenya for structural purposes to hold two or more poles, rafters or even trusses together in the absence of nails. The young bark is therefore used while fencing, thatching roofs, building the superstructure of houses etc. When the wattle tree bark gets older, it is sold to several companies who process it and extract tannin, a phenolic compound that is used to tan leather from hides and skins. When used in the construction of traditional «ceiling boards» that act as crop storage facilities, the bark mellows with age and wood smoke.

In a number of communities in Kenya, where black wattle trees are grown bacterial throat infections and infected wounds are traditionally treated using the juice extracted after chewing mellowed wattle bark from the traditional «ceiling board». These traditional «ceiling boards» act as the best storage facilities for crop harvests by minimizing on post-harvest crop losses due to pest and disease infestation.

The seedbed for black wattle is prepared to a very fine tilt because of the small seed size and the fact that direct seeding is the most economical method of planting black wattle. To break seed dormancy and ensure more even germination, the seed are put in water that is brought to boil and then left to cool under ambient conditions overnight. This procedure is called parboiling. Seeding is done either by drilling in rows or by broadcasting. Optimal plant population densities are achieved when seeding is done by drilling. The agro-ecological zone, the end use of the trees and its products, determines spacing. It is common practice to use a much higher seeding rate than finally required, and then keep on thinning periodically as the saplings grow until the required plant density is reached. The thinnings have found many uses in places where black wattle trees are grown. The thinnings are used for weaving baskets; making fish traps; making traditional doors and windows among the poor; in the construction of houses, granaries and toilets; for construction purposes as fence droppers; as stakes for horticultural crops like tomatoes etc.

Trees start being pruned once they are 6-10 ft tall dependent on intended final use. Pruning leads to good quality bark and structural wood. The fre quency of pruning depends on intended final use of the trees and tree products but it is common to prune once a year.

Once a black wattle plantation is established, it is normally unnecessary to re-establish the plantation again by re-seeding. The seed dormancy, long viability of the seed and ease of regeneration of black wattle guarantees that there are successive waves of seedlings to re-establish the plantation. For this reason, it is common to find farmers with black wattle plantations that have several clearly discernable generations of trees growing in one plantation as a multi-canopy.

With the increasing human population and pressure on land, a number of farmers and East African Tanning Extracts and Tanning Company (EATEC) have been researching on using *Acacia mearnsii* as an agroforestry species with various degrees of success. Available evidence suggests that the black wattle tree, when it is still young, is a good agroforestry tree and can coexist well with other trees and crops with little or no deleterious effects on the production and productivity of the other trees or crops, especially with good management practices. When the wattle tree grows to a diameter beyond 10 inches it starts to affect the production and productivity of trees or crops growing near it.

As a rule of thumb, black wattle trees are stripped of bark when they have achieved a diameter of 5-7 inches and are about 5-9 years old. This is when the bark has the highest tannin content of 42-47%. The tannin content of the bark fluctuates according to the age and size of the tree, the season, soil type, amount of ferrous iron in the soil, soil drainage, and handling of the bark before extraction of the tannin.

The tannin content is highest in the bark when the black wattle trees are 5-7 inches in diameter and are 5-8 years old. The tannin content is also highest in those trees grown in deep, well-drained volcanic soils that have low ferrous iron content. The time of harvesting determines the quality and quantity of tannin. During the dry season, tannin acts as a source of food for the plant and is found more concentrated in the leaves than during the wet season. During the wet season, tannin reaches the highest concentration in the bark. To get the highest quantity tannin it is therefore recommended to strip the bark when the tannin content is highest, i.e. during the wet season.

Tannin quality is determined by the level of oxidation of the Phenolic compounds that make the bulk of the tannin extract: the less the oxidation, the higher the quality of the tannin extract. Oxidation also affects the coloration of the tannin extract, given on a technical scale of reds, which decreases the leather quality during treatment. Oxidation of tannin within the bark increases with increase in age, tree diameter, and ferrous content of the soil and poor drainage, exposure to the sun and compression of the wet bark. Handling of the bark after harvesting therefore determines the quality of the tannin. For the best quality tannin, the bark should be stripped during the wet season when the trees are at the right development stage (diameter 5-7 inches; age 5-8 years) and the tannin extracted immediately, before undergoing oxidation. The bark is then technically referred to as green bark. Drying green bark results in stick bark. Stick bark has low quality tannin because the tannin has undergone oxidation.

Wattle bark is made up of tannin, non-tannin, and moisture. Wattle bark from trees younger than 5 years has much more non-tannin than tannin. Tannin from trees older than 9 years has undergone oxidation and is of poor quality. Within the tree bark, oxidation of tannin is highest near the base of the tree and least towards the shoot. Oxidation of tannin makes the bark turn dark or gray in color, instead of the normal green. Drying of stick bark under direct sunlight or heat increases oxidation of tannin, and it is therefore recommended to dry stick bark in the shade.

4. MARKETING

In the wattle bark trade, one could process stick bark, bale it and sell it, or process wattle bark, extract tannin and sell it. Sale is normally by export. There are only two companies in East Africa, which extract tannin, namely EATEC and KTE. These are business concerns that are profit driven. Two other concerns, the Sotik Wattle Bark and the MWB, were the only two other concerns dealing with wattle bark within East Africa. These were owned by farmers and dealt only with processing stick bark and marketing it. The Sotik Wattle Bark Company has since gone under and it is only the MWB that remains as the sole flagship of black wattle farmers.

Tannin extract is traditionally produced from black wattle trees but there is now stiff competition from tannin extract produced from Chestnut and Quebracho trees from South America. Locally there is stiff competition with synthetic tannin. It is this synthetic tannin that is used by all shoe-making companies in Kenya. Synthetic tannin is made using sulphur, which adversely affects the environment, the ozone layer and users. Synthetic tannin produces poorer quality leather than natural tannin and it is much cheaper than tannin extract from black wattle, which makes the best quality leather.

There are three conventional types or grades of tannin extracts from black wattle. These are:

1. Solid mimosa extract.
2. Green Sulphited mimosa powder, commonly called GS powder.
3. Spray dried ordinary mimosa powder.

Solid mimosa extract is the cheapest and poorest quality tannin extract. It is made out of stick bark alone. Its demand worldwide appears to be waning and the profit margin is lower than the other two types of tannin extract. Spray dried mimosa powder is the next cheapest tannin extract. It is of higher quality than the solid mimosa extract. Spray dried mimosa powder is made out of stick bark, green bark or a combination of the two. GS powder is the highest quality tannin extract, has the highest demand and profit margin. GS powder is made only from green bark. The green bark that is used to make GS powder must be processed within a maximum of 24 hours after being stripped.

The four companies dealing with wattle bark have traditionally exported to India, Pakistan and Bangladesh mainly. Ethiopia and Egypt are potential countries to earmark for exporting to. Europe and the Americas mainly use synthetic tannin or tannin extracted from trees other than black wattle. The main source of tannin for Europe and America is South America.

The MWB has always supplied the Indian market that had been identified and used by the original plant management. Local or African markets could be targeted only if the MWB can get the capacity to extract tannin. In this case, Tanzania, Ethiopia, Uganda, Somalia, Sudan, Zambia and Zaire could become lucrative markets.

The main competitors to EATEC, KTE and the MWB are South Africa and Zimbabwe in Africa and Brazil and Chile in South America. Kenya and Zimbabwe produce the best quality tannin in the world but this targets the discerning market. Local competitors to the MWB are EATEC and KTE who are the only ones that extract tannin.

5. ENVIRONMENTAL ISSUES

Traditionally, the inhabitants of these districts were mixed farmers who practiced a form of shifting cultivation. Hunting and collection of honey were widespread to substitute livestock production. The land was left fallow, normally for 3-5 years, sometimes for more than fifteen years after only 1-2 years of farming. This ensured that the land use was sustainable and the environment, together with its flora and fauna, was conserved. The long period that the land was left fallow ensured that there was adequate regeneration and rejuvenation of the genetic resources. No single species of either flora or fauna was under threat of over-exploitation or extinction. The soil was well covered and maintained its fertility and characteristics. The Akamba inhabitants peacefully co-existed with their environment, harvesting out of nature's bounties only for sustenance. Land was communally owned and the Kamba population was mobile.

The advent of land adjudication localized the population, which started putting pressure on the land and the natural environment. As the population increased, the period that the land was left fallow continued to decline and pressure continued to bear on the exploitation of the flora and fauna for sustenance. Rejuvenation and regeneration of the flora and fauna was becoming more and more incomplete before being exploited for sustenance by the population.

The introduction of a monetary economy introduced a profit motive in the exploitation of the resources of the environment. This led to the opening up of much more land. The enforcement of poll and hut taxes further acerbated the land opening up demand.

The soil in the Machakos and Makueni district is of medium to low fertility, fragile and often shallow. It requires several years, on average 3-5 years, of being left fallow after 2-3 years of tillage so as to regain its fertility, according to some farmers interviewed.

The increased opening up of new land, the almost continuos tillage of the soil, the effects of overgrazing and overstocking, the increase of the carrying capacity of the land for wildlife as well as livestock had their toll on the environment. Soil erosion increased as more land cover disappeared and surface water ran-off increased and soil water seepage decreased. The decreased soil water seepage led to decreased productivity of the soil which in turn led to decreased yields for the inhabitants' sustenance and budding commerce. Over-hunting of game for meat, hides, skins, ivory, horn and other game trophies set in. Over-exploitation of specific tree species for woodcarving and other traditional purposes and herbs for medicinal and other uses also set in.

By the early 1940s, the environment was so degraded as to necessitate the colonial government to take urgent measures to reverse the trend. Natural resources were being over-exploited for immediate benefit with little worry for long term negative consequences. Over-exploitation of the soil led to decreased farm yields and poverty. Excessive soil erosion led to severe gullying. Over-hunting led to dwindling numbers of wildlife. A number of birds, small herbivores, elephants, rhinos and snakes started disappearing.

All these factors led to increased poverty, which could only be alleviated through opening up more land.

The colonial government gazetted the Tsavo National Park and Amboseli National Park to protect the threatened wildlife populations. It also enforced draconian measures to conserve the soil. Soil conservation became a form of forced labour for all residents. The full force of the law was used to dig deep cut-off drains as well as bench terraces. Stabilization of soil conservation structures using *cenchrus ceriaris, eragrostis superba,* and later 'makarikari' grass were enforced by village headmen, chiefs and the colonial administration using whips and other more physical means. Other environmental conservation activities like the prudent use of natural forest plantations were also enforced by the use of legal force. Poachers were sometimes shot on sight, forced to do soil conservation work or imprisoned.

The independent government enforced some of these strategies to conserve the environment for a while before it became apparent that the strategy was not working and needed to be changed to gentler and individually convincing methods.

6. MAJOR CONSTRAINTS TO FULL EXPLOITATION OF BLACK WATTLE TREES

The major constraints to the full exploitation of black wattle trees for income generation and environmental conservation in Machakos and Makueni Districts are related to:

- Low capitalization of the MWB factory at Konza and its operations;
- Inadequate technical and human resource capacity of the MWB
- Widespread poverty, economic marginalization, sporadic famine and drought within the two districts;
- Poor attitudes of the local population to economic ventures;
- Resistance to environmental conservation due to stigmatization through bad experiences in the past; and
- Inadequate marketing of stick bark and its low returns as compared to tannin extract.

7. CONCLUSIONS AND RECOMMENDATIONS

Acacia mearnsii offers a unique opportunity for income generation and environmental conservation in Machakos and Makueni districts. This conclusion has been reached comparing the black wattle farmers in Uasin Gishu and Nandi districts in Kenya. On this basis, the following recommendations are made:

- To increase the share capital of the share holders of the MWB;
- To raise the level of capitalization and turn round the financial position of the MWB;
- Increase the shareholding, if necessary, through farmers supplying stick bark to the MWB factory;
- To raise awareness of the current status of affairs MWB amongst share holders and the action that need to be undertaken to revive its operations;
- To increase the number of shareholders;
- To search for and acquire necessary funding to jerk the MWB into operation;
- To run the affairs of the MWB as a profit making business and not just as a service delivery;
- To improve the technical and human resource of the management and staff of the MWB as required in marketing, business management and administration;
- To undertake such measures as are necessary to add value to the products e.g. extraction of tannin;
- To reverse the deteriorating financial position and operations of the MWB;
- To start a reserve fund which acts as a buffer fund that can be used to regularly part pay farmers once they have delivered their produce;
- To undertake awareness raising measures that will increase the use of wattle bark for environmental protection and income generation.

The United Nations Convention to Combat Desertification, CCD, offers a major opportunity to the MWB through the provision for a National Desertification Fund (NDF) that will channel resources from national level rapidly to populations affected by drought and desertification. The NDF in Kenya has been established in Kenya as the Community Development Trust Fund (CDTF). The MWB will do well to access funding from the CDTF to assist in overcoming the financial constraints it faces.

ACKNOWLEDGEMENTS

I am deeply indebted and very grateful to all those who participated in this research: men and women in the then famine stricken districts of Machakos, Uasin Gishu and Makueni. They gave willingly of their time to explain the activities they were involved in and expressed their opinions.

These opinions were about the activities of the Machakos Wattle Bark Cooperative Society (MWB), East African Tanning Extracts Company (EATEC) and Kenya Tanning Extracts Company (KTE). I am also indebted to staff of the MWB, Claymo, MOCD, EATEC and MOALDM who made time available for interviews and travel to the field. I owe a lot of gratitude to Mr. David Musya who selflessly gave of his time to accompany and direct me to all the places I visited in Machakos and Makueni. My sincere gratitude also goes to the unnamed staff of Claymo Company Limited, EATEC and KTE who gave sensitive and valuable information freely so that some starving wattle tree growing Kenyans in Machakos and Makueni may be availed an opportunity to lead dignified and self-supporting lives. Last but not least, Mr. Maitha of MOCD and Messrs. Alex Kirui and Joseph Mwangangi of Technoserve Kenya for their guidance, logistical support and genuine concern and support for local development initiatives.

LIST OF ACRONYMS

ASAL	Arid and Semi Arid Lands
CBK	Cooperative Bank of Kenya
CBO	Community Based Organization
DDC	District Development Plan
EATEC	East African Tanning Extract Company Limited
GS	Green Sulphited Mimosa Powder
KTE	Kenya Tanning Extract Company Limited
KVA	Kilo-Volt-Amperes
MWB	Machakos Wattle Bark Farmers Society
MENR	Ministry of Environment and Natural Resources
MOALDM	Ministry of Agriculture, Livestock Development and Marketing
MOCD	Ministry of Cooperative Development
AGM	Annual General Meeting

REFERENCES

1. Action plan proposal for 1997 and beyond for MWB
2. MWB Minutes of general delegates meeting held on 17/1/97
3. MWB Chairman's report for the year ended 1996
4. MWB Chairman's report for the year ended 1979
5. MWB Minutes file
6. MWB Miscellaneous office documents

7. Technoserve report on meeting held on February 13, 1997 in Machakos district regarding the Machakos Environment Project and Disaster Relief/Development Assistance.

POTENTIAL OF AUSTRALIAN ACACIAS FOR HUMAN FOOD

Adoption of seeds of Australian acacias as a new food in Niger Republic.

Tony Rinaudo
World Vision, AMEEE team, IIP, 1 Vision drive, East Burwood Vic 3151, Australia, Tel: 613-9287-2309, Fax: 613-9287-2377, email: rinaudot@wva.org.au

1. BACKGROUND TO RECENT INTEREST IN ACACIAS FOR HUMAN FOOD

There is a sense of increasing alarm over Africa's decreasing ability to feed itself. From 1969-1992 the hungry population of Africa doubled from 103 million to 215 million. FAO projects that the number of undernourished Africans could rise to 265 million or 30% of the regions population by 2010. The per capita food consumption among hungry people declined from 1,490 calories to 1,470 calories per day, far below the minimum requirement of 2,350 calories (Hunger, 1997).

The Maradi Integrated Development Project (MIDP) working in Niger Republic, West Africa, has been helping farmers to evaluate their problems and work towards solving them, using local resources and people, since 1975. Between 1984 and 1999 the author Tony Rinaudo was involved in five famine relief interventions conducted through the infrastructure of MIDP. Famines have occurred throughout the history of Niger. What is alarming is their increasing frequency and scale. Within one season, farmers have had to replant their millet crops 5-8 times due to drought, windstorms and pests. Reasons for hunger are many and complex. No simple, single solution is likely to reverse this situation in the near future any time soon, however, growing edible-seeded Australian acacias can help address some of the

Combating Desertification with Plants, edited by Dov Pasternak and Arnold Schlissel.
Kluwer Academic/Plenum Publishers, New York, 2001.

causes of famine by diversifying the range of crops grown and thus reducing the risk of total crop failure. This paper's focus is on experiences of introducing acacias into Niger Republic and its aim is to promote acacias as an overlooked food crop of high potential, particularly in sub-tropical dry-zoneArid and semi arid regions.

Since European settlement in Australia, the food potential of acacias has gone largely unappreciated. Early explorers and settlers had little enthusiasm for native foods. Unfortunately, even Aboriginal populations living in many outback regions today are not fully utilizing these foods, thus valuable indigenous knowledge is being lost. In fact, if cooked appropriately, the seeds of acacias are not only very tasty they are quite nutritious, being high in protein (17-25%), fat (4-16%) and carbohydrate (30-40%) (Brand & Cherikoff, 1985).

This apathy largely persisted until recently when native foods became popularized by personalities such as the colorful "Bush Tucker Man", Major Les Hiddens, and through results of research into the nutritional attributes of the seeds by Brand, Cherikoff and others. Scientists such as Dr. Chris Harwood and Dr. Lex Thomson in Australia's CSIRO Forestry and Forest Products Division saw the potential of acacias as a food source in developing countries in the semiarid tropics, and have been researching and promoting this possibility steadily for over a decade.

In Australia today the use of ground and roasted Acacia seed as a gourmet food flavouring and ingredient is a rapidly growing but still small-scale industry. A number of individuals, and commercial and research groups are actively investigating the potential of Australian acacias for human food (Maslin et al.,1998).

Of the acacia species tested in Niger, so far *A. colei* and *A. elachantha* (formerly known as *A. cowleana*) have stood out in terms of ease of propagation, rapid growth, survival under field conditions and seed production, thus showing the most potential for development as a new food crop. They have good agronomic, harvesting and food characteristics. Analysis of *A. colei* seed flour indicated that it has good nutritional value and that known toxic and anti-nutritional factors were absent or at levels below those that would cause any concern (Harwood *et al.* 1999).

Details of botanical descriptions, characteristics, natural occurrence etc. of edible Australian acacias are available in a number of references (Doran and Turnbull 1997, Thomson 1992, Maslin *et al.* 1998).

Appendix I describes advantages, limitations and recommendations for *A. colei* and *A. elachantha* based on field experience and observations in Niger. Other species under investigation that warrant further research include *A. torulosa, A. coriacea var. sericophylla, A. ampliceps, A. victoriae* and *A. tumida.*

2. GRADUAL ADOPTION OF ACACIAS AS A NEW FOOD CROP

The author's interest in acacias came out of the realisation after severe famine in 1984, that hunger could occur in any year as long as farmers relied on annual crops in the uncertain climate of Niger. The ACIAR-sponsored conference "Australian Acacias in Developing Countries" held in Gympie in 1986, a subsequent visit by Australian forester, Lex Thomson in 1989 and the encouragement and support of the Australian Tree Centre generated the interest needed to begin acacia species and food trials in Niger. While results of these trials have been very positive, there has not been an overwhelming response on the part of the people, rather, acceptance has been sporadic and gradual. However, in light of the long period of time it took for the western world to accept potatoes and tomatoes (Appendix II), progress has been surprisingly rapid! Collaboration between the Australian Tree Seed Centre, MIDP, local authorities and the farmers themselves has resulted in successful evaluation of acacias. An understanding of their potential has been gained and optimum silvicultural methods have been largely determined. It is up to the farmers now to accept or reject the findings.

Year after year, Nigerien farmers are perplexed as they see their millet crop fail or yield inadequate harvests. They do not know where to turn as there are few options.

Existing strategies include:

- Exodus. In some districts 90% of working males seek work in neighbouring countries during the 8-month dry season.
- Clearing new land. However there is virtually no new or even fallowed rested land to cultivate and with few nutrients being returned to the soil, land average millet yields are as low as 250 - 300 kg/ha in "good" years.
- Cutting and selling the few remaining trees for firewood to supplement income. Even tree stumps are uprooted and sold in times of severe hunger.
- Selling farmland.
- Searching for and testing new crops. However, amongst traditional food crops grown around the world none have been found which perform better than millet and sorghum under traditional farming methods.

Acacia colei (then known as *A. holosericea*) was planted quite widely in southern Niger during the late 1980s by forestry agencies, as a fast-growing small tree for windbreaks and firewood production, and its potential to produce seed had already been established. Species trials were commenced in 1992 to determine whether there were other species also suited to local conditions, which might be superior to *A. colei*. In 1994 famine struck again and so as part of the food for work program, farmers were required to raise and plant out acacia trees on their own land in order to receive grain handouts. Along with this, acacia food demonstrations were conducted in key villages to introduce people to the new food. Well over 100,000 acacia trees have been planted in the district as a result of food for work programs.

Despite this good start, some farmers pulled acacias up several years later, in anger that the project did not buy acacia seed from them. Despite all the discussion and training aside, they had seen acacias as a potential cash crop and had missed the point about the importance of a backup food crop for when their staple crop fails.

A human volunteer trial in 1995 (Adewusi et al, 1998) followed up and complemented earlier chemical and animal testing conducted by CSIRO and showed *A. colei* to be safe, and gave acacias a clean bill as far as toxicity and anti-nutritional factors were concerned. This was a milestone, as being a completely new food, there was an element of doubt about the safety of acacias. The experiment received official approval and almost unanimous backing from government agencies (health, agriculture, forestry, state government) and traditional leaders. The result of the trial, that people fed diets incorporating 25% acacia seed flour put on more weight than those on traditional "control" diets and experienced no ill effects, was not lost on the local population.

Meanwhile, two charismatic farmers (Barro Mahaman and Dan Mani Idi) had become unofficial acacia promoters through their sheer inventiveness and enthusiasm. Upon hearing that "acacia coffee" was marketed in Australia, both experimented and perfected home grown ways of making "coffee" (a palatable hot drink can be prepared using ground, roasted acacia seed flour. It does not contain caffeine). It was an instant success, being much cheaper than imported coffee, and stocks rarely lasted one week after production. Both men made up new acacia recipes and consumed acacia foods regularly, introducing it to their neighbours as they did so. It is interesting that both also saw the potential of acacias for restoring the health and body weight of emaciated infants. They prepared "Bitamin" packets (a mixture of acacia, corn, bean and peanut flour), which they gave away for free at first and later sold to mothers who by then were convinced of the value of the product!

Noticing that goats ate the fallen seed of acacias, some farmers began collecting seed and feeding it as a supplement to their animals during the 8-month dry season when fodder is scarce and of poor quality. and animals generally lose body weight. Individual farmers have even begun purchasing acacia seeds for this purpose as a more effective substitute to bran as a dry season supplement.

In the village of Garin Sale there is an *Acacia colei* plantation on an eroded hillside. Each year at harvest time the chief proclaims an open day for anyone in the village to go and harvest acacia seed. The chief claims that every family participates and all see the value of acacia for food. They eat it because they like it, not just because of hunger.

On a visit in March 1999, heavily bearing trees growing on sandy farmland were inspected. Because of the size of the trees and large amount of seed (estimated at between 6 – 10 kg per tree) the chief was asked to measure the yield. However, when he went to collect the seed, he found that the village children had already come and harvested it! They had roasted it, removed the seed coat, pounded the seed to flour and then they added salt and ate it, sharing some with their parents. This was a bad day for science, but a very good day for the gradual adoption of acacias into the regular diet!

In most villages in the MIDP program however, people are not yet eating the seed regularly. In many cases seed was harvested and stored away in the false expectation that the project would buy it. In 1998, food shortages were severe and out of desperation as much as anything, some people decided to eat the seed themselves. They had heard many times of the food value of acacias and some had tasted the prepared foods. To their surprise, the foods actually tasted good when they prepared it themselves. This is especially so when compared with traditional famine foods such as bran, various grasses, the treated seed of a poisonous berry (*Boscia senegalensis*) and even the foul meat of animals that perished from hunger. It is hoped that experiences like this will help popularise acacia in the local diet.

Because of the widespread "food for work" plantations, many farmers have a daily reminder of the potential of these trees. In 1998, despite good rains, both millet in control plots and millet grown between rows of *A. elachantha* barely produced any seed. This was a result of a late start to the rains and insect attacks. The acacias were not affected by this and produced good quantities of wood and edible seed. Repeated demonstrations like this must have an effect of reminding people of the opportunity to adopt a new food.

MIDP has been able to share its experience with acacias right across Niger through invitations of staff to other areas and visits by farmer groups to Maradi. One such exchange was to the Lake Chad area in eastern Niger where farming is extremely precarious. Rainfall is only 100 – 200 mm per year and very sporadic. There are many pests including monkeys, birds and insects that cause heavy losses. Typically 7 in 10 years are bad (hungry) years. Keen to increase food security, one project manager tried to raise 10,000 *A. colei* seedlings There was almost total failure as plants germinated but then yellowed and died in the pots. It was realized later that the alkaline, potash rich soils were unsuitable for *A. colei*. In 1997, 130 *A. ampliceps* were planted out. These did well but goats broke into the gardens and decimated most of them. In one of those providential turning points, just two trees were not eaten. After all that effort the project leader says he was ready to give up on acacias but for these two trees which not only survived but also thrived. At 1 years they were 3 – 4 meters tall and setting seed.

Farmers in the area were asked why they were still planting the acacias even though the food for work program was over. They said that they were tired of being hungry. They had tasted the acacia food being demonstrated in 1996 and were motivated enough to persevere, despite many set backs. Raising trees in this region has proven even more difficult than in the Maradi area. Rainfall is lower, temperatures are higher, incidence of insect attack is greater and grazing pressure is excessive. The low areas where tree planting is possible are subject to flooding during good years.

In addition to physical problems there is considerable social pressure working against anyone who dares to be different. In the Maradi area, herders look down on any type of farming and Fulanis actually despise anyone in their tribe who would work with trees like their traditional Hausa enemies. In contrast it is ludicrous to a Hausa millet farmer to think of spoiling his cropland by planting trees on it. In both regions, farmers would have to have extremely good understanding of the advantages of growing acacias in order to be motivated and committed enough to be successful.

Acacia seed has not entered the formal market place yet, but there are increasing instances of farmers buying seed from each other informally for human and animal food. From time to time women have sold popular deep fried bean cakes made with

15 – 25% acacia flour. Stories like these may be the essential first steps towards widespread acceptance of acacias.

From the start, project management was very aware that all the incentive and promotion was coming from white foreigners who most likely didn't eat acacia seeds themselves, except on demonstration days! In 1998 the Australian government sponsored a cultural exchange visit. Rosie Nangala and Kay Napaljarri, two Aboriginal women from Central Australia who retained traditional skills of collecting, preparing and eating acacia seed travelled to Niger to share their knowledge on various acacia species and how they prepared them for eating. To the Niger people who met them, this showed that acacias were not just a white man's "thing", but black people like themselves had discovered this food and used it for thousands of years.

It is interesting to note that Rosie and Kay were equally impressed with the variety of ways in which their foods were incorporated into the local diet. Their uses of the acacias had remained static, and perhaps partly because of that their knowledge was being lost. In Africa, acacia seed flour was not promoted in totally new recipes, it was incorporated into established local dishes developed by local people by blending with staple millet and sorghum flours and experimenting to get the right taste and texture.

Two interesting exchanges were noted during the visit. Some urban based dignitaries and public servants seemed to miss seeing the back up and supplementary crop role acacias could play for farmers in rural areas. Their oft-repeated question was: " do white Australians eat this and are acacias available as packaged products in your supermarkets? "They were interested in the samples of commercial gourmet acacia products from Australia. It appears that in trying to promote a new food, prestige and status areis as important as functionality and taste!

The second exchange was with some forestry agents. In most Forestry Department circles in Niger, *A. colei* (known locally as "holo" from its former name *Acacia holosericea*) has been written off **as a tree** because of its short lifespan (only 3-4 years in close spaced plantations, and certainly no more than 10 years under ideal conditions).

When challenged about this, Dr Chris Harwood responded that if the Australian acacias were compared with other commonly planted tree species such as Neem and *Eucalyptus camaldulensis*, they were inferior as a species for tree plantations to produce firewood, timber and, windbreaks, because of their short life and small adult size. However, if acacias were thought of **as a short-lived perennial crop** yielding edible seeds and firewood, providing a temporary windbreak for nearby crops and restoring exhausted farmland through their nitrogen fixing capability and production of leaf litter, they could prove an attractive component of farming systems.

Nobody gets frustrated with pigeon peas, which have a lifespan of 2-3 years, because they are seen primarily as crop plants. Pigeon peas live longer than most crops, but nobody expects them to behave as long-lived trees. Acacias should be treated as a crop, giving them the due attention (spacing, fertilizing, cultivation, pruning etc.) required by other crops. The forestry agents were very interested to hear this different perspective on acacias.

3. CONCLUSION

Australian acacias have been successfully evaluated and show high potential as a new food crop in Niger Republic and beyond. An understanding of their potential has been gained and optimum silvicultural methods have been largely determined. Within Niger it is up to the farmers now to accept or reject the findings.

Enough is known about the potential of edible Australian acacias to warrant introduction, testing and promotion in other areas. While food security is tied to a small selection of mostly annual crops growing under agro-climatic conditions that are biased against them, the spectre of famine will never be distant.

Millions of third world farmers have no access to the usual green revolution inputs and they are farming on infertile and degraded lands under adverse climatic conditions. If there is to be a green revolution for them, it will have to be by way of plants which thrive under such conditions, yield well and require minimal inputs. There is much to be gained and little to lose in testing the potential of Australian acacias for food in such new areas.

ACKNOWLEDGMENTS

The Australian Aboriginal people, whose traditional knowledge was freely shared and made this work possible.

Australian Tree Seed Centre Staff - for their hard work , encouragement, advice and follow up of the acacia species trials, nutritional analysis, animal trials and human volunteer trials and especially Dr. Chris Harwood for invaluable editing advice for this article.

CIDA which supports the ongoing work of SIMs Maradi Integrated Development Program.

ECHO for permission to use material from ECHO Notes article "Edible Australian Acacias, A "New" – "Old" Food for semi Arid Zones.

IPALAC for sponsoring Tony Rinaudo's participation in the conference "Combating Desertification with Plants.

SIM (Society for International Ministries) International, SIM Australia and SIM Canada and the many individuals who support this work through SIM.

World Vision Australia for encouraging Tony Rinaudo go to the IPALAC conference and allowing time for preparation of this document.

ACRONYMS

ACIAR. Australian Centre for International Agricultural Research.
CIDA. Canadian International Development Agency.
CSIRO. Commonwealth Scientific and Industrial Research Organisation.
ECHO. Educational Concerns for Hunger Organisation
IPALAC. International Program for Arid Land Crops.
MIDP. Maradi Integrated Development Project.
SIM. Society for International Ministries.
WV. World Vision.

REFERENCES

1. Adewusi, S.R.A., Falade, O.S., Oyedapo, B.O. & Harwood, C.E. (1998). *Acacia colei seed as human food: a dietary trial.* Human Biology. (submitted)
2. Brand, J.C., Cherikoff, V. (1985). *The nutritional composition of Australian Aboriginal food plants of the desert regions.* In: Wickens, G.E, Goodin, J.R. & Field, J.V. (Eds) Plants for Arid Lands, pp 53-68. Proceedings of the Kew International Conference on Economic Plants for Arid Lands held in the Jodrell Laboratory, Royal Botanic Gardens, Kew, England, 23 -27 July 1984. George Allen & Unwin, London.
3. Devitt, J. (1992). *Acacias: a traditional aboriginal food source in Central Australia*
4. Doran, J.C. and Turnbull, J.W (Eds). (1997). *Australian Trees and Shrubs: species for Land rehabilitation and farm planting in the tropics.* (Australian Centre for International Agricultural Research): CSIRO, Forestry and Forest Products, Australian Tree Seed Centre, Canberra.
5. Harwood, C. Pers comm. Australian Tree Seed Centre, CSIRO Forestry and Forest Products, PO Box 4008, Queen Victoria Terrace, ACT 2600, Australia.
6. Harwood, C.E., Rinaudo, T. & Adewusi, S. *Developing Australian acacia seeds as a human food for the Sahel*, in Unasylva, Vol 50 No. 196, 1999/1 An international journal of forestry and forest industries. FAO.
7. Hunger, (1997). *What Governments can do.* Seventh Annual Report on the state of World Hunger. Bread for the World Institute, 1100 Wayne Avenue, Suite 1000. Silverspring. MD 20910, USA.
8. Maslin, B.R, Thomson, L.A.J, McDonald M.W and Hamilton-Brown, S. (1998). *Edible Wattle Seeds of Southern Australia. A Review of Species for Use in Semi-Arid Regions.* CSIRO and Department of Conservation and Land Management. CSIRO Australia.
9. Ruskin. F.R, (Ed) (1989) *Lost Crops of the Incas.* National Academy Press, Washington D.C.
10. Thomson, L.A.J. (1992). *Australia's subtropical dry-zone Acacia species with human food potential.* House, A.P.N. & Harwood, C. E. (Eds.), Australian Dry-Zone Acacias for Human Food, pp. 37-53. Canberra: Australian Tree Seed Centre, CSIRO Division of Forestry. 145 pp.
11. Thomson, L. Forest Resources Division, Forestry Department, FAO. pers comm.

4. APPENDIX I

4.1 Acacia colei var. *ileocarpa*

Particularly in marginal areas *A. colei* should be seen as a multi purpose shrub which not only provides food for humans and animals but large amounts of firewood and light construction timber, organic matter and protection from winds. *A. colei* has high seed yield potential if minimum requirements of site selection, spacing, cultivation, pruning and fertilising are met. Regular cultivation of newly planted acacias has been shown to be far more critical for survival and healthy and rapid growth than any fertiliser regime. Likewise, placement of trees along water flow lines or digging micro water catchments greatly increases growth, vigour and seed yield compared to all other treatments.

Advantages of *A. colei* include:

• High yields of nutritious seed that can be stored for many years.

• Ease of nursery propagation.

• High survival rate on a range of infertile sites from sands to hard pans, making them useful for land restoration. Farmers in Niger who dedicated exhausted, non-productive fields to acacias for 4 – 6 years were able to resume millet cropping once the trees were harvested or died out.

• Rapid early growth rate.

• Hardiness and drought resistance.

• Fixation of atmospheric nitrogen and therefore improvement of soil fertility for other crops.

• The ability to lower wind speeds and reduce sandblasting of adjacent crops, making it a valuable windbreak species.

• The production of large quantities of wood which is an excellent fuel and readily marketable. Through harvesting acacia wood, pressure on native stands is reduced.

• The production of leaf litter which can be used for mulching and to increase the typically low organic matter content level of dry zone soils.

• Resistance to insect attack, including termites and locusts.

• Leaves on the tree are usually not attractive to livestock, thus fencing or guarding for establishment is usually not necessary.

• Use of the dried leaves of A. colei as low-grade animal fodder.

Limitations of *A. colei* include:

• Short life span, ranging from 6 - 12 years. Though a consistent seed harvest could be ensured by planting a number of replacement trees each year, a large paradigm shift is required by traditional farmers to plan ahead in this way. Forestry agents have discounted A. colei after seeing high mortality rates from the third year after planting. This is most common however in close spaced (4 m x 4 m) plantations.

• The extensive shallow root system of A. colei competes with crops planted near by, particularly in low rainfall years. This factor has scared many farmers from planting *A. colei* on cropland. Planting should be concentrated on farm borders, exhausted abandoned fields and areas not suitable for cropping. Trees can also be planted as shade trees in home compounds where there is less competition for light and water and there are concentrations of animal wastes. Very fast growth rates and high seed yields have been realised in home compounds.

• *Acacia colei* is prone to wind damage as the plant gets older or when severely water stressed. This can be largely overcome by heavy pruning every 2 years, which also stimulates seed production.

• Under Niger's conditions, direct seeding in the field has not been successful, but it has worked in coastal areas of Senegal[1] where rainfall is higher. Thus, propagation has been limited to the expensive and labour-intensive process of raising trees in a nursery. In Niger where water tables can be as low as 40 – 100m and the eight-month dry season is harsh, farmers have not readily embraced the work of raising seedlings in nurseries.

4.1.1 Growing Requirements of *A. colei*

4.1.1.1 Climate

A. colei withstands hot (45+ degrees C), dry conditions being able to set seed with only 200 mm of rain. Heights of 3 - 4 m are attained within 2 - 3 years on infertile, sandy soils with 200-500 mm rainfall and dry seasons of up to 8 months. In drier areas (i.e. receiving less than 350 mm of rainfall) it does best, planted along flow lines, where run off water passes.

4.1.1.2 Soils

A. colei grows on a wide variety of soils from pure sand to heavy clay. It tends to not do well on shallow lateritic soils, especially when rainfall is limited but thrives on wasteland such as hard pans and eroded sites where regular crops cannot be grown.

4.1.1.3 Planting distance

Acacia colei has an extensive shallow root system and will be severely stressed if planted too close together. When severe moisture stress occurs, closely planted *A. colei* fails to grow or set seed. The current planting distance recommended for Niger is 10 m apart. *A. colei* has been observed doing well at closer spacing in Northern Nigeria where rainfall is greater than 450 mm. Potential competition should be considered when deciding how close to plant acacias to traditional crop plants. Farmers will not be interested in trees which reduce yields of their main food crops.

4.1.1.4 Fertilising

Acacia colei responded more favorably to light applications of rock phosphate than NPK or Superphosphate in the first year of trials. The effect of nodulating *A. colei* with nitrogen fixing rhizobia is currently being tested and certainly during the rainy season there is a positive response. It is critical to use viable rhizobia which hasn't been exposed to heat. Inoculated acacias in Asia (Harwood, pers comm) have shown significant growth rate increases.

4.1.1.5 Cultivation

Trees will not set seed if weed competition is too great. It is critical to give two to three cultivations in the rainy season, especially in drier climates. Cultivation in the dry season may also be helpful as flowering and seed set occurs after the rainy season.

4.1.1.6 Pruning

Branches of *A. colei* are brittle when stressed and strong winds have been known to completely uproot trees of age three years and older. However *A. colei* responds positively to pruning at the onset of the rainy season. Trees should be pruned at breast height at the onset of the rainy season once they reach 33 months of age or when loss of vigor or die back occurs. Re-growth is enhanced when a number of smaller (finger diameter) branches are left on the pruned trees. After the first pruning, vigorous new branches will grow to 2 - 3 m in just six months and flowering and seed set can be heavy on these new branches. Pruning also helps reduce die back and wind damage. Pruning trials are in progress, and have shown that *A. colei* can withstand at least two cycles of heavy pruning with most trees remaining vigorous to age 6 years.

Pruning at the onset of rains also reduces competition with crops while the trees replace shoot and leaf cover. Some farmers have planted *A. colei* at 10 x 10 m spacing on worn out farms and found that, because of the trees presence, winds have deposited a layer of (relatively) fertile silt on the site. They expect the site to be fertile enough to return to cropping when the trees finally die out or are removed.

4.1.2 Seed Handling

4.1.2.1 Harvesting Seed

A. colei is an early and heavy seed bearer. Two and three year old trees produced up to 6 kilos of seed with less than 400 mm of rain while two-year-old trees produced up to 3.5 kgs of seed with only 296 mm of rainfall.

The seeds ripen in March (in Niger) when no other non-irrigated crops are growing. The clusters of bean like seed pods can easily be harvested by hand either immediately before drying or for some time afterwards. Ripening occurs within a 4-week period. Coiled-podded *A. colei var. ileocarpa* does not shatter and the seeds are firmly held in the pod by a thread-like aril. It is well worth the effort of seeking this variety as seed of curved-podded *A. colei var colei* ripen sequentially in Niger[2] and drop to the ground within a day or two of the pod turning brown. Unless the pods are picked green, heavy losses of seed are incurred.

Protective glasses should be worn while harvesting. The seeds cause irritation to the eyes on contact. Once the seeds are dried, they can be removed from the pod by threshing with a mortar and pestle. Threshing should be done on a windy day as the dust from pods is irritating to nose and throat, causing sneezing and running nose.

In their native habitat, trees are easily harvested by placing a sheet on the ground and beating the branches with a stick. The extended ripening period in some species and provenances of acacias seems to be more of a problem in Niger than in Australia.

4.1.2.2 Grain storage

Seeds may be stored for many years without deterioration, however the aril is susceptible to weevil attack. Periodic winnowing reduces the weevil population. The weevils do not damage the grain unless their numbers are allowed to build up to the point where their wastes cause spoilage.

4.1.2.3 Food processing

The seeds are washed, dried and ground to flour. As much of the black seed coat as possible is removed by sieving through a fine mesh sieve. The flour is then used in traditional recipes, being mixed with the flour that would be used for a particular recipe (e.g. wheat flour for pancakes, millet flour for traditional porridge etc.) at a rate of no more than 25% acacia flour to 75% of other ingredients. The acacia flour has proven to be very acceptable as it is tasty and versatile, being usable in many different dishes including traditional porridge, bean cakes and pancakes. The seed can also be roasted and ground to make a substitute coffee.

4.2 Acacia elachantha

A. elachantha has similar advantages and limitations to *A. colei*. The biggest difference is the need to pick *A. elachantha* pods before they ripen to prevent seed shattering and loss.

4.2.1 Advantages of *A. elachantha*

Like *A. colei, A. elachantha* is a fast growing, early and heavy seed bearer. In drier years A. elachantha has out yielded *A. colei* in Niger. It produces large amounts of firewood and light duty poles. *A. Elachantha* has a more erect form and a lighter canopy than *A. colei* suggesting it would be a better choice for crop windbreaks and farm border planting.

4.2.2 Limitations of *A. elachantha*:

In Niger, seeds ripen over an extended period. As more pods ripen each day, seeds fall to the ground. The tall growth habit makes harvesting for poor farmers difficult. However, CSIRO scientists pointed out that seeds could be harvested for food when the pods just begin to turn brown, thus reducing the number of harvest "passes" necessary, and eliminating the shattering problem.

If trees are pruned regularly, the supple new branches can be pulled down for seed harvesting without breaking.

4.3 Other Species

As there are some 50 different Acacia species with food potential, it is well worth testing as many as possible. Even two provenances of the same species may perform quite differently at the same site. A species might be written off on one trial site with say, sandy soil, and yet thrive on a heavy clay soil. Thus trials should also be replicated over a number of sites.

Acacia torulosa appears to have greater drought tolerance than *A. colei. Acacia ampliceps* and *A. victoriae* have done well on alkaline, salt affected soils where *A. colei* has failed totally.

Acacia tumida has not performed well in Niger but has excellent characteristics. It warrants further testing in other regions.

Acacia coriacea var. sericophylla is slower growing and produces less biomass than other species but it is extremely hardy and long lived and still is a favourite food of various Aboriginal groups.

4.4 Recommendations

1. There is considerable yield variation between specimens of otherwise uniform *A. colei* under apparently identical environmental conditions. Investigation into genetic causes of yield difference could greatly boost yield potential of this species.

2. Since farmers are first and foremost interested in yields of their traditional food crops varying agro-forestry cropping systems using acacias and field crops should be tested.

3. So far the focus of trials has been on *A. colei* and *A. elachantha*. Other species of high potential also require in depth research to determine their potential and constraints.

4.5 Footnotes

1. Research in Maradi has demonstrated that direct seeding has very limited or no potential there. Even so the technique has worked for *A. colei* at a near-coastal site in Senegal and in northern Nigeria and it should not be discounted, without trial, in milder areas where the species might be grown (Harwood, pers comm).

2. The observations relating to seed ripening/shattering in the two main varieties of A. colei are of critical practical importance in the Sahel but diverge from what is observed in northern Australia. There the most common experience is that the whole crop ripens rather uniformly and can be harvested in "one" hit. In northern Australia seed is lost fairly rapidly after reaching full maturity from bushes of both podded types of colei eg. within 1-2 weeks in hot, windy weather. Pods of both types can be hand picked in the green state, perhaps up to a couple of weeks before pods turn brown. Seeds should be fully formed, hard and with yellow aril (Lex Thomson, pers comm.).

5. APPENDIX. II.

When you read the history of the adoption of such common and widely accepted crops as the tomato and potato it is not so surprising that Australian acacias are only gradually finding their way into the diets of people in new lands.

Today the tomato is one of the top 30 food crops in the world but its popularity is only a relatively recent phenomenon. Introduced into Europe in 1523 it was regarded with suspicion and as a curiosity for over a century. It was widely believed that only by prolonged cooking could it be rendered safe for eating. It came into intense production in the U.S. only about the time of World War 1 and only in the last 20 years has its popularity soared to make it number one vegetable in China. (Ruskin, 1989)

The potato has had a similar history. People in Europe believed it to be dirty and sinister, causing leprosy, syphilis and scrofula. Even though peasants in Prussia regularly faced hunger, threats and decrees by Frederick the Great, his father, grandfather and great grandfather could not convince them to plant the potato. It was not until the late 1700's that

Europeans finally became serious potato eaters even to the exclusion of most other vegetables (Ruskin, 1989). Today, the modern potato is the world's fourth largest food crop.

Ruskin (1989) attributes the turnaround in popularity of the potato to certain "crop champions"– individuals of vision who dedicated themselves to the crop's cause. One of the most famous was Antoine Parmentier. After convincing Louis XVI of the potato's qualities, he tricked French peasants into thinking potatoes were fit only for royalty. As a result, the people pilfered the king's potato fields and the plant quickly ended up in gardens all over France. Perhaps acacias need similar promoters to show the world their value.

THE EFFECT OF SHADE ON INITIAL GROWTH, DEVELOPMENT AND OCCURRENCES OF LEAF DISEASES ON WILD INDIGENOUS FRUIT TREES

Sclerocarya birrea ssp. Caffra and Strychnos cocculoides in shade netting conditions

Stanley M. Mateke

Veld Products Research & Development, P. O. Box 2020. Gaborone, Botswana, Tel: 267 347 047; Fax: 267 347 363. E-mail: veldprod@info.bw

1. INTRODUCTION

In rural areas of Botswana there are two needs facing the people living there. Firstly, there is the need to generate food and household security, and secondly, the need to combat the increasing environmental degradation around villages and settlements. Botswana is a semiarid to arid country with annual rainfall as low as 250 mm in some areas. Drought is a regular occurrence. Two-thirds of the country is covered by the deep Kalahari sands, and is almost completely absent of surface water. The majority of the population lives in the eastern part of the country, where conditions are not as harsh as the central and western areas.

Although agriculture only contributes 4% to Botswana's GDP, it remains an important source of food, income and employment for the majority of people living in the rural areas. Almost all farmers in Botswana are traditional, practising mixed subsistence farming, with individually managed arable fields and communal grazing of livestock. Many rural households (49% - 1995; NDP8) do not own cattle and are engaged in subsistence crop production. Nearly 50% of these resource-poor farmers are female-headed households.

Combating Desertification with Plants, edited by Dov Pasternak and Arnold Schlissel.
Kluwer Academic/Plenum Publishers, New York, 2001.

In some areas of the Central and the Kweneng Districts, particularly those areas bordering the arid central and western areas of the country, people practice subsistence farming adjacent to woodlands and *forested areas*[1]. But, with low inputs, low soil fertility, low and erratic rainfall and high evapo-transpiration rates of arable production is low and crop failure is common. The need for food and household security forces people to be highly dependent on the "free" forest resources in their environment.

It is widely recognised that at local and national level woodlands and natural resources in general contribute to the improvement of food security at household level, to meeting rural subsistence needs, to generating income, to agricultural productivity and to environmental protection. Although woodlands benefit all members of society, the poor are the sector most dependent on natural resources. Thus a strategy for the integration of agriculture and the sustainable management and utilisation of forest resources offers the greatest potential for alleviating poverty and protecting forested areas in rural areas.

Sclerocarya birrea and *Strychnos cocculoides* grow in the sandy veld of Botswana. Both species produce traditionally important wild fruit, show promise for domestication and genetic improvement, and seem to have great commercial potential. Traditionally, the varied uses of the *S. birrea* nut range from its role as a dietary staple for the Basarwa (Bushmen) groups, to its use by traditional healers as "medicine" in their divining dice (Taylor, *et al*, 1995). Leaves of *S. birrea* are very popular amongst browsers. In drought years with poor grazing during the spring and summer, Basarwa will lop off branches of male trees to feed livestock (Taylor, 1988). The bark is used traditionally for the treatment of ulcers and malaria.

Domestication would necessitate high survival and improved growth rates of young plants in nurseries. No information has previously been published concerning the performance of the seedlings of these indigenous fruit trees when raised under shade cloth as compared to direct sunlight. This experiment was therefore designed to show the effect of shade on the growth and development of shoots and roots of young seedlings of *S. birrea* and *S. cocculoides* and their resistance to powdery mildew leaf disease at nursery in the semiarid conditions of Botswana.

[1] _In general, Botswana's vegetation ranges from open to dense savannah. Dry forests occur in the northern, wetter area of Chobe where the dominant hardwoods are Pterocarpus angolensis and Baikaea plurijuga. South of this, on the hardveld, the vegetation is dominated by Colophospermum mopane, various Combretum species and Terminalia sericea. In the central, western and southern parts of Botswana, the dominant trees are Acacia species.

2. MATERIALS AND METHODS

2.1 Selection from Wild

Identification of superior phenotypes in the wild was done by conducting competitions amongst school children and marginalised groups in remote areas to find those tree species which produce the biggest and sweetest fruit. The search for superior phenotypes was conducted as and when the fruit ripened. Letters to primary schools and publicity posters were placed in villages located in areas where high concentrations have been reported within Botswana (Taylor, 1983).

The provenances that produced most, biggest and sweetest fruit, looking healthiest in appearance to the eye were the ones selected as baseline materials for germplasm collection and use in propagation investigations. Variation in the wild ranges from height of trees, branching characteristics, time of fruiting to fruit size.

For our research purposes, the following criteria was used to identify superior phenotypes:

- *Sclerocarya birrea;* 100 fruits should weigh at least 4kg, tree to have > 1500 healthy fruit, sweet juice with at least 12.0 °Brix count and pH >3.5.
- *Strychnos cocculoides;* 10 fruits should weigh at least 1.6kg, tree to have > 300 healthy fruit, sweet juice with at least 12.0 °Brix count and pH >3.5.
- Both species' nutritional and mineral compositions were determined from the Botswana Food Technology laboratories, at Kanye.

2.2 Shade Net Studies

Several seeds from healthy fruits from the selected wild superior phenotypes of *S. birrea* and *S. cocculoides* were germinated in a green house under 50% shade netting 12 months later. The ground was irrigated regularly to keep soil moisture around seeds at filled capacity. Good germination of seeds was achieved. Then, three groups of the new seedlings were raised under 50% shade, 25% shade and 0% shade (direct sunlight), respectively, for six consecutive summer months.

Each group of seedlings contained 16 replications of each species in a completely randomized design.

The experiment began in November 1996 and continued until April 1997. Records of plant survival, basal stem diameter and height of seedlings were kept at the start of the experiment. At six months, records of survival, basal stem diameter, height, dry shoot weight and dry root weight were taken again.

2.3 Diseases Surveys

On the same seedlings mentioned above, the degree of occurrence of leaf diseases (observed by naked eye) on each seedling was recorded. Each seedling was examined on a scale of 0 to 3 (0=no disease, 1=slight, 2=50% of leaves disease covered, and 3=complete disease cover on all leaves) over the weeks for 6 months.

The data generated in these studies were subjected to a statistical analysis at the Department of support services at the University of Botswana's College of Agriculture.

3. RESULTS

3.1 Selection from Wild

Ten best entries of superior phenotypes were selected from villages around Botswana where the species occur naturally in the wild. These were selected on the basis of size of fruit and relative production potential determined from physical counts of fruit at the time of competition according to criterion set for these studies. Most entries came from central Botswana most probably due to best infrastructure such as roads and telephones that enabled the teachers or parents of the pupils contact us in time for the researchers to find the trees still having fruit on the trees. The selected trees were those that produced bigger fruit than the average known sizes in reference to recorded information by Taylor and his colleagues (Taylor *et al.* 1995)

The search for superior phenotypes was conducted as and when the fruit ripened. The weights of fruit and its components are shown in *Table 1*.

Table 1. Fruit component composition of the selected superior phenotypes

Species	Av. # fruits/tree	Av. fruit wt (g)	Av. Seed wt. (g)	Av. Pulp + juice wt (g)	pH	°Brix
S. birrea	11905.9	52.0	17.4	14.9	4.6	13.4
S. cocculoides	551.6	197.8	3.6	83.6	3.7	20.0

In both species, the biggest fruit producing tree does not necessarily produce the sweetest fruit. The weight of pulp and seed was generally greater in bigger fruit as compared to smaller fruit (Data not included in this paper). The mineral and nutritional compositions of the fruits are comparable or in some cases superior to most commonly imported, into Botswana, tropical fruits (*Tables 2 & 3*).

Table 2. Nutritional composition of fruits of selected superior phenotypes

Species	Mean nutritional composition of fruits from selected trees/species (g/100g of edible potion, dry matter)						
	Moisture	Fibre	Protein	Fat	Total Carb.	Vit. C mg	Edible portion (%)
S. cocculoides	84.1	0.9	3.8	4.7	91	38	51
S. birrea	74.0	1.2	8.0	15.7	75	127	51
Orange	86.0	0.3	0.8	0.0	13	43	78
Peach	85.0	0.5	0.8	0.0	13	10	33

Table 3. Mineral composition of fruits from selected superior phenotypes

Species	Mean composition fruits from selected trees/species		
	Ca (µg/100g)	Fe (µg/100g)	Energy (kJ)
S. birrea	20	0.5	225
S. cocculoides	19	0.2	308
Orange	30	0.5	222
Peach	8	0.5	234

Rural communities have more access to indigenous fruits than the exotics that are not only expensive but are also far (in cities) from their living areas. It seems more desirable to develop techniques that would favour mass production of seedlings of some of these indigenous fruit tree species and supply them to the rural masses.

3.2 Shade Net Studies

At six months after germination, survival was 100% for all the treatments. Both *S. birrea* and *S. cocculoides* seedlings grew faster under 50% shade than those under 25% or 0% shade (*Table 1*).

Stem basal diameters were significantly ($P < 0.001$) thicker for *S. birrea* in direct sunlight than for those under shade. *S. cocculoides* showed no signficant difference ($P < 0.5$) in stem thickness between seedlings that were kept under shade and those which received direct sunlight (*Table 4*).

S. birrea and *S. cocculoides* which were raised under 0% shade (direct sunlight) developed the greatest amount of total underground biomass (roots) (*Table 4*).

The wet weights of the roots were 25% and 54% heavier than the dry root weights for *S. birrea* and *S. cocculoides* respectively. No significant differences were observed in shoot weight.

Table 4. The performance of S. birrea and S. cocculoides seedlings under different shades of light in semiarid Botswana.

Species/ % shade	Survival %	Height (cm)	Stem diameter (mm)	Shoot wt. (g)	Root weight (g)
Sclerocarya birrea					
0	100	20.6	9.6	11.6	89.9
25	100	29.3	8.3	11.4	72.0
50	100	37.8	7.5	11.0	67.0
LSD	ns	2.1	1.0	1.6	4.3
Strychnos cocculoides					
0	100	11.2	3.3	1.4	5.5
25	100	12.9	3.7	1.0	3.1
50	100	14.9	3.1	1.4	2.5
LSD	ns	2.1	1.0	1.6	4.3

3.3 Diseases Surveys

After 6 months of experimentation, Powdery mildew was the main leaf disease that was easily identified as it covered the leaves of susceptible seedlings. Incidences of high Powdery mildew (more 50% of plants' leaves) was greatest in *S. birrea* seedlings that were raised under 50% shade net and no infection of the disease was noticed in seedlings that were under direct sunlight (*Table 5*). Slight infection was seen in plants in the 25% shade.

Table 5. The relative infection of Powdery mildew on *S. birrea* and *S. cocculoides* seedlings under different shades of light in semiarid Botswana.

Species/ % shade	Survival of seedling	Start of experiment	6 months later
Sclerocarya birrea			
0	100	0	0
25	100	0	0.83
50	100	0	2.81
Strychnos cocculoides			
0	100	0	0
25	100	0	0
50	100	0	0

4. DISCUSSION

The use of primary school pupils in the selection of superior phenotypes from the wild was a very successful method. Pupils know the best fruit producing trees in their areas. However, due to poor road systems in the Ngamiland region it was not easy to reach the San people that live far from main roads. The areas bordering Botswana and Namibia on the north-western Botswana still need to be visited. The San people are still largely gatherers in the Ngamiland's remote areas.

S. birrea and *S. cocculoides* seedlings showed improved growth when raised under shade netting. *S. birrea* seedlings in particular demonstrated greater height in shoot growth.

However, development under shade may begin to compromise the ability of seedlings to resist drought. *S. birrea* seedlings which were raised under direct sunlight developed thicker basal stem diameters. These results indicate that *S. birrea* seedlings will start adapting to water stress at very early stages of plant growth, perhaps explaining the species' drought tolerance. In addition, seedlings of both *S. birrea* and *S. cocculoides* developed more substantial root systems when raised in direct sunlight. This also may enable the seedlings to resist drought more easily.

Powdery mildew did not attack *S. cocculoides*, a sign that this species may be resistant to the disease.

The results of this experiment lead to some practical recommendations for the propagation of *S. birrea* and *S. cocculoides* in nursery conditions. In order to ensure adequate root development, as well as shoot growth, seedlings of both species should be raised under 25% shade. However, seedlings which are being raised not for grafting onto as rootstock can be raised under 50% shade as height of seedlings increases faster in 50% shade than 25% or no shade conditions.

To reduce on Powdery mildew incidences on *S. birrea*, seedlings may be raised in 25% shading and then be transferred to direct light whenever sight indications of the disease are observed.

ACKNOWLEDGEMENTS

I am grateful to staff of VPR&D for technical help in data collection and administrative support during our studies.

Many thanks to Mrs B. Sebolai of BCA for analyses of data and Dr. Omo of Botswana Food Technology Centre for nutritional and mineral analyses of the fruits.

Funding was provided by CED-FINNIDA.

REFERENCES

1. Taylor, F.W., Butterworth, K.J., & Mateke, S.M. (1995). "*The Importance of Indigenous Fruit Trees in Semi-Arid Areas of Southern and Eastern Africa.*" African Academy of Sciences Second Roundtable Discussion on Non-Wood/Timber Products, Pretoria, RSA.
2. Taylor, F.W. et al (1988). "*An outline of some potential projects for Rural Employment Creation.*" Prepared for the Bophuthatswana National Development Corporation, by Development Consultancies.
3. Taylor, F.W., (1983). "*The Potential for Commercial Utilisation of Veld Products in Botswana.*" Vol. 1.

AFFORESTATION OF SEMIARID AREAS OF TANZANIA: FOCUSING ON THE ROOT COMPARTMENT

Esron Munyanziza
Faculty of Forestry and Nature Conservation,Sokoine University of Agriculture,P.O.BOX 3009 Morogoro, Tanzania, Tel; 255-564648/3718/3259, Fax: 255-654865/4648/3511-4, email: e_munyanziza@yahoo.com

1. INTRODUCTION

The process of land degradation or desertification is often (if not always) initiated by the removal of original vegetation (Munyanziza *et al.*, 1997). The most logical step towards combating desertification is the restoration of plant cover. Understanding the site characteristics and the natural plant ecological strategies is a key towards successful rehabilitation of degraded or degrading lands. The performance of plants is generally a function of their efficiency in exploiting the atmosphere and the soil system (Munyanziza and Oldeman, 1996). Early days of a developing plant start by the development of the root system. The subsequent survival and growth of such a plant hinge upon its fitting in the soil system (Munyanziza, 1994).

Tanzania which looses about 500,000ha of its woodlands or forests every year has launched a big campaign for reforestation (Ministry of Natural Resources and Tourism, 1998). The areas of focus are the marginal or bare lands deprived of their native vegetation and hence prone to erosion. The removal of native ecosystems renders the site more difficult for any subsequent endeavour for reforestation since micro-organisms which contribute to the system productivity are also affected (Munyanziza, 1999). Screening of plants with peculiarities based on their ecological strategies becomes a necessity. The present study focuses on key strategies of some of

Combating Desertification with Plants, edited by Dov Pasternak and Arnold Schlissel.
Kluwer Academic/Plenum Publishers, New York, 2001.

the main trees of the miombo woodlands. The focus is on the root compartment. The relationship between the shoot and the root is highlighted.

2. MATERIALS AND METHODS

2.1 Symbiotic Status of Canopy Trees

The survey of symbiotic status in canopy trees of Caesalpiniaceae or Papilionaceae (*Table 1*) was carried out in Morogoro, Tanzania. The survey was based on root sampling or mushroom association in the field. Both adult trees and wildlings were covered by this survey. Over 50 seedlings or trees were assessed.

Nodulation was visually assessed while mycorrhizal association was confirmed by combining morphological features of fine roots, microscopic laboratory work and the systematic occurrence of mushrooms around a particular tree species (Hogberg, 1982; Hogberg and Piearce, 1986; Alexander and Hogberg, 1986; Munyanziza, 1994).

2.2 Root and Shoot Architecture and their Dynamics

Dynamics in shoot and root was monitored in *P. Angolensis* during the rain and the dry season. While mere visual observations on phenology and the branching habit were used to assess the above ground part, the assessment of the below ground part of the plant involved careful excavation of seedlings followed by a close observation of the root system in terms of nodulation and the pattern of branching.

2.3 Relationship Between Taproot and Shoot

The upper part of the taproot of 34 wildlings of *P. angolensis* was exposed. The diameter of this taproot was measured at 1.5 cm soil depth. The relationship between shoot height and the taproot was analysed using the SPSS programme. In another trial 35 wildlings with a range of taproot top diameters were transferred to the nursery where they were supplied with ample water. Six months later height measurements were made. The relationship between height and taproot top diameter was again computed.

In the last set of seedlings two sub-sets were made: taproots 1.5 cm long or 5.5cm long. These were raised in the nursery with ample water supply. Height growth in each case was monitored.

A test was implemented to look at the variations as attributed to taproot length.

3. RESULTS AND DISCUSSION

3.1 Symbiotic Status of Canopy Trees of The Miombo Woodlands

All the trees or seedlings investigated in the field formed at least one type of symbiosis (*Table 1*).

Table 1. Miombo covered by this study: symbiotic status

Species	Family	Symbiotic status
Afzelia quanzensis	Caesalpiniaceae	Estomycorrhizal
Brachystegia boehmii	Caesalpiniaceae	Ectomycorrhizal
Bracystegia spiciformis	Caesalpiniaceae	Ectomycorrhizal
Brachystegia microphylla	Caesalpiniaceae	Ectomycorrhizal
Julbernardia globiflora	Caesalpiniaceae	Ectomycorrhizal
Pterocarpus angolensis	Papilionaceae	VAM, nodulates

Arbuscular mycorrhizae and nodules were observed in *P. angolensis*. All the Caesalpiniaceae investigated formed ectomycorrhizae. None formed nodules. Root symbiosis is the key strategy adopted by plants growing in hostile conditions (Mikola, 1980). Earlier studies carried out in Tanzania (Hogberg and Nylund, 1981) indicated that indeed Caesalpiniaceae form ectomycorrhizae. Nodulation observed in *Afzelia quanzensis* by Hogberg and Nylund (1981), however, was observed in none of the hundreds of seedlings of Caesalpiniaceae assessed in the field or in the nursery in this study or others (Munyanziza, 1994; Munyanziza and Kuyper, 1995).

Ectomycorrhizal occurrence in Caesalpiniaceae was reflected both underground by prominent, very hairy ectomycorrhizae and by a high diversity of ectomycorrhizal mushrooms aboveground. They were formed by fungi belonging to several genera namely, *Amanita, Boletus, Cantharellus, Cortinarius, Inocybe, Lactarius, Pisolithus, Russula, Scleroderma*. Over 100 species of ectomycorrhizal mushrooms were observed. Occurrence of mushrooms in the miombo ecozone is widespread (Pegler and Piearce, 1980; Rammeloo and Welleyn, 1993). There is no doubt that they play a role in nutrient recycling, ectomycorrhizae being the major if not the sole absorbing organs in miombo ectomycorrhizal species.

3.2 *P. angolensis:* Root and Shoot Architecture and Dynamics under Natural Conditions

The shoot was seasonal. During the dry season all the living biomass was hidden underground (Munyanziza and Oldeman, 1996). The leaves were shed, the stem and branches died back to the ground. Repeated die back of the shoot was reflected in the number of dead shoots or marks on the short surviving stem. Thus one could estimate the age of the plant by counting the number of marks. This gives but an indication since (a) in some years of exceptionally good weather conditions the shoot may survive; (b) browsing may induce multiple shoots; (c) some dead shoots may be blown away by the wind or the passing animals.

There is no doubt that the architecture of plants of arid zone is partly influenced by various stresses.

The shoot reappeared, few weeks prior to or during the short rain season of November to January and remained active throughout the long rains of March to May and some weeks later. It is a period when carbon assimilation, translocation and storage take place.

Shoot dynamics aboveground was somehow similar to that observed underground. The root compartment was dominated by a single, robust, long carrot-shaped taproot (Munyanziza, 1994). This was a continuously expanding organ which was (1) the sole source of energy required to reinitiate growth prior to the short rains and (2) an extension of the plant to the ground water.

Lateral roots behaved like shoots in that they (1) were active during the rain season (2) contributed to taproot expansion (3) were temporal (4) died back during the dry season and regrew in the subsequent rain season. It is noteworthy that as the size of the seasonal shoot increased with the size of the taproot so did that of seasonal laterals. While shoot die back has been reported in other studies (Boaler, 1966; Ernst, 1988, Chidumayo, 1991), the present is the first report on lateral root die back.

Under similar conditions, the phenomenon of die back is such that a shoot produced in year 2 is bigger and taller than that produced in year 1. The dynamics in laterals assume similar pattern: a lateral emerges, dies back during the dry season, regrows in the next growing season becoming longer, dies back during the following dry season, etc.

Shoot die back is often thought to stretch over 10 years after which the seedling grows into a pole in one growing season. There is a possibility that the laterals have acquired a permanent status at this stage.

3.3 Relationship between Shoot size and the Taproot Size

There was a positive relationship between taproot top diameter and shoot height in seedlings growing in the field (R= 0.85) or those raised in the nursery (R= 0.89). Similarly, heights of seedlings having a long taproot were significantly bigger than those with short taproots ($P < 0.05$; *table 2*). The taproot is the sole source of energy to reinitiate growth prior to or during the rain season (Chidumayo, 1991; Munyanziza and Oldeman, 1996).

Table 2: Mean shoot elongation (cm) in nursery seedlings of *P. angolensis* as affected by the length of the taproot (n=6)*

Taproot length	Shoot height	SE
1.5	8.0	2.1
5.5.	16.6	1.7

* = Highly significant ($P < 0.05$)

That is why the bigger the store of reserves the bigger the initial height growth in a new season. This may not always be the case especially in two consecutive seasons, the first being very favourable the following being markedly dry (*Table 3*). As *Table 3* indicates indeed seedlings of 1996 were shorter than those of 1995 though the latter had a bigger taproot, this expanding yearly (Munyanziza and Oldeman, 1996).

The focus of the plant on the root compartment and the positive relationship between the size of the taproot and its growth and survival give us an indication of where we should focus when we are preparing seedlings for reforestation of arid zones.

Table 3. Mean shoot elongation (cm) in seedlings of *P. angolensis* in two consecutive years as recorded in the field in Morogoro, Tanzania (n=21)*

Year	Mean shoot length	SE
1995	145.4	19.5
1996	107.0	12.2

* = Highly significant ($P < 0.05$)

3.4 Preparation of Nursery Stock for Arid Zones: Focusing on the Root Compartment

As demonstrated and discussed above, the taproot of plants of arid zones is the plant focus, the entire life in seedlings being hidden in it during the dry periods. The preparation of seedlings for arid zone should focus on the root compartment. As noted the magnitude of the shoot produced during the rain season was a function of the tap root size. The question is, can we enhance in the very nursery subsequent seedling growth and survival in the field?

Three nursery operations affect directly the root compartment and hence seedling growth. These operations include root pruning, inoculation, fertilisation and certainly irrigation (Rook, 1971; Gosh et. al., 1974; Chamshama and Hall, 1987; Munyanziza and Oldeman, 1995). The function of root pruning is to enhance seedlings attributes which will increase survival chances. In non miombo trees root pruning's function is to promote a fibrous root system (Chamshama and Hall 1987). Fine roots however, as mentioned above are ephemeral organs. In *P. angolensis*, pruned seedlings had a very stunted growth with a significantly thin taproot (Munyanziza and Oldeman, 1995). Pruning the taproot is to confine it in the pot or in the upper layer of the seedbed where the chances of suffering from scorching by the sun or bush fires are high. As observed above (*Table 2*), the shorter the taproot in seedlings for outplanting, the smaller the shoot to be produced. It is important to mention that the longer the taproot the earlier will be seedling access to the ground water. Pruning of the taproot therefore in species like *P. angolensis* whose bare root seedlings have a high survival rate should be discouraged. Instead taproot development should be encouraged by keeping it undisturbed, by irrigating at the nursery stage and by fertilizing where possible and or by inoculating with the appropriate symbionts (Munyanziza and Oldeman, 1996).

It should be remembered, however, that excessive fertilisation with inorganic fertilisers may depress symbiosis (Aziz and Habte, 1989; Munyanziza and Oldeman, 1995). The point is to strike the balance between temporal growth and symbiosis. It is also to be borne in mind that the effect of fertilization on symbiosis may be temporal.

4. CONCLUSIONS

The growing seedling of arid zones bear marks imparted by the harsh weather and edaphic conditions. In miombo trees studied these marks include shoot and lateral root die back. The function of these ephemeral organs is to mobilise all resources gathered or manufactured during the short favourable rainy period for the build up of the permanent organ, the taproot. The functions of the taproot are to preserve life during the dry season, to reinitiate growth prior to or during the rainy period, to extend the plant to the ground water and certainly to act as an achorage for the tree. The build up of the taproot is the plant focus and is achieved through the work of mycorrhizae and N fixing bacteria. Under natural conditions a positive relationship between the size of the taproot and the shoot was observed. In the nursery improved water conditions and fertilisation make it possible to speed up the taproot development and hence the shoot, and to preserve life both underground and above ground during the dry seasons. Proper nursery techniques which focus on the root system are recommended for nursery stock prepared for afforestation of arid zones

ACKNOWLEDGEMENTS

This is part of the studies generously sponsored by the International Foundation for Science that I gratefully acknowledge.

REFERENCES

1. Alexander, I.J.and Hogberg, P. (1986). *Ectomycorrhizas of tropical angiospermous trees*. New Phythologist 102: 541-549.
2. Aziz, T. and Habte, M. (1989). *Influence of inorganic N on mycorrhizal activity, nodulation and growth of Leucaena leucocephala in an oxisol subjected to simulated erosion*. Comm. Soil Sci. Pl. Anal. 20: 239-251.
3. Boaler, S.B. (1966). *The ecology of Pterocarpus angolensis D.C. in Tanzania*. Overseas Research Publications 12. H.M.S.O., London.
4. Chamshama, S.A.O.and Hall, J.B. (1987). *Effects of nursery treatments on Eucalyptus camaldulensis field establishment and early growth at Mafiga, Morogoro, Tanzania*. For Ecol. Manage. 21: 91-108.
5. Chidumayo, E. N. (1991). *Seedling development of the miombo woodland tree Julbernardia globiflora*. J. Veg. Sci. 2: 21-26.
6. Ernst , W.H.O. (1988). *Seed and seedling ecology of Brachystegia spiciformis, a predominent tree component in miombo woodlands in South Central Africa*. For. Ecol. Manage. 25: 195-210
7. Gosh, R.C., Singh, R.P. and Sharma, K.K. (1974). *Standardisation of nursery technique of tropical pines III. Root pruning of P. patula*, Schiede and Deppe seedlings in nursery- a study. Ind. For. 100: 544-547.
8. Hogberg, P. (1982). *Mycorrhizal associations in some woodlands and forest trees and shrubs in Tanzania*. New Phytologist 92:407-415.
9. Hogberg, P. and Nylund, J.E. (1981). *Ectomycorrhizae in coastal miombo woodland of Tanzania*. Plant Soil 63: 283-289.
10. Hogberg, P. and Piearce, G.D. (1986). *Mycorrhizas in Zambian trees in relation to host taxonomy, vegetation types and successional patterns*. J. Ecol. 74: 775-785.
11. Mikola, P. (1980). *Mycorrhizae across the frontiers*. In: mikola, P. (ed). Tropical mycorrhizal research. Clarendon Press, Oxford. pp. 3-10.
12. Ministry of Natural Resources and Tourism (1998). *The National Forest Policy*. The united Republic of Tanzania. Dar Es Salaam.
13. Munyanziza, E. (1994). *Miombo trees and mycorrhizae. Ecological strategies, a basis for afforestation*. PhD thesis, Wageningen Agricultural University, the Netherlands. 193pp.

14. Munyanziza, E. (1999). *Agroforestry systems, trees species composition and its effect on ectomycorrhizal fungal diversity in Africa.* In: Bagyaraj, D.J., A. Verma, K.K. Khanna and H.K. Kehri. pp. 91-103.
15. Munyanziza, E., Kehri, H.K. and Bagyaraj, D.J. (1997). *Agricultural intensification, soil biodiversity and agro-ecosystem function in the tropics: the role of mycorrhiza in crops and trees.* Applied Soil Ecology 6: 77-85.
16. Munyanziza, E. and Kuyper, Th.W. (1995). *Ectomycorrhizal synthesis on seedlings of Afzelia quanzensis Welw. using various types of inoculum.* Mycorrhiza 5: 283-287.
17. Munyanziza, E. and Oldeman, R.A.A. (1995). *Pterocarpus angolensis D.C: field survival strategies, growth, root pruning and fertilization in the nursery.* Fertilizer Research 40: 235-242.
18. Munyanziza, E. and Oldeman, R.A.A. (1996). *Miombo trees: ecological strategies, silviculture and management.* Ambio 25: 454-458.
19. Pegler, D.N.and Piearce, G.N. (1980). *The edible mushrooms of Zambia.* Kew Bull. 35: 475-491.
20. Rammeloo, J. and Welleyn, R. (1993). *The edible fungi of Africa south of the Sahara: a literature survey.* Scripta Bot. Belg. 5: 1-62.
21. Rook, D.A. (1971). *Effect of undercutting and wrenching on growth of P. radiata D. Don seedlings.* J. Appl. Ecol. 8: 477-490.

INTRODUCTION OF CACTI AS VEGETABLE CROPS FOR ISRAEL

Avinoam Nerd[1], Eli Afflalo[2] and Yosef Mizrahi[1, 2]
[1]The Institutes for Applied Research, [2]Department of Life Science, Ben Gurion University of the Negev, P.O.Box 653 Beer- Sheva, 84105, Israel, Tel: 972-77-6461966, Fax: 972-7-6472984, email: aavi@bgumail.bgu.ac.il

1. INTRODUCTION

The stems of several cactus species native to Mexico are picked at the younger tender stage and consumed as a vegetable known as nopalitos; either fresh, cooked or pickled. Nopalitos are collected from wild stands, home gardens, or plantations which are planted on a wide scale in the center of the country (Flores-Valdez, 1995; Mizrahi et. al., 1997; Pimienta-Barrios, 1993). The nutritional composition of nopalitos is similar to that of common green vegetables, e.g. lettuce or spinach (Mizrahi et al., 1997). The main cultivated species are cactus pears bearing flattened succulent stem joints (cladodes), such as *Opuntia ficus-indica* and *Nopalea cochenillifera*, for which cultivation systems, marketing and storage practices have been developed (Cantwell et al., 1992; Florez-Valdez, 1995; Mick, 1992; Mizrahi et al., 1997; Nerd et al., 1997). Because of their potential for high biomass production in warm climates, the low water demand for dry matter production and the high tolerance to drought, vegetable cactus pears appear to be ideal candidates for cultivation in arid countries (Mizrahi et al., 1997; Nobel, 1988).

With a view to commercial cultivation of vegetable cacti in Israel, which suffers from water shortage, a field experiment was carried out to determine the productivity of selected clones of *O. ficus-indica* and *N. cochenillifera*.

Combating Desertification with Plants, edited by Dov Pasternak and Arnold Schlissel.
Kluwer Academic/Plenum Publishers, New York, 2001.

Additionally, the performance of *N. cochenillifera* as a vegetable crop was studied in a commercial plantation.

2. COMPARISON BETWEEN SELECTED CACTUS PEARS IN GROWTH AND CO_2 UPTAKE

2.1 Plant material and methodology

The study was carried out in Beer- Sheva (northern Israeli Negev desert) on two clones of *O. ficus-indica*, "Milpa-Alta", widely cultivated as a vegetable in central Mexico, and "Ofer", the major fruit cultivar in Israel, as well as on a single clone of *N. cochenillifera*, # 1308, selected by Texas A&M University at Kingsville, Texas (Mick, 1992). Rooted cladodes were planted in March 1994 in 3 blocks, 7 at each replication, with a spacing of 1 m within the row and 4 m between rows. The soil was loamy loess, and annual rainfall (winter) averaged 180 mm for the period of the study, 1994-96. The plants were drip irrigated to keep the soil wet with water containing NPK fertilizer. Max/min temperatures averaged 17/5 ^{0}C in the coldest month (January) and 35/17 ^{0}C in the hottest month (August) (*Fig. 1*).

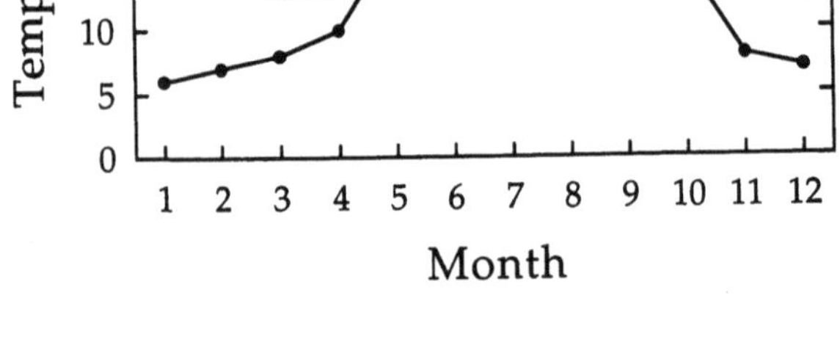

Figure 1. Average maximum and minimum air temperatures measured in Beer-Sheva (1994-95)

Flower and cladode number and cladode area were determined periodically in five plants of each replication. Cladode biomass was estimated by multiplying cladode area by unit area mass. Cladode samples were oven dried at 70^0 C for assessing cladode moisture. CO_2 uptake of terminal mature cladodes was monitored with a Li-Cor LI-6200 portable photosynthesis system on selected days in summer and autumn over 24 h periods.

2.2 Phenology, cladode production and CO_2 uptake

No substantial difference was found between the two investigated opuntias with respect to the measured parameters, but both differed significantly from *N. cochenillifera*. *O. ficus-indica* produced vegetative (cladode) and reproductive (floral) buds in a major flush during spring (April) and a minor flush in autumn (August) whereas *N. cochenillifera* produced vegetative buds continuously throughout the warm season (March-October) and reproductive buds throughout the cold season (November-February).

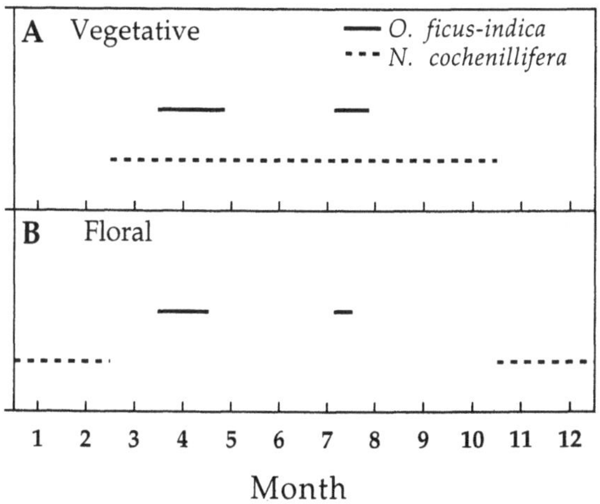

Figure 2. Times of cladode (A) and floral (B) bud emergence in Beer- Sheva

These results are consistent with our findings that low winter temperatures induce vegetative and reproductive budbreak in *O. ficus -indica* in regions with a subtropical climate (Nerd and Mizrahi, 1995). However in *N. cochenillifera*, daylength may be the determining factor in budbreak, with long days inducing cladode production and short days inducing flowering.

Surface area growth measured in the summer showed a sigmoidal pattern for cladodes of both species (Fig. 3). The time elapsed from cladode emergence to the end of the rapid growth phase and attainment of full size was about 30 and 50 days respectively for *N. cochenillifera* and about 50 and 90 days respectively for *O. ficus-indica*. The cladodes of both are consumable (tender) and ready for harvest as a vegetable during the rapid growth stage which lasts until the small rudimentary leaves which cover the young cladode are shed.

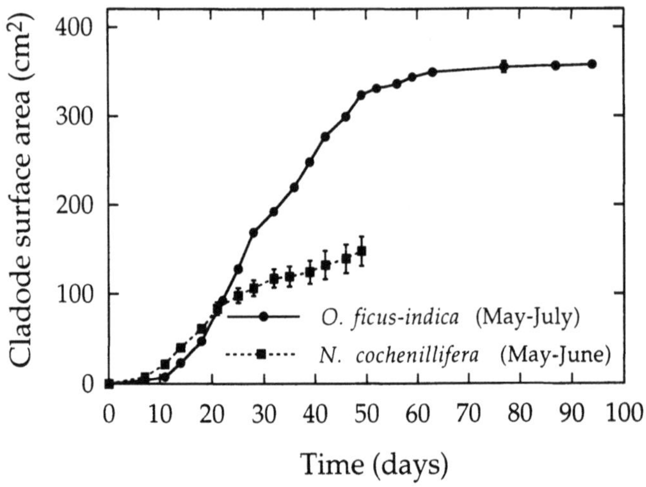

Figure 3. Time course of the expansion of new cladodes in summer (Beer- Sheva). Values are means ± SE (n=5)

The two species also differed in the characteristics of their cladodes; those of *N. cochenillifera* are free of spines, unlike *O. ficus-indica* whose cladodes bear spines that have to be removed before marketing. In addition, mature cladodes of *O. ficus-indica* are thicker and their mass per unit area, ~29 kg per m^2, is two folds higher.

In both species the increase in surface area and dry mass of the cladodes occurred during the warm season (*Fig. 4*). The rate of increase was much higher in *N. cochenillifera* and cladode dry mass per plant for that species at the end of the first and the second year was 1.4 kg and 8.5 kg respectively or about 2.5 times higher than for *O. ficus-indica*. However, cladode surface area, which represents the photosynthetic area, was 5-7 times higher for *N. cochenillifera* at the end of the second year and therefore *O. ficus-indica* should be regarded as more efficient in dry matter production.

In both species the mature cladodes exhibited Crassulacean acid metabolism (CAM), i.e. CO_2 uptake occurred primarily during the night. Daily CO_2 uptake was higher in *O. ficus-indica* in both summer and autumn (713 and 460 mmol m^{-2} respectively, vs. 528 and 361 mmol m^{-2} respectively in *N. cochenillifera*). However, productivity of *N. cochenillifera* was significantly higher, conformable with its longer period of cladode emergence. This finding taken together with the absence of spines led us to select *N. cochenillifera* for further evaluation as a commercial vegetable crop.

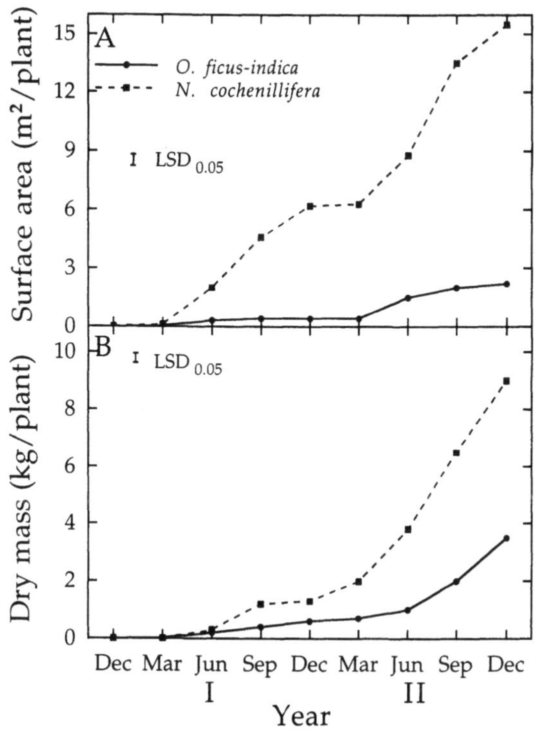

Figure 4. Cladodes growth per plant in the first and second years; surface area (A) and fresh mass (B). Values are means for 5 plants

3. COMMERCIAL PRODUCTION OF *NOPALEA COCHENILLIFERA*

A plantation of *N. cochenillifera*, clone # 1308, was established at Moshav Sede-Yaakov, which is located in the north of Israel in an area with an annual rainfall of 600 mm. Cladode cuttings were planted in March 1996 in heavy soil, along the tops of 0.8 m-wide soil-beds spaced 1.0 m apart; two cladodes were planted per 1.0 m row (~5500 cladodes /ha). A poultry manure (20 tons/ ha) was supplied prior to planting and 100 mm water was supplied to establish the plants; no water was applied later on. At the end of the first year the average height and width of the plants was 82 and 170 cm respectively and estimated cladode fresh mass production was 240 ton/ha. Harvesting of vegetable cladodes 10-15 cm length started in the second year. Cladodes were picked at 3-4 weeks intervals during the annual growth period. At each harvest date cladodes harvested from 4 m rows in three replications, were weighed for determination of yields. The harvest period stretched from May to November, peak in July-September when the average monthly yield was around ~10 kg/m^2. Total annual yield in the second year was 50 kg/m^2, estimated as 290 t/ha (taking in account the space between soil beds). This yield is three times greater than that reported for rain-fed (>600 mm) plantations of vegetable opuntias in Mexico (Flore-Valdez, 1995).

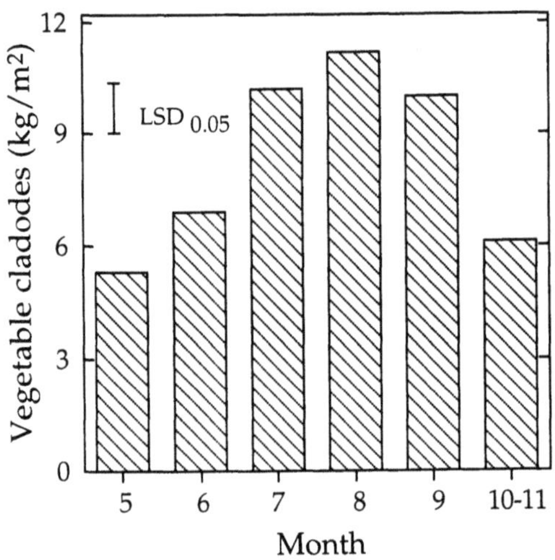

Figure 5. Monthly vegetable cladode production in the second year for a non-irrigated N. cochenillefera vegetable orchard planted in the north of Israel.

The experiments with cultivation were successful, as reflected in the high yields of quality vegetable cladodes, obtained under conditions of sustainable agriculture (no irrigation or pest and disease control). However, at present the future crop is in doubt owing to weak demand. As for other new crops, extensive promotion to generate appreciation for this novel vegetable in local and other markets is required.

REFERENCES

1. Cantwell, M., A. Rodriguez-Felix, and Robles-Contreras, F. (1992.) *Postharvest physiology of prickly pear cactus stem.* Scientia Hortic., 50: 1-9.
2. Flores-Valdez, C. A. (1994). "*Nopalitos production, processing and marketing*". p. 92-99. In: Agro-ecology cultivation and uses of cactus pears. FAO, Rome, 1995.
3. Mick, R. J. (1992). *Growing and marketing the nopalito variety 1308 for year around in Texas.* p. 7-9. In: P. Felker and J.R. Moss (eds), Proc. 3rd Annual Texas Prickly Pear Council, Kingsville Texas.
4. Mizrahi, Y., A. Nerd and Nobel, P. S. (1997). *Cacti as crops.* Hort. Rev. 291-319.

5. Nerd, A., and Mizrahi, Y. 1995. *The effect of low winter temperatures on bud break in Opuntia ficus-indica.* Adv. Hort. Sci. 9: 188-191.
6. Nerd, A., Dumoutier M. and Mizrahi, Y. (1997). *Properties and post harvest behavior of the vegetable cactus Nopalea cochenillifera.* Postharvest Biol. Tech. 10: 135-143.
7. Nobel, P. S. (1988). *Environmental biology of agaves and cacti.* Cambridge Univ. Press, Cambridge, MA.
8. Pimienta-Barrios, E. (1993). *Vegetable cactus (Opuntia).* p. 177-192. In: J. T. Williams (ed.), Pulses and vegetables. Chapman & Hall, London.

UPSCALING THE EXPERIMENTAL PLANTING OF *ARGANIA SPINOSA* AT KIBBUTZ KETURA, ISRAEL

Elaine M. Solowey
Arava Institute of Environmental Studies, Kibbutz Ketura, 88840 Israel, email: emsolow@shani.net

1. INTRODUCTION

The agricultural and economic base of the southern Negev and the Arava valley is unhealthily narrow, based largely on a few veteran enterprises such as date orchards and many water-wasting and environmentally stressful crops such as melons and out of season vegetables.

To broaden the agricultural base, save water and be more sensitive to the negative effects of agriculture on the fragile desert ecosystem, sustainable projects must be introduced, including the cultivation of industrial crops in the context of organic orchardry/agroforestry.

Organic orchardry for this region should be based on the use of animal manure and agricultural wastes from established projects, thrifty application of water by drip irrigation and the planting of trees suitable for arid and saline lands.

2. UPSCALING THE ARGANIA PROJECT

In 1985 *Argania spinosa* , a nut tree from Morocco was introduced to Kibbutz Ketura in the Arava through the Fruit Tree Introduction Program for Arid and Saline Lands directed by Professor Yossi Mizrachi and Dr.

Combating Desertification with Plants, edited by Dov Pasternak and Arnold Schlissel.
Kluwer Academic/Plenum Publishers, New York, 2001.

Avinoam Nerd of Ben Gurion University of the Negev. The Argania trees adapted well to the extreme climate and saline water. With light irrigation the trees produce two falls of nuts per year. There is a great deal of variability among the trees of the 1985 planting. Several trees produce over 50 kilos of nuts per tree per year which is far beyond the amount produced by wild trees (the low yield of wild trees has been one of the obstacles in the development of Argania as a crop).

In 1996 efforts were made to upscale the Argania introduction project into an orchard format to investigate the possibility of growing Argania as a commercial crop.

2.1 The Orchard Site

The site chosen is an area that had formerly been planted in pomelo trees. A date orchard provides a western windbreak and the area is sheltered from both the north and the south. The soil is sandy with some pockets of chalk and clay but few stones.

2.2 Germplasm

Four exceptional trees with high yields from the initial planting were chosen and cuttings were taken from them to produce saplings for the new orchard. Propagating Argania by cuttings is difficult as conventionl methods do not produce viable saplings. In order to propagate these four superior trees it was necessary to make a large number of cuttings of mature wood, bundle them perpendicularly for 48 hours then plant them in a trench filled with peat. Only forty percent of the cuttings were viable. Even with limited and careful applications of water sixty percent of the cuttings rotted or did not form normal root structures.

Young trees sprouted from the seeds of promising plants were also including in the new orchard. Moroccan genotypes from Ben-Gurion University of the Negev's plant introduction project were planted in the southern-most row of the twenty-row orchard. The BGU plant introduction project also contributed genotypes which had done well at an introduction site in Ramat Hanegev but had been pulled up due to lack of funds. Researchers from the introduction project managed to save some live wood and brought some saplings from this source to the new orchard.

2.3 Fusarium

Many Argania trees are vulnerable to Fusarium oxy. Therefore selection for fusarium resistant trees is important for the health and survival of future plantings. Exposure of the seedling trees to fusarium-carrying soil resulted in a 15 % rate of infection. Infected saplings were discarded in the nursery and only trees which appeared healthy were planted in the new orchard.

Despite this early screening 67 trees in the new orchard became infected and 60 trees died. Seven trees appeared to recover indicating that there may be degrees of fusarium resistance in the Argania population rather than a straight immunity/vulnerability equation. Checking the seven recovered trees to see whether they had the disease or other conditions which resemble fusarium's pattern browning of foliage would help answer this question.

Overall, 21% of the germplasm for the new planting was lost to fusarium.

2.4 Salt Tolerance

Argania trees appear to be generally tolerant of Ketura's heavily mineralized water. Only seventeen trees showed signs of salt sensitivity. All were locally produced seedlings. Three of these trees died shortly after transplant and were replaced with other seedling trees from the nursery.

2.5 Orchard Format

The orchard was planted with an eye to mechanical harvesting of Argania nuts, consequently there is twice as much distance between rows as there is between trees. This 5 x 21/2 meter format should leave room for a vacuum harvester (of the general type which is being used in Jojoba harvesting) to enter between the rows to collect the nuts.

The irrigation system was assembled from recycled materials. A 50 mm water main attaches to a standard three inch irrigation head. The 50 mm main divides into four 32 mm. extensions to which the 20 mm drip lines are attached by the "collar and rider" system. There are two to four litres per hour calibrated drip fixtures by each tree. The new orchard is irrigated twice weekly from May until October and once weekly from November to April. Adult Argania trees in the initial planting (1985) receive 12 cubic meters of water per year. The young trees in the orchard format receive between three and six cubic meters of water per tree per year.

2.6 Harvesting

Trees are shaped and pruned upward from a tender age to leave sufficient room for nut collecting. This is accomplished by the use of hand tools with young trees but may have to be done by hydraulic pruning equipment when dealing with a mature orchard because of the extreme toughness of the Argania wood. The Argania fruit has proved vulnerable to the med-fly therefore it may not be practical to leave the nuts to cure on the ground as was the practice in the initial planting. Speedy collection of the fallen nuts and curing in deep trays covered with plastic eliminate this problem. So far no other pest problems have appeared.

2.7 Uses

The flesh surrounding the Argania nut has value as animal feed. Tray dried Argania nuts can be cracked out easily and the dried material can be fed to livestock. Argania forage and timber will probably have little economic value in the Arava.

The oil from the Argania nuts could be the basis for a small-scale production plant as the pure oil is a valuable medicinal and culinary commodity. The Argania nutshells could be incorporated into chipboard or button board.

There is great variability in the size and oil content of the Argania nuts. Some nuts contain one kernel. Others contain two or three small kernels. At some point it will be necessary to determine which type of nuts yield the most oil so that future screening may be done for the desired characteristics.

2.8 Sustainability

Raising Argania trees instead of conventional cultivars makes better use of available land and water. Since machinery will only be used to prune the trees annually and harvest the nuts twice a year, the fragile desert soil will not be compacted.

Moderate use of organic soil amendments in conjunction with sparing drip irrigation does not contribute to the salinization of the soil or pollute through chemical run-off.

A truly sustainable system of desert agriculture does not exist yet but the Argania project is much closer to the desired level of sustainability than any other arboreal cultivar.

3. CONCLUSION

The Argania tree shows great promise as an orchard tree in the Arava valley because of its hardiness, salt tolerance, and low water requirements. Argania trees should now be planted in Israel's barren south in the context of organic agroforestry for the purposes of land reclamation, windbreaks and other anti-desertification strategies.

REFERENCES

1. Julia F. Morton and Gilbert Voss, (1987). *The Argan Tree, A Desert Source of Edible Oil*, Economic Botany 41-20.
2. Leaky, Temu, Melnyk and Vantomme, (1996). *Domestication and Commercialization of Non Timber Forest Products in Agroforestry Systems*, ICRAF
3. Nerd, Eteshola, Borowyand, Mizrachi, (1994). *Growth and Oil Production of Argan in the Negev Desert of Israel*, Industrial Crops and Products.
4. Y. Monnier, (1989). *Les Problemes Actuels de L'Arganeraie Marocaine*, Revue Forestiere Francaise.
5. *Non- Wood Forest Products*. Proceedings of the International Conference held in Nairobi, Kenya, Febuary 1996. Published by FAO 1997

MUNGBEAN (*Vigna radiata L.*), A CHOICE CROP FOR IMPROVEMENT OF HUMAN AND SOIL HEALTH IN SOUTHERN AFRICA

M.L. Chadha
AVRDC-Africa Regional Program, P.O. Box 10, Duluti, Arusha, Tanzania,
Tel: 255-57-4232/4270/8491, Fax: 255-57-4232/4270, email: avrdc-arp@cybernet.co.tz

1. INTRODUCTION

Droughts are frequent and severe in many countries of sub-Saharan Africa (SSA) and have a devastating impact on their people and economies. The extreme vulnerability of rainfall in the arid and semiarid areas of the continent and the poor capacity of most African soil to retain moisture result in almost 60% of SSA being vulnerable to drought and 30% being extremely vulnerable. Since the 1960s, rainfall in parts of the Sahel and Southern Africa has also been significantly below the norms of the previous 30 years. Moreover the El Nino effect has led to more focus on the impact of drought in SSA.

Conventional wisdom suggests that poorer, low-income countries in SSA are likely to be the most severely affected by drought. However, an initial examination of Gross Domestic Product (GDP) aggregates for countries that have suffered intense drought, and a review of response by international and bilateral agencies to the 1991-92 Southern African drought, suggested a more complex relationship between the effects for drought and the structure of a country's economy. The overall impact of drought may be particularly great because of the relative importance of the agricultural sector.

The first round of the agricultural development programs focused on cereals and totally neglected other food crops, such as legumes, which were

Combating Desertification with Plants, edited by Dov Pasternak and Arnold Schlissel.
Kluwer Academic/Plenum Publishers, New York, 2001.

traditionally an integral part of cereal-based cropping systems. Cereal-based cropping systems are more prone to the devastating effects of droughts. Pulses were relegated to relatively unproductive areas. This led to expansion of intensive cereal-based systems, and virtual monocropping.

Questions have long been raised about the sustainability of such systems, due to soil degradation and increasing difficulty controlling insects and disease. Now, mounting evidence suggests a slowdown in productivity growth and resource base degradation-yields from long-term experiment plots are declining, yields in farmers' fields are stagnating despite growing input use, and total factor productivity is declining. Meanwhile, lack of pulse research resulted in increased prices, which resulted in protein deficiency and malnutrition among low-income groups in most of the SSA.

In such a scenario, Asian Vegetable Research and Development Center chose in 1971 to undertake intensive research aimed at improving the lowly mungbean.

2. WHY MUNGBEAN?

Mungbean is one of the important grain legumes of the arid and semiarid tropics. Mungbean is providing protein-rich (24%protein) nutrition to the poor (*Table 1*).

Table 1. Nutritive value of mungbean (calculated on the basis of 100g dry weight of edible portion)

Item	Whole grain
Moisture (g)	10.4
Protein (g)	24.0
Fat (g)	1.3
Minerals (g)	3.5
Fibres (g)	4.1
Carbohydrates (g)	56.7
Energy (K cal)	334.0
Calcium (mg)	124.0
Phosphorus (mg)	326.0
Iron (mg)	7.3
Vitamin A I.U.(mg)	158.0
Thiamine (mg)	0.47
Riboflavin (mg)	0.39
Niacin (mg)	2.1

Source: Singh and Chaabra 1988

Its short duration (60 to 65 days) facilitates diversifying cropping systems by making soils more productive. Mungbean is particularly important to resource poor farmers due to its low requirements for water (*Table 2*) and other inputs (*Table 3*).

Table 2. Soil Moisture use and water-use efficiency of mungbean (Field experiment, 1985)

Genotype	Seasonal moisture use (cm)	Water-use efficiency (kg/ha-cm) on the basis of		
		seed yield	above ground	Total biological yield
K851	26.7	49.9	168.5	197.2
TI-18	26.6	81.4	137.8	150.4

Source: Phool Singh et al. 1988.

Further, this crop has the ability to restore soil fertility through the symbiotic nitrogen fixation. Results from a two-year rotation with Mungbean in Thailand showed 58 to 107 kg N per ha per crop (Firth et al. 1973). Other reports (Mortan et al. 1982, Sinha 1976) also estimated fixation by mungbean to range from 50 to 100 kg N per ha. A report from Mahatma Phule Agricultural University, Maharashtra, in India that mungbean fixes 38.67 Kg of nitrogen per ha which is very useful to the next crop to be grown.

Table 3. Effects of different rates of inputs on mungbean planted after rice at Phatthalung, Thailand, 1985

Input	Treatments	Seed yield (kg/ha)
Insecticide	Not sprayed	622
	Spray every 2 weeks	988
Fungicide	Not sprayed	550
	Spray every 2 weeks	727
Fertilizer	Not applied	668
	47-20-38 kg/ha	1094
Planting	Broadcast	948
Method	Hill planting 50x20 cm, 2 plants/hill	1284

Source: Paisan Laosuwan et al. 1988

Nutritionally mungbean pulse is especially good when combined with cereals, as the relative deficiency of sulfur amino acids in mungbean is compensated for by the surplus in cereals, while the relative deficiency of lysine in the cereals in likewise compensated for by the relative surplus in mungbean (*Table 4*).

Table 4. Protein content and quality of mungbean-rice porridge at different formulations.

Formulation Mungbean		Rice	Protein content (%)	Lysine	Sulphur containing protein (mg/g)	Chemical score (%)
0	:	100	7.50	36	37.0	62
10	:	90	9.25	41	35.3	71
20	:	80	11.00	46	33.6	79
30	:	70	12.75	51	31.9	88
40	:	60	14.50	56	30.2	97
50	:	50	16.25	61	28.5	100
100	:	0	25.00	85	20.2	81

Source: Chao-Yih Chen et al. 1988

More than a dozen different food preparations of mungbean are known in India. Some food preparations of mungbean, as the mungbean split pulse cooked with rice, are especially suitable for children, elderly people and invalids. In China, mungbean is used to prepare mungbean soup, mungbean-rice porridge, bean sprouts, starch noodles, mungbean cake and other products. In Thailand, mungbeans are used as food products at both the household and the industrial level.

Mungbean sprouts cooked alone or mixed with meat or other vegetables provide a fresh and naturally nutritious dish. Mungbean sprout production simply involves germination, which can be done indoors throughout the year. It does not require soil and sunlight. Just 1 kg of mungbean can give 6 to 8 kg of mungbean sprouts. It means we can multiply 6 to 8 times food availability.

Mungbean is widely grown in Asia but not very common in Africa. Mungbean contains additional iron, which is deficient in diets of Southern Africa. Past research revealed that just by combining mungbean with tomatoes in a meal can increase mungbean's iron bioavailability. Mungbean might provide an attractive, economically viable answer to serious problem of iron deficiency in developing countries. Thus mungbean provides true benefits to farmers as well as consumers in drought prone areas of Southern Africa.

3. LIMITATIONS WITH MUNGBEAN CROP

Many cultivated mungbean varieties are photoperiod sensitive, low yielding (with a potential of about 700 kg/ha at experiment stations and less than 300 kg/ha farmers' fields), asynchronous and maturing in 95-115 days. They have spreading growth, small pods (5-6 cm), and small seeds (20-25 g/1000 seeds), which are mostly green and lacking in luster. Their strong response to day length forces farmers to postpone sowing, which delays the sowing of following cereal crops. They are susceptible to mungbean yellow mosaic virus (MYMV), Cercospora leaf spot (CLS), and powdery mildew (PM).

The long duration and unsynchronized maturity of these unimproved varieties created strong competition with other crops for land, inputs and labor. Low yield threw mungbean out of this competition. What's more, its susceptibility to diseases made it a risky crop to grow and its dull color made it unattractive to consumers. By the mid-1970s, the crop was relegated to a very limited area outside the cereal-based system.

4. ROLE OF AVRDC IN MUNGBEAN IMPROVEMENT AND ITS UTILIZATION IN DIFFERENT CROPPING SYSTEMS

Recognizing the critical need for sustainability of cropping systems and the importance of quality protein in human nutrition, the AVRDC researchers took the challenge to overcome mungbean's many constraints. They focussed their research to make mungbean cultivation an economically viable option for use in a wide range of cropping systems, so that its soil enhancing ability and its nutritional benefits could be put to use.

AVRDC researchers set out to create a mungbean with a maximum yield potential of 4 t/ha at maturity of 70 days (57.1 kg/ha/day) under optimum environmental and cultural conditions. The improved mungbean would have 20 pods/plant, 10 seeds/pod, and seed weight of 60g/1000 seeds in a plant population of 335,000 plant/ha.

The plant would have to have rapid seedling growth, dark green lanceolate leaflets, a strong erect non-branching stem, a vigorous root system, abundant nodulation, a vegetable reproductive ratio of 4:3, extended photosynthesis, high and efficient translocation of photosynthesis.

Better distribution of the sink would require a high harvest index with a large number of nodes and short internodes, multiple peduncles at each node with abundant pods per cluster, synchronized flowering and ripening, non-shattering pods, its pods above the leaf canopy, large number of seeds per pod, and large seeds. For yield stability, resistance to diseases and insects, tolerance to temperature and moisture stress and resistance to seed weathering was needed.

Innovation is a product of imagination. Crop improvement in mungbean began with an imagined ideal plant and performance targets. Collection, characterization, and evaluation of germplasm formed the foundation of the mungbean improvement program.

Some 5615 accessions have been collected, and most have been evaluated for agronomic character, nutritional components, and resistance or tolerance to major pests and diseases. At AVRDC, some 6460 crosses have been made; and up to 300,000 plants were screened each season. More than 90,000 seed packets have been distributed to 127 countries.

The work continues. Each year, AVRDC's elite lines and superior accessions from different countries are evaluated for adaptability in the International Mungbean Nursery (IMN)

Identifying germplasm with desirable characteristics was the first step. From germplasm collected, accessions were identified with insensitivity to photoperiod and temperature, resistance to CLS and PM, and resistance or tolerance to bean flies, bruchids, and pod borers.

MYMV is the most serious constraints in mungbean production in South Asia. To contain the spread of the virus, which is transmitted by the white fly *(Bemisia tabacci Genn),* AVRDC and NARS in South Asia agreed in 1992 to collaborative through a regional network. The object of the network was to develop and screen mungbean populations for MYMV resistance and other important plant characters. This Program has resulted in identifying lines resistant to MYMV and CLS, high yielding, large seeded and, uniform in maturity and early in duration.

The first international MYMV nursery, including 17 resistant lines originating from India, Pakistan, and Sri Lanka, was sent to more than 20 collaborations in south Asia and neighboring countries.

To introduce mungbean lines, and evaluate and select the best lines for specific different locations in Africa, AVRDC started its initial work at its Africa Regional Program. Using the AVRDC germplasm line, VC 1482E, an improved variety by name *Mmelegi* has been released in Botswana. Similarly , using V 1380(1), an improved variety by name *Imaca (Imara)* was released in Tanzania.

At present work is in progress and 14 lines are have been tested for two years both for yield and disease severity (*Table 5*). Lines VC 6148(50-12), VC 6372 (45-8) and VC 6153 (B-20P) showed good level of field resistance to the MYMV disease.

The best line will be promoted to farmers through on-farm demonstration trials.

Table 5. Yield of selected mungbean lines in a preliminary yield trial, Madiira Research and Training Station, AVRDC- Africa Regional Program during 1998 and 1999.

Line Accessions	Yield (in t /ha) 1998	1999	MYMV Disease incidence value (%)
NM 92	0.88	0.85	31
Kanti (check)	1.18	0.85	41
VC 6144 (B12)	1.35	1.26	53
VC 6173 (B-10)	1.14	2.33	3
VC 6173 (B-6)	1.25	0.92	23
VC 6148 (50-12)	1.39	1.96	35
VC 6173 (B-33)	0.86	1.31	19
VC 6379 (23-2-1)	1.17	1.09	60
VC 6367 (44-55-2)	0.97	1.20	12
VC 6173 (B-14)	1.08	0.99	27
VC 6175 (41-13-6)	NA	0.89	19
VC 6372 (45-8)	1.40	1.77	8
VC 6153 (B-20P)	0.92	1.77	10
VC 6173 (B-10b)	1.14	1.26	33

Source: AVRDC-ARP Data

5. RESULTS OBTAINED AND CONCLUSIONS

Three decades of AVRDC collaborative research has produced tangible results reflecting the following improvements in mungbean lines:

- Uniform maturity reducing the harvesting costs
- Resistance to MYMV, CLS, and other diseases.
- Reduced photoperiod sensitivity.
- Improved plant type with pods above the crop canopy
- Increased seed weight- 70g/1000 seeds.
- Larger and shinier seeds
- Early maturity - 60 days compared to more than 100 days for the unimproved cultivars.
- High yield-up to 2t/ha, compared to about 300 kg/ha for the unimproved cultivars grown in farmers' fields.
- Resistance to bruchids, an important storage pest.
- With the above-improved traits, the demand for the improved mungbean varieties developed by AVRDC has increased.

- 66 improved mungbean breeding lines have been released in 22 countries (see table in appendix).
- A total of 1.2 million ha worldwide are planted to mungbean derived from AVRDC germplasm.

6. SUMMARY

The crop improvement research on mungbean by AVRDC researchers has paid dividends and mungbean is delivering nutrition to the poor, diversifying cropping patterns, improving sustainability of cereal-based cropping systems by making soils more productive, and providing true benefits to farmers as well as consumers. Thus, mungbean is a choice crop for combating desertification and to improve human and soil health in Southern Africa.

AVRDC Africa Regional Program has initiated research on mungbean and is already identified with promising material ready for on-farm testing at strategic locations in Africa to know the potential of mungbean production and its acceptance in different parts of Africa. We hope that its low requirement of inputs, especially water, its high productivity, and a low management requirement will make it successful in African continent. Mungbean's high iron and protein availability make it especially valuable in places where these nutrients are deficient in the human diet.

REFERENCES

1. Ali Mubarik and Samson C.S. Tsou. (1999). *Promising Best Practices for Sustainable use of Medicinal and Indigenous Food Plants in Developing Countries. Case Study* (Unpublished).
2. Anon. (1982). *Roles of Legumes in Nitrogen Economy of Crops*. Mahatma Phule Agricultural University, Parbhani, Maharashtra, India.
3. Chao-Yih Chen, Samson C.S. Tsou and His-Hua Wang. (1988). *Utilization patterns of Mungbean in the Chinese Diet.*. In S.Shanmugasundaram, ed. Second Int. Mungbean Symp. Proc. AVRDC, Shanhua, Tainan. Pages 498 to507
4. Firth, P., H. Thitipoca, S. Suthpradit, R. Wwetselaar and D.F. Beech. (1973). *Nitrogen balance in the Central Plain of Thailand.* Soil Biol. Biochem. 5:41-46.
5. Paisan Laosuwan, Piti Sripana and Autapol Thoubsomsri (1988). *Research on Mungbean as Intercrop with Young Rubber and as a Sequential Crop with Rice.* In S.Shanmugasundaram, ed. Second Int. Mungbean Symp. Proc. AVRDC, Shanhua, Tainan. Pages 392-397.
6. Phool Singh, D.P. Singh, A. Kumar, B.D.Chaudhary and S.K. Thakural. (1988). *Response of Mungbean – Blackgram Hybrids under Water Stress Conditions. In* S.Shanmugasundaram, ed. Second Int. Mungbean Symp. Proc. AVRDC, Shanhua, Tainan. Pages 263-271.

7. Mortan, J.F., R.E. Smith and J.M. Poehlman. (1982). *The Mungbean.* Department of Agronomy and Soils, University of Puerto Rico, Mayaguez. 136 p.
8. Sinha, S.K. (1976). *Food Legumes., Distribution, Adaptability, and Biology of Yield.* FAO, Rome. 124 p
9. Singh, V.P., A. Chhabra and R.P.S. Kharb. (1988). *Production and Utilization of Mungbean In India.* In S.Shanmugasundaram, ed. Second Int. Mungbean Symp. Proc. AVRDC, Shanhua, Tainan. Pages486-497.

MANY PROBLEMS AND SOME SOLUTIONS TO THE COMMERCIALISATION OF NTFPs IN BOTSWANA

Frank W. Taylor
Veld Products Research & Development, PO Box 2020, Gaborone Botswana,
Tel/Fax: 267-347047, email: FrankTaylor@botsnet.bw

1. BACKGROUND

Botswana is a land-locked country of 584,000 square kilometres – almost exactly the same size as Kenya – but with a population of about 1.5 million. It is semiarid, with rainfall in the south being 200 mm., increasing to 800 mm in the extreme north. 70% of the country is covered by the Kalahari sandveld – the Kalahari Desert. The land is divided into tribal land, state land, wildlife reserves and freehold farms (the latter being less than 10% of the country). With such a small population in such a large country, everyone has access to tribal land to build a home in their village and have lands for agriculture. The area around each village is used for communal grazing but if a person has many cattle he may have a distant area set aside for him to drill a borehole for water and graze his cattle there.

Non-Timber Forest Products (NTFPs) play an important, but often un-recognised, role both in the rural and the national economies; for example, in times of drought the rural poor depend heavily on wild resources to sustain them. It is roughly estimated that the total value of NTFPs to the national economy is around US$20M; these include commercial products as well as those which can be considered as import substitution such as fuel wood, wild foods and medicines, building materials, etc. As far as income generation is concerned, it is estimated that about 20,000 people, mainly women and

Combating Desertification with Plants, edited by Dov Pasternak and Arnold Schlissel.
Kluwer Academic/Plenum Publishers, New York, 2001.

children, are involved in the harvesting and processing of NTFPs, the street value of which is thought about US$12M. There are no accurate statistics in this regard.

Historically, the management of NTFPs for subsistence had been with the tribal authorities that, in the colonial era, were emasculated by the colonial government which took over jurisdiction of the management of all natural resources (amongst other things). After many years of fruitless attempts to enforce natural resource management on the rural people, government's policy changed to encourage them to develop their own management systems so as to ensure sustainable utilisation of the resources.

People living in the communal (tribal) land areas always have had free access to NTFPs in **any** communal area in the country. In the past people harvested only what they needed from the wild in order to meet their subsistence needs such as foods, medicines, building materials, etc. Due to the low population density and vast harvesting areas there never was any pressure on the NTFP resource base. People had their traditional ways of managing their resources to ensure their sustainability: in particular the rare medicinal plants had to be harvested with care so as not to kill them because the active ingredients are in the roots and /or bark.

Until the late 1970s this situation was not a problem as there had been almost no commercialisation of NTFPs other than the sale of produce on the informal market – but even that was very limited as Botswana was unusual in never having had traditional markets as found in most other African countries. When the commercialisation of NTFPs became a reality the problems started on both the ecological and social fronts – hence this paper and the need to look at the causes and means to overcome them. The major problem areas are looked at and suggestions made on possible solutions. These are discussed below.

It is important to note that there are two categories of NTFPs which evidently are not threatened by over-harvesting, namely, wild fruits and most dried florist materials. However, it may be argued that if **all** the fruit of a certain species are marketed then there will be no regeneration in the long run. In practice there has been a decrease of the numbers of most, if not all, indigenous fruit trees around villages, mostly due to the grazing of goats and cattle which destroy the small seedlings before they are more than 20 cm high. They never recover.

2. COMMERCIALISATION

The commercialisation of NTFPs has brought quick benefits to thousands of rural people but in many cases the dividends are short-lived as the resources are over-exploited within one or two decades, if not sooner. In the case of Botswana the effects of commercialisation of NTFPs on the natural resource base in the 1970s caught government authorities unawares as at that time these products were regarded as having little significance. Not until about 10 years ago did the government start to recognise that NTFPs were an important means of subsistence for a substantial number of rural people.

Below is a brief history of the commercialisation and over-exploitation of some NTFPs in Botswana, starting with a short case study on an important medicinal plant.

2.1 Grapple Plant

This NTFP merits a short case study to show what happens when a widely available NTFP is commercialised and the way the different stakeholders reacted to the scenarios as they developed. The resource in question is *Harpagophytum procumbens* (otherwise known as the Grapple plant). The tuber-like secondary root system on this plant has medicinal properties which international pharmaceutical companies market widely as an effective remedy for arthritis and other ailments.

The plant is endemic in the Kalahari sandveld which occurs throughout most of Botswana as well as parts of Namibia, South Africa and Angola. It first came to the notice of German pharmaceutical companies in about 1925 after an arthritic German farmer in Namibia had a miraculous healing when a local herbalist encouraged him to drink an infusion made from the root. Tests in Germany proved the roots to be highly effective in the treatment of this ailment and the demand for the roots gradually increased over the years. Today, about 620 tons of dried roots are harvested annually in Namibia and Botswana.

For many years there was little concern about the long-term future of the resource base particularly as many regarded the plant as a pest because its large seedpod with vicious hooks often cause livestock to go lame – hence its other name of Devil's Claw. Namibian villagers eagerly harvested the plant as for many it was their only source of income. There was no thought about management practices to ensure the sustainability of the resource. The digging up of the secondary root system, which contained the active ingredients, usually killed the primary root and, thereby, the plant.

The fact that after some years of harvesting the plant became virtually extinct around some villages did not worry the merchants who just went to other villages to encourage the people to dig up the plant. This situation continued for many years until the merchants found that they had to travel great distances to find dried roots to buy. In about 1970 they started to come into Botswana to buy the roots as access was easy and the Botswana government took little notice – the need for phytosanitary certificates was ignored by all parties. It was not till 1976 that word got back to the Botswana government that the people in the settlements along the border were complaining that plants had disappeared along with their important source of income. Government immediately took a somewhat novel step to attempt to ensure the sustainability of the plant resource: they knew it was a perennial with vines that dried up and dispersed in the wind soon after the seed had set, usually about April each year. Therefore they decided to control the harvesting to a limited period after April each year – but by permit only. This meant that it would be more difficult for the harvesters to find where the plants were, thereby ensuring that not all the plants would be harvested in any given area.

In practice this simple control generally has proved to be true. Harvesting permits were issued to individuals. The front of the permits gave permission to the bearer to harvest a specific quantity of roots within a specified three month period which lay between April and September. On the reverse of the permits the harvesters were instructed to dig around the tap root without disturbing it and just dig up the lateral secondary root system. The instructions also stated that the holes had to be filled in.

Most of the harvesters were the poorest of the poor and, by definition, illiterate. Government took an additional precaution and hired a consultant to advise on how to further ensure the sustainability of the resource. Without studying the plant distribution, nor consulting the harvesters, the consultant recommend that the areas around each village should be divided into four quadrants: north, south, east and west, with each quadrant being harvested in turn each year, thereby allowing the plants three years to re-grow their secondary root systems.

The villagers accepted the permit system but in practice ignored the instruction to harvest by quadrant because they knew it was highly impractical as the plant never is evenly distributed around each village or settlement. As for the injunction not to disturb the tap root when digging up the secondary root system, this initially was generally ignored as it is much less effort to dig up the tap root in order to find the secondary roots – at any rate government rules often are regarded as being there to be disobeyed, and if one can get away with it so much the better!

It was not till 1985 that a local NGO took the initiative to buy the dried root at a far higher price than the pittance paid by the outside buyers. It was only then harvesters really started to take an active interest in the management of the resource and to harvest with greater care. In 1994 the government realised that imposed management systems do not work in the long run and brought out an enlightened policy that rural communities should develop their own strategies to **sustainably** utilise and manage their natural resources. In 1996 another local NGO started a programme with some rural communities to help them to take up government's new policy on management of their resources, the first step being to accurately estimate their potential harvest in a given area, divide total weight between the number of harvesters and then ask for harvest permits accordingly. As the government officer was an observer during this exercise, he had no difficulty in accepting the proposed figures.

Government officers now have become ultra cautious in the issuing of harvesting permits and are reluctant to issue these permits in new areas until such time as proven management methods have been developed to ensure a sustainable resource base.

2.2 Mopane Worms

About 50 years ago the first NTFP to be commercialised was the Mopane Worm *(Imbrasia belina)*. The worm is a large edible caterpillar that feeds on the Mopane tree (*Colophospermum mopane*) which usually occurs in extensive, pure stands. The worms are harvested, cooked and dried before being sold. Merchants sold the worm on the South African formal markets which today will take all that can be produced, with prices going up almost every year. Presently there are about 10,000 harvesters earning about US$1 million in a good season. However the harvests can vary considerably – from a glut to almost zero - and concern is being expressed that the resource base is being eroded through over-harvesting. Research is underway to resolve this issue and to develop a management programme to ensure sustainable harvests.

2.3 Baskets

The commercialisation of hand-woven baskets using the *Hyphaene petersiana* (Ilala palm) started in the late 1960s, and it is estimated that by 1980 a couple of thousand women were involved, mainly in the north western area of the Okavango Delta.

The pressure on the resource base became acute: instead of harvesting the leaves in the traditional way, i.e. every third leaf that appears, people started harvesting every leaf the moment it fully emerged. No plant can survive such a harvesting regime, and in 1981 the basket weavers approached government for help. A consultant quickly identified the problem and the women soon started to plant their own trees in their yards thereby increasing their source of supply. The situation now has improved to some degree but it is by no means solved, probably because the individual weavers have not planted enough palm trees to meet their needs.

2.4 Wild Silk Industry

An amateur entomologist discovered that in north eastern Botswana there were huge supplies of cocoons of the Wild Silk Moth (*Gonometa rufobrunnea*) from which a good quality of wild silk can be produced.

European companies expressed much interest in obtaining the carded raw silk and it was estimated that it would be a profitable business if cocoons were harvested within a 50km radius of the processing plant. About US$1M was invested in a factory to process the cocoons into carded raw silk.

The company took the precaution of hiring the services of a consultant in entomology to study the life cycle of the moth to ensure that the harvesting would be sustainable. The harvesters were instructed to harvest all cocoons that they could find whether they had pupated (hatched) or not.

The consultant went into the field to study the affect of harvesting the cocoons on the overall moth population in the harvest area, and his verdict was that only about 10% of the cocoons were being harvested and therefore the overall moth population was not being threatened.

In the first year about 30 tons of cocoons were purchased; in the second year 154 tons; in the third year 375 tons; and in the forth year only about 30 tons. It would seem evident from the rapid decline in the harvest that the moth population had been decimated through over-harvesting.

The consultant blamed other factors for the cause but it would seem self-evident that the harvesting of over 500 tons of (mostly?) live cocoons to be pulped and processed would involve millions of moths – in fact, much of the breeding resource base.

It is possible that the consultant visited areas where the harvesting was inefficient and only 10% of the resource was harvested, and that he then assumed that the harvesting was uniform.

Further research by the consultant outside the harvest areas did reveal severely depressed worm populations in some seasons. The exact cause(s) could not be pinpointed

The financial backers decided not to put more funds into the project and it finally closed. Now, about 10 years later, the moth populations are reported to almost be back to normal.

3. PROBLEM AREAS AND POSSIBLE SOLUTIONS

This paper now explores the problem areas and makes suggestions on possible solutions.

3.1 Access to Resources

Problem: Because there is uncontrolled access to communal resources, people from any part of the country can go into any communal area to harvest NTFPs. For example, this problem becomes apparent during most Mopane Worm harvests when large numbers of people disperse in the Mopane forests to search for the caterpillars. If they are in short supply people almost come to blows, with the local residents feeling very angry at the intrusion of outsiders. Similar scenarios may be experienced with other NTFPs such as wild foods, thatching grass, etc.

Solution: Government should enact legislation to allow rural communities to have control over access to their NTFPs.

3.2 Community-Based Natural Resource Management (CBNRM)

Problem: While the government policy on encouraging CBNRM is laudable, there is very little to encourage people to become actively involved, especially when outsiders have access to "their" resources. In addition, people are not easily motivated to participate in CBNRM if the benefits are relatively low and from only a few seasonal products. When only a few products are exploited heavily, the resource base is put under severe pressure.

Solution: A very wide range of NTFPs using diverse natural resources should be developed so as to allow people the opportunity to generate income **throughout** the year and at the same time reduce the current narrower pressure on the resource base.

3.3 Identification of NTFPs of Economic Importance

Problem: There are many resources which are unexploited due to ignorance of their commercial potential. The main reason for this is that such NTFPs are for the export market about which local people are totally ignorant. Such products could include florist materials, ornamental plants, essential oils, gums, resins, etc.

Solution: The services of a consultant should be engaged to undertake a survey of the wild resources to identify potential NTFPs.

3.4 Value of NTFPs to National Economy

Problem: As mentioned above, there has been no accurate assessment of the value of NTFPs to the national economy with the result that there is no government policy that is particularly concerned with these resources other than from a conservation point of view. If their full value to the economy were known then more could be done to enhance income generation through protection and supportive measures.

Solution: Undertake a survey to assess the value of NTFPs to the national economy.

The government presently is planning such a survey with the involvement of stakeholders.

3.5 Access to Markets

Problem: In general the commercialisation of most NTFPs is severely hindered as people do not have easy access to markets which will pay a fair price. International markets want volume, quality control and delivery on time which conditions 99.9% of the producers cannot meet on their own.

Solution: Establish an NGO with its members being drawn from producer groups and other grass roots NGOs with the objective of establishing a national buying and marketing network.

3.6 International Lobby Against Harvesting Wild Medicinal Plants

Problem: The "Greens" movement is acting as a pressure group to persuade the pharmaceutical industry not to buy herbal medicines harvested in the wild as the resources are being wiped out in many countries. Also CITES is considering listing *Harpagophytum procumbens* as a Category #2 endangered species. In addition there is a rumour that the European Union may bring in regulations that will only allow the importation of *H. procumbens* if the harpagosid contents are above a specified level. As the harpagosid content cannot be guaranteed from plants in the wild, this major market will be closed.

Solution: Domestication of *H. procumbens* will have to be undertaken using superior genetic material well above the minimum specifications required. This also will enhance the value of the product.

3.7 Environmental Degradation

Problem: Environmental degradation is happening around every village and settlement in the country without exception. This means that the NTFP resource base is being severely reduced, wild fruits being one example.

Solution: Domestication of indigenous plants of economic value should be undertaken (two NGOs have started along this path but considerably more needs to be done). Concurrent with this it is recommended that exotic arid land plants with economic potential should be screened for adaptability to local conditions – the IPALAC programme is an ideal vehicle for this. The CBNRM programme, if fully taken up by the communities, also should have a significant impact on reducing the environmental degradation.

3.8 Who is Responsible in Government?

Problem: There is not one department in government that is responsible for NTFPs: there is the Agricultural Resources Board (ARB) which is responsible for seeing to it that the natural resources are not over-exploited. They also deal with the issue of harvest permits for *H. procumbens* and, among other things, they are very occupied with controlling the extensive bush fires that sweep across the country every dry season. They are under-staffed and their hands are full. Also in the Ministry of Agriculture, the Department of Forestry and Range Ecology is directly involved in NTFPs for obvious reasons. Then there is the National Conservation Strategy Committee which is a government body seemingly operating on its own. The Ministry of Commerce & Industry also has an interest in NTFPs through the commercialisation of the products.

Solution: The ARB be given full responsibility for the policy and other matters pertaining to NTFPs.

3.9 Pandora's Box

Problem: The ARB, which is the main government body responsible for the sustainable use of NTFPs, is faced with a problem: producer groups, NGOs and individuals are identifying new NTFPs to exploit commercially, some even coming up with proposed management methods to ensure the sustainability of the resource. While the ARB would like to encourage income generation for the rural poor, at the same time it is wary of putting the resources at risk just in case there is mismanagement and the resources are wiped out. Also they do not have the manpower to police the management of the resources that presently are being harvested, let alone any expansion of these activities. At present they are reluctant to allow the harvesting of *H. procumbens* in new areas even though there may be an abundance of plants there. Proposed new projects are being put on hold while, at the same time, they are agonising over the fact that there are poor people out there who desperately need money to feed their families. They realise that it could take a considerable amount of time, effort and money to develop management systems which will ensure the sustainability of each individual resource being exploited.

Solution: There is no easy, quick-fix solution. The ARB presently is trying to develop a practical *modus operandi* to screen all applications for NTFP exploitation. This is a good step in the right direction but they face a problem of finding suitably qualified and/or experienced people to be involved. They will be faced with making decisions on how to control the harvesting of NTFPs while the management systems are being worked out. One possible solution may be to issue permits for harvesting conservative amounts of material – as with *H. procumbens* - until a clear picture emerges of the most suitable management system for each resource. Of course this will mean that the ARB will have to get capable personnel to undertake the implementation and policing of the programme.

An alternative solution may be for government to contract with a suitably experienced NGO to undertake the entire programme, which will include:

- Product identification and development,
- Assisting the people to develop their own sustainable management systems.
- Monitoring the impact on the resource base with the people, and adjusting the management systems as and when required.
- Reporting to government on a regular basis on all aspects of the work.

If government does this then it has gone a long way to overcome its problem of policing the environmental impact of commercialisation of NTFPs.

4. CONCLUSION

It is important that every effort be made to sustainably exploit NTFPs to the benefit of rural people without endangering the resource base. This requires an awareness of the potential pitfalls and the means to overcome them.

There is no doubt that there are a considerable number of potential NTFPs which have yet to be identified and exploited in any given area, but the temptation to identify and exploit them must be tempered by the careful planning and execution of project. Possibly the most important aspect is that there should be a continuous monitoring of the resource base by the communities themselves in collaboration with government or with an NGO appointed by government to undertake this role.

FIGHTING DESERTIFICATION BY PRODUCTION AND USE OF JATROPHA OIL

Experience of a Systematic Approach Applied by a GTZ-Supported Project in Mali, West Africa

Reinhard K. Henning
Rothkreuz 11, D - 88138 Weissensberg, Germany, Tel.: 49 8389 984 129, Fax: 49 8389 984 128, e-mail: Henning@bagani.de, internet: www.jatropha.org

1. CENTRAL HYPOTHESIS

The Jatropha System creates a positive reciprocity between energy production and environment/food production: i. e. the more energy Jatropha hedges produce, the more food crops are protected from animals and erosion. Also additional income is created, mainly for women.

2. INTRODUCTION

Mali is a landlocked country in the middle of West Africa, just at the southern edge of the Sahara desert. The country's average annual rainfall ranges from 200 mm in the north to 1200 mm in the south. For generations, farmers have protected their gardens with hedges of *Jatropha curcas*, or physic nut, which is not eaten by animals and thus protects the foodcrops as a living fence.

Jatropha curcas is a plant of Latin American origin which is now widespread throughout arid and semiarid tropical regions of the world. A member of the Euphorbiaceae family, it is a drought-resistant perennial, living up to 50 years and growing on marginal soils. A close relative to the castor plant, its oil has the same medical properties. Jatropha seeds contain about 35% of non-edible oil. The production of seeds is about 0.8 kg per meter of hedge per year, with an oil yield of 0,17 l [1]. Currently, Mali has about 10.000 km of Jatropha hedges with a growth rate of 2.000 km per year, which represents a potential of 1.700.000 liters of oil per year. The average length of these hedges, in those areas of Mali where they are most prevalent, is between 2 and 15 km per village, with a maximum of up to 40 km per village [2]. These areas, which have been evaluated as part of this project, include the:

- Kati area, 180 km. west of Bamako (average 15 km of hedges per village)
- Kangaba area, 100 km southwest of Bamako (av. 7 km of hedges)
- Fana area, 120 km east of Bamako (av. 2 km of hedges)
- Ouelessebougou area, 80 km south of Bamako (av. 6 km of hedges)
- Bougouni area, 200 km south of Bamako (av. 4 km of hedges).

Jatropha curcas is generally well known among the populations of Mali and has long been recognized as a plant of many uses. If carefully planted, Jatropha hedges not only protect gardens from hungry livestock but also reduce damage and erosion from wind and water. Traditionally the seeds were harvested by women and used for medical treatments and local soap production.

As far back as at the end of the 1930's the oil's potential as a fuel source was also recognized [3]. Currently, it can be used to substitute for the "gazoil" mixture used in the Indian type diesel engines that drive grain mills and water pumps in rural areas of Mali. The high-quality oil extracted by engine-driven expellers or by manual Bielenberg-ram-presses or the sediment of the oil purification process may be used for larger-scale soap making in rural areas, giving local women the chance to gain additional income and thus strengthen their economic position. The press-cake as another extraction by-product can be used as a high-grade organic fertilizer.

The Jatropha activities were initiated in Mali in 1987 by German Technical Assistance (GTZ) in the framework of a renewable energy programme. The Jatropha Project itself started 1993 and ended in 1997. It worked to combine the above mentioned and other factors into the "Jatropha System". This system focuses not simply on the use of Jatropha oil as fuel, but rather on the use of this oil as a crucial element to activate a circular system combining ecological, economic, and income-generating effects, the latter specifically for women [4].

Thus, the Jatropha system promotes four main aspects of development, which combine to help assure a sustainable way of life for village farmers and the land that supports them:

- Erosion Control and Soil Improvement
- Promotion of Women

- Poverty Reduction
- Renewable Energy

In the rural areas in Mali, Lister-type engines are used to drive grain mills and water-pumps. These inexpensive pre-combustion chamber diesel engines of Indian origin require only the addition of a fuel filter to be able to run on pure Jatropha oil, thus eliminating the need for gazoil entirely. Furthermore, at maximal load conditions the Jatropha oil gives even better results than gazoil because of its high oxygen content [5]. Based on tests conducted by the Jatropha Project, the oil can also be successfully used as a lubricant in these engines[6].

In equivalent terms, the energy needed to produce Jatropha oil in mechanical presses amounts to about 10% of the oil obtained. Because Jatropha oil can be produced inexpensively [7], it can also be sold at prices lower than gazoil's official price at the petrol stations. Even more important than the price is the possibility of local energy production, because of the periodic unavailability of gazoil in the rural areas caused by lack of road access during rainy season.

The technology for using natural pure Jatropha oil as substitute for paraffin oil for lamps and cookers is not yet available. Different research centers are working on it.

3. EROSION CONTROL AND SOIL IMPROVEMENT

Jatropha "living fences" in Mali not only control unwanted animal access to the fields; they also reduce wind erosion and, if planted parallel to slopes to fix small earth or stone dams, they help control water erosion. The plant's roots grow close to the ground surface, anchoring the soil like miniature dikes or earthen bunds. These dikes effectively slow surface runoff during intensive downpours, which are common, thus causing more water to penetrate into the soil and boosting harvests.

The press cake which remains after oil extraction by the expellers is a very good organic fertilizer, with mineral composition comparable to that of chicken manure. This has great value for agriculture in the Sahelian countries, since soils there are rapidly depleted of humus and chemical fertilizers are very expensive.

The Malian cotton-growing company, CMDT (Compagnie Malienne de Dévelopement Textile), uses Jatropha hedges to assure a program of improved fallow: the cotton fields are protected with Jatropha hedges to keep out cattle, while the fields are sown with legumes to improve soil fertility.

4. PROMOTION OF WOMEN

Many government and non-government organizations provide rural Malian women with engine-driven grain mills to ease their work of food preparation. However, these grain mills need external resources of fuel, lubrication oil, spare parts and maintenance. Consequently, the introduction of such a grain mill tends to lead to an impoverishment of the village because of the cash required both to buy and to transport these external resources to the village. By using locally produced Jatropha oil as fuel and lubrication oil, some of this cash outflow from the village can be stopped.

Traditionally, rural women used *Jatropha curcas* for medicine (seeds as a laxative, latex to stop bleeding and against infections, leaves against malaria) and for soap production. The traditional soap-making process is very labor-intensive, producing small amounts of relatively poor-quality soap.

Table 1: Economy of soap production in Mali (prices in US$, 500 FCFA = 1 US$)

Description	Quantity	Unity	Price per unity US$	Amount in US$
Inputs				
Seeds (give 3 l of oil with handpress)	12	kg	0,1	1,20
Caustic soda	0,5	kg	1,2	0,60
Labour (4 h for pressing, 1 h soap production)	5	h	0,2	1,00
Depreciation/maintenance (5 years, 10 t/a, 240,-)		US$/kg	0,02	0,24
Total expenses				**3,04**
Revenues				
Presscake	9	kg	0,03	0,27
White soap	28	pieces (170 g)	0,15	4,20
Total revenues				**4,47**
Net Profit				**1,43**
Profit per liter of oil			**0,48**	
Profit per kg of soap			**0,31**	
Price per kg of soap			**0,89**	

When Jatropha oil is used, either alone or in combination with other local plant oils such as shea butter, larger amounts of a more refined soap are produced. The women can easily sell this soap in local markets and nearby towns, increasing their possibilities of earning income with local resources.

Some details of the economy of soap production with the means of the Bielenberg hand press are shown in the above table Poverty Reduction.

By promoting the integrated utilization of the Jatropha plant, the Jatropha System can provide direct financial benefits to the rural economy. To illustrate this with a rough calculation, assume the average village of the pilot area has 15 km of Jatropha hedges, which represents 12 tons of seeds.

These 12 tons of seeds may generate 1.800 US$ of cash income when the oil is extracted and the products are sold:

• 9.000 kg of presscake for 0.03 =	270.- US$
• 2.400 liters of oil for 0.60 =	1.440.- US$
• 600 kg of sediment for 0.15 =	90.- US$
Total	**1.800.- US$**

If we take the real example of an entrepreneur in a small village near Bamako, who buys the seeds for soap production and hires people for the production process (extraction with Bielenberg ram press, soap production, see table above), the cash income for the village population, including the entrepreneur, amounts to 3.630 US$:

• 12.000 kg of seeds for 0.10	1.200.- US$
• 5.000 hours of labor for 0.20	1.000.- US$
• profit of the entrepreneur	1.430.- US$
Total	**3.630.- US$**

If these figures are extrapolated to Jatropha plantations, a profit in the range of cotton farming is within reach.

The "Jatropha System" also helps reduce poverty by:

- Reducing crop losses caused by wandering livestock or wind damage;
- Increasing rainfall infiltration, resulting in less work/irrigation water needed for local gardens;
- Increasing soil fertility by use of presscake as fertilizer;
- Increasing use of inexpensive local resources rather than expensive external resources;
- Reducing disputes between farmers and livestock owners regarding crop damage, as well as among farmers themselves regarding the boundaries of their fields;
- Providing local jobs, lessening the need for local villagers to migrate to cities to find employment.

Because of its economic value the rural people are planting new Jatropha hedges in a large extent. In Kita, one of the pilot regions of the Jatropha project, the average length of hedges went up from 5 km to 15 km in the last 8 years.

5. CONCLUSIONS

The results of the Jatropha Project to date show that the chances of this system being successfully implemented are high, provided that a cautious approach is taken. Above all, care must be taken to ensure that women retain their traditional responsibilities for harvesting and processing the seeds.

Furthermore, Mali is a typical Sahelian country; its large geographic expanse and climatic variations mirror the ecological conditions found throughout the Sahel. Because of this, the efforts already being made in Mali to derive value from oil-bearing plants can be taken as representative and used to elaborate a "concept for production and use of plant oils as fuel" that is valid for the Sahel region as a whole, and even for other African countries.

To summarize, the Jatropha system is characterized by the many positive ecological, energetic and economic aspects which are attached with the commercial exploitation of this plant. The more this plant is exploited, the better for the environment and for food production.

REFERENCES

1. Henning, R.K. (1994). Produktion und Nutzung von Pflanzenöl als Kraftstoff in Entwicklungsländern. In: VDI-Berichte Nr. 1126, , 215 - 229
2. Henning, R.K. (1996). 3. Fachlicher Zwischenbericht zum Projekt: Produktion und Nutzung von Pflanzenöl als Kraftstoff; unpublished project report, Projet Pourghère,GTZ.
3. Kone, S. (1988). Les activités précédentes sur le Pourghère au Mali, unpublished project report, Programme Spécial Energie Mali.
4. Henning, R.K. (1995). Klaus v. Mitzlaff, Produktion und Nutzung von Purgieröl als Kraftstoff und Rohstoff für die lokale Seifenherstellung im Sahel. In Witzenhäuser Hochschulwochen.
5. Bielenberg, C. (1994). Personal communication.
6. Metzler, R. (1996). Plant Oil as Fuel and Lubrication Oil, unpublished project report, Project Pourghère.
7. Wiemer, H.J. (1995). Rapport de mission: Etablissement d'un système de suivi et évaluation des effets du projet,unpublished project report, Projet Pourghère.
8. Samake, F. (1996). Valorisation du tourteau de Pourghère comme engrais sur le coton, unpublished project report, Projet Pourghère.

Part III

LAND REHABILITATION

COMBATING DESERTIFICATION WITH SEABUCKTHORN

Rongsen Lu
Chengdu Institute of Biology, Chinese Academy of Sciences, P.O.Box 416, Chengdu, Sichuan, China , Tel / Fax: 0086-28-7326017, email: mangotree@163.net

1. BACKGROUND

The environment is a common concern all over the world, as it is the basis for the existence of human beings. Around the world, 900 million people in more than 100 countries are threatened by loss of productive lands and increasing food shortage due to scarcity of rainfall and the desertification process. The Institute of World Resources has estimated that over the last 45 years, about 1.2 billion hectares of land, or about 11 percent of the earth's surface, has been badly degraded, particularly in Africa and Asia, although in other continents as well. It has been also estimated that desertification has been responsible for over 42 billion dollars in land and property loss annually in recent years, with the prospects for the situation worsening in the future unless strong global commitment is made to address the root causes of the destructive desertification process taking place.

The destructive nature of desertification, and its relation to poverty and human suffering, has been taken up in various international fora over the years. Most notably the need to combat desertification was discussed and agreed upon at the first United Nations Conference on the Human Environment in Stockholm, Sweden, in 1972, and was given prominence in its Declaration on the Human Environment. Desertification was again given major attention at the United Nations Conference on Environment and Development in Rio de Janeiro in June 1992, and specific measures to control desertification were outlined in the Global Agenda 21 Strategy for

Combating Desertification with Plants, edited by Dov Pasternak and Arnold Schlissel.
Kluwer Academic/Plenum Publishers, New York, 2001.

Sustainable Development agreed upon by Heads of States attending that Conference. Most recently in June 1994, countries of the world ratified a UN Global Convention on the Prevention of Desertification, which outlines specific responsibilities of countries in helping to control desertification process around the world.

China is among the countries most severely affected by desertification, with one third of its surface area covered by deserts and a loss of about 1,250 square kilometers of agricultural lands annually during the 1980s due to wind and water erosion. This rate of farming land loss is up from about 1,560 square kilometers annually during the 1960s and 1970s. This increasing desertification of arable lands in China has taken place despite very considerable efforts to control desert encroachment and degradation of soil conditions. Over the last four decades China has planted over 100,000 square kilometers of shelter forests in desert areas, bringing 10 percent of the desert areas under control, and reclaiming 1,300,000 square kilometers of agricultural lands.

Over the last two decades, a major element in China's strategy to control desertification process has been the planting of seabuckthorn (*Hippophae* L.) in desert border areas suffering from severe soil and water losses. Seabuckthorn has proved able to hold and build back top soil in such areas, and thereby to help to improve the productive capacity of fragile ecosystems. Seabuckthorn plantations have demonstrated their ability to facilitate the new development of sustainable agriculture, forestry, and livestock raising and the improvement of farming family incomes.

In the areas of northwest China, including the Loess Plateau region, where poverty is widespread and which is threatened by desertification, seabuckthorn production and harvesting of their berries have proved to be effective in generating employment and income, and thereby helping to alleviate poverty.

2. BIOLOGICAL FEATURES OF SEABUCKTHORN AND ITS USES

Seabuckthorn belongs to the genus *Hippophae*, family Elaeagnaceae, which includes six species and nine subspecies and is widely distributed in many countries, mainly in Eurasia between 9°W-122°E longitude and 27°-68°N latitude. In terms of altitude, it grows in a variety of ecological niches ranging from seaside (Baltic Sea) to 5,200 m above sea level (Qumulangma Mountain). Seabuckthorn can grow well under the following climatic conditions: monthly average temperature of the hottest month 15°C-25°C, maximum radiation on clear days in the period of vigorous growth 23.5-26.5 kcal/cm^2, and annual rainfall no less than 300-350 mm.

Seabuckthorn can stand extreme ground temperatures from -40°C to +40°C. Some species *(H. tibetana* and *H. neurocarpa*) can grow in the area with annual average temperature of 0°C. Some species (*H.rhamnoides subsp.turkestanica*) can resist the highest summer temperature of 40°C, and the species (*H. rhamnides* subsp.*turkestanica* and *H. rham.* subsp. *mongolica*) can tolerate 1.1% of salt in the soil. The hardiest species, *H. rhamnoides* subsp. *sinensis*, can adapt to very harsh, poor and arid soil conditions like arenaceous rock areas in Inner Mongolia, China.

Seabuckthorn is effective in controlling soil erosion and water loss. From comparison tests, seabuckthorn covered areas have 75% less soil loss by water erosion, 90% less runoff, and 80% less wind erosion than the barren lands. Seabuckthorn is a nitrogen fixing plant. One *mu* (1/15 of a hectare) of 10-year-old seabuckthorn forest can fix 12 kg of nitrogen annually, which is equivalent of 25 kg of urea fertilizer.

In 1985-1998, a number of seabuckthorn forests were established in the north, north-west and north-east of China. Each year, about 1.0 million *mu* of seabuckthorn plantations were established, which greatly reduced soil erosion and water loss in those areas.

Recent practice has shown that seabuckthorn is also a powerful weapon for reclaiming lands that have suffered desertification. There are three major areas of desertification distributed along the Great Wall in the Loess Plateau in China. They are the Mouwusu Sand Land, the Kubqi Desert, and the Hedong Desert. The total area of land subject to desertification is about 118,000 km^2, while lands threatened by desert total about 200,000 km^2. Each year the desert moves south 3 m, entraining a land loss of 50 km^2. Since the middle of the 1980s, seabuckthorn has been introduced to stop desert encroachments in Yulin, Jingbian and Huanxian which are located in the south of the Mouwusu Sand Land.

The results showed that seabuckthorn was capable for forming closing canopies in the sand lands, the average height being about 3 m and the diameter of the trunk above ground about 7 cm. The seabuckthorn forest between sand dunes could not only form dominant populations, but also send out suckers from roots to the tops of the dunes. The density of suckers reached 2-3 plants/m^2. Two-year-old plant suckers are 0.5 m in height and 0.6 cm in diameter. Due to the strong ability to develop suckers, a certain quantity of seabuckthorn plants could form a chessboard-like sucker network. Sucker networks can effectively stop the movement of sand dunes.

Owing to its unique biological features, seabuckthorn has played a very important role in controlling soil erosion and water loss in the area of desert encroachment in northern China. During the recent decade, seabuckthorn development has been tried in even harsher areas like western Liaoning province, northern Shanxi province, north-western Shanxi province, and the arenaceous rock areas of Inner Mongolia. In the 1990s, the seabuckthorn plantation areas increased by one million *mu* annually. Now, the total area of natural seabuckthorn forests and plantations has reached about 15 million *mu* in the whole of China.

3. ECONOMIC BENEFITS OF SEABUCKTHORN

Seabuckthorn is a multipurpose plant whose fruits, seeds, leaves and bark can be exploited. In particular, its fruit contains very rich nutrients and bio-active substances (both in its juice and oil) which have been widely used in making food, beverages, medicines, and cosmetics.

It is estimated that one ton of seabuckthorn fruit can be made into several primary products with a total value of about 17,860 RMB Yuan (8.3 Yuan equals to 1 $US), including seabuckthorn oil (5,400 Yuan), total flavonoids (11,200 Yuan) and fruit juice (1,260 Yuan). Farmers can earn 1,000-1,200 RMB Yuan by selling a ton of fresh fruit. If the seabuckthorn oil and flavonoids could be further processed into final products, the total value would be 5-10 times that of the raw materials.

The leaves of seabuckthorn are rich in protein and fat and their nutritional value is similar to that of alfalfa, so the leaves of seabuckthorn are very good fodder. Studies show that a four-year-old seabuckthorn forest can be grazed by animals. The leaf production of 7 *mu* of seabuckthorn forest is sufficient to support one sheep. Once the seabuckthorn forest is established, it can be sustainably used for a long period.

The tender leaves of seabuckthorn can be used for making tea and have been marketed for this purpose in many cities in China. The branches and trunks of seabuckthorn are a good fuel wood with heat value of 4,788 calorie/gram, which is appropriately equal to the heat value of crude coal. This is very important for areas where vegetation is scarce and there is a lack of alternative energy resources.

By the end of 1998 there were about 100 seabuckthorn processing factories in the whole country, preparing 200 products including soft drinks, food, health products, medicines, cosmetics, fodder and pigments, with a total value of about 0.5 billion RMB Yuan annually.

Since new China was established in 1949, many valuable plant species such as willow, pine, *Canagona, Amorpha, Sophora, Camphlotropis* and *Melilotus* have been recommended for planting in the desertification areas. But they have been slow to gain popularity. One reason is that the farmer had little interest in planting the recommended species because he could not derive benefits from them. Farmers were willing to accept seabuckthorn because they could get various economic benefits from seabuckthorn plantations, as mentioned previously.

4. SOME SUCCESSFUL CASES OF SEABUCKTHORN CULTIVATION

Encouraged by successful examples in the Loess Plateau , people believe that seabuckthorn is a unique, precious plant which has tremendous ecological, economic and social benefits, and that it is a powerful tool with which to harness the lands threatened by desertification. The decision-makers of several provincial governments resolved that seabuckthorn should be utilized to combat desertification in their own provinces, with the result that establishment of seabuckthorn plantations has taken place on a large scale in northern China. Below we mention several successful cases.

4.1 Yikezhao Prefecture, Inner Mongolia

In the arenaceous rock areas in Yikezhao Prefecture, Inner Mongolian Autonomous Region, the vegetation cover is so severely damaged that sand and clay bedrock is entirely exposed over large areas, and large stretches of land have become deserted, being known as "the moon on the earth", and "environmental cancer". Soil loss/km^2/year reaches over 30,000 tons, with the highest loss being over 60,000 tons, amounting to over 2 cm of surface soil lost every year. The region produces coarse sediment, with severe wind erosion and water erosion affecting the middle reaches of the Yellow River. The serious water and soil erosion here is not only a great threat to the lower reaches of the Yellow River, but also causes much hardship to the local people. For a long time, degradation of land quality and diminution of land quantity have resulted in a situation where the local people have no land to cultivate and no grassland on which to graze their animals. They have lost the basic preconditions for existence and have become "environmental refugees".

In order to reclaim the severely deteriorated environment, develop animal husbandry, and raise the living standard of the people, the government of Yikezhao Prefecture has insisted on the principle of comprehensive reclamation, with priority given to biological measures, and has put forward the call to "plant trees, plant grasses and plant peashrub." In the area covered with loess, peashrub (*Caragana korshinskii*) and astragalus (*Astragalus adsurgens*) planted on a large area have played an important role in preventing wind erosion and in fixing the sand to conserve soil and water and boost livestock development. But in the arenaceous rock area, it is very difficult for peashrub with its deep root system to survive and to grow, and even where it survives, it grows poorly. In five years, peashrub grows less than 40 cm in height. Among tree species, white elm, mountain apricot and small leaf poplar were planted in such areas without any success. It seems, then, that measures involving planting of trees and grasses cannot bear fruitful results in the harsh arenaceous rock area. Seabuckthorn, however, with its well-developed root systems, nitrogen-fixing ability to enrich soil, strong emergence of root suckers, and tolerance of drought and poor soil fertility, is well equipped to thrive in the arenaceous rock area.

Since 1986, Yikezhao Prefecture has been carrying out a seabuckthorn afforestation demonstration. The experiments were conducted in Henitu, Husatu, Hantaichuan and Deshengxi, among others, with satisfactory results. Seabuckthorn plants planted along gully slopes, elongated mounds, gully bottoms and riverbanks were successful. Early-planted seabuckthorn on an area of 6.67 ha grew vigorously, plants being sturdy, with an average height of 1.5 m and rich fruits, and forming a canopy layer. The local people were able to see for themselves the green canopy, the vigor of the plants, and the hope they represent for control of water and soil erosion in their region. These features of seabuckthorn also drew attention and support from governments at various levels.

After the national workshop on seabuckthorn in 1990, the Water Conservancy Committee of the Yellow River increased its investment in Yikezhao Prefecture to accelerate the speed of seabuckthorn afforestation. The planned area in the project is 700 km^2, including Zhunger county, Dalat county and Dongsheng city and 8 townships. Largely owing to the enthusiasm of the local people, strict management, and practical technical measures, 20,000 ha of seabuckthorn forests were completed to form a rather large-sized demonstration area.

Since 1992, the local people and government officers have realized that seabuckthorn is their "green hope". Seabuckthorn planting has thus become a mass campaign. In Zhungar county, for example, from 1987 to 1998, through integrated harnessing (mainly growing seabuckthorn), vegetation coverage has been increased to 61%, water loss and soil erosion have declined by 70%, and the erosion modulus has been reduced from 40,000 to 5,000 tons/km^2/year. By the end of 1998, about 53,000 hectares of seabuckthorn forest had been established in Yikezhao Prefecture, Inner Mongolia. Out of the total seabuckthorn forest, about 10,000 hectares of seabuckthorn forest are bearing fruit, and it is estimated that 75,000 tons of fruit could be harvested per year.

4.2 Janping County, Liaoning Province

Jianping County is located in west Liaoning Province. With an annual rainfall of 400 mm, the climate falls in the semiarid category with monsoon climate. The rainfall in June-August accounts for 73% of the yearly total. The frost free period is short. The average elevation within the county is 600 m above sea level. There are three kinds of landforms, i.e. the rocky low mountains, rocky hills, and flat lands along the river valleys.

According to the historical record, several hundred years ago, Jianping County had green mountains and clean and limpid water, as well as thick and vigorous forests. By the 1950s, however, prolonged indiscriminate felling, overutilization of land, and overgrazing had left only 200 ha of residual forest i.e., the forest coverage rate was only 1.9%. The destruction of the plant cover resulted in severe deterioration of the environment. Such disasters as floods, droughts, wind and hailstone occurred frequently, leading to further intensification of water and soil erosion. The eroded area in Jianping county occupies as much as 39,300 ha, accounting for 81% of the total land. There are still 67,000 ha of sand dunes and deserted land. Accordingly, the worsened natural environment seriously constrains economic development. Jianping County is one of the poverty-stricken counties in China.

At the end of the 1950s, seabuckthorn was introduced from northwest China into the county to control water and soil erosion and improve the environment. Seabuckthorn, first introduced purely as an ecological measure, is now yielding dual benefits - ecological and economic. Planting seabuckthorn is proving an important tool for promoting rural economic development.

Now that the value of seabuckthorn in developing the rural economy has been established, a new surge of seabuckthorn afforestation has sprung up in Jianping county, with 6,700 ha being planted annually. At present, the size of the artificial seabuckthorn forest has reached 66,700 ha. Of these, 55,000 ha bear fruit (annual yield: 19,680 tons of fruit, i.e., production rate of 0.36 t / ha). Thus Jianping county has become the county with the biggest artificial seabuckthorn forest in China. It is also the largest stretch of artificial seabuckthorn forest in the world.

The planting of seabuckthorn over a large area in Jianping county has not only reversed the deterioration in the environment, it has also brought huge socio-economic benefits. At the same time, rich experience has accumulated.

4.3 Wugou Township, Zhenyuan County, Gansu Province

Zhenyuan County in the Loess Plateau is also a poverty-stricken county, with a degraded natural environment and low agricultural production level. Wugou Township is one of the poorest townships in the county. Over a period of 30 years, a forest with a total area of more than 2,300 ha was planted there. However, owing to the irrational selection of tree species - pine, black locust, popular, and white elm - only 400 ha of the forest remains, and the trees are growing quite poorly. In the middle of 1980s, with support from the Seabuckthorn Office of the Water Conservancy Committee of the Yellow River, Wugou township planted 4,000 ha of pure seabuckthorn forest, then, in 1990, a mixed forest with other tree species. Together they account for 27% of the total area in the township. As a result of this successful plantation effort, the bare mountains and waste gullies have become green.

The large-scale seabuckthorn forest has produced remarkable economic returns, yielding practical benefits for the people in Wugou township. To begin with, the hazards caused by water and soil erosion have been greatly reduced. Based on preliminary observations and measurements, it seems that the four-year-old canopy of the seabuckthorn forest is very thick and vigorous, while beneath it grass species grow in abundance. And its litter layer is 3-4 cm in thickness, so that a vertical protective structure is being formed. Secondly, the problems of lack of animal feed, fertilizer and fuel, which has been affecting the development of agro-production for a long time, have been solved. Without using crop residue for fuel, the cereal yield has been increased by a wide margin so that the problem of eating well and dressing warmly is being solved progressively. Zhenyuan County has extended the experiences obtained from Wugou Township to the other townships. In 5 years (1993-1997) the county established 13,000 ha of seabuckthorn forest, which makes the total area of seabuckthorn forest up to 27,000 ha.

5. PROSPECTS IN THE FUTURE

Since seabuckthorn has proved to be a powerful weapon for combating desertification, the central government of China (the State Council) has established a long-term project called "National Plan for Afforestation Using Seabuckthorn". According to this plan, a huge tract of desertified land with a total area of 6,476,667 hectares (mainly in the north, northwest China) will be covered by seabuckthorn forest in 23 years (1998-2020). In other words, the area of seabuckthorn will be increased fivefold (it was 1,300,000 hectares in 1997). By the end of 2020, the ecological environment of the target area will be greatly improved due to the establishment of seabuckthorn forest.

In conclusion, the exploitation and utilization of seabuckthorn is of great significance in combating desertification, increasing the opportunity of employment, and alleviating poverty in rural and mountain areas of many countries. Seabuckthorn is not only being used in the 20th century - it is also the plant of the next century.

REFERENCES

1. Pan Ruiling et al. (1989). *The Distribution Characters of Seabuckthoen and its Rresearch Progress in China.* Proceedings of International Synposium on Seabuckthorn, Xi'an, China.

2. Sun Zhenhua et al. (1995). *Exploitation and Utilitation of Seabuckthorn (H.Rhamnoides L.) in China*, North-West University Press.
3. Lu Rongsen (1992). *Seabuckthorn A Multipurpose Plant Species for Fragile Mountains*, ICIMOD Occasional Paper No. 20, Kathmandu, Nepal.

REVERSING DESERTIFICATION WITH TAMARIX IN CHINA

Tamarix and its Utilization in Xinjiang Desert Zone

Wenjiang Liu and Mingting Liu

Xinjiang Institute of Ecology and Geography, Chinese Academy of Sciences, 40-3 South Beijing Road, Urumqi, Xinjiang 830011, P. R. of China, Tel: 86-991-3835642, Fax: 86-991-3835459, email: camel_83001@yahoo.com

There are about 90 species of Tamarix in the world, mainly distributed in Asia and North Africa. Other Tamarix shrubs and trees occur in the arid and semiarid regions of Europe and Western Coast of South Africa.

1. GENERAL INTRODUCTION OF TAMARIX IN CHINA

China Flora mentions 18 species and one subspecies of Tamarix, but the book written by Prof. LIU Mingting records 20 species. The discrepancy is due to the addition of *Tamarix aphylla* and *T. korolkovii* and omission of *T. laxa* var *palystachya*.. Most Tamarix species in China are found in the Northwestern desert regions, Inner Mongolia and North China, the provinces with the largest distributions being Xinjiang, Gansu, Inner Mongolia, Qinghai and Ningxia.

The population in Xinjiang is the most extensive in terms of both area (54 million mu) and species (16) and accounts for 80% of the Tamarix species in China. The climate, soil and hydrologic conditions of Xinjiang provide appropriate environments for Tamarix immigration, accounting for the fact that Xinjiang is the richest region for this species. *Table 1* shows the species of Tamarix in China and their distribution.

Combating Desertification with Plants, edited by Dov Pasternak and Arnold Schlissel.
Kluwer Academic/Plenum Publishers, New York, 2001.

Table 1. The species of *Tamarix* in China and their distribution.

No.	Species	Distribution								Height (m)	Flowering Period
		Xinjiang	Inner Mongolia	Ningxia	Gansu	Shaanxi	Qinghai	Tibet	other provinces		
1	*Tamarix elongata*	+	+	+	+		+			1-3(5)	3/3April - May
2	*T. laxa*	+	+	+	+	+	+			1-2(3)	3/3March- 1/3April
3	*T. laxa var. palystachya*[1]	+	+	+	+	+	+			1-2(3)	3/3March- 1/3April
4	*T. androssowii*	+	+	+	+					2-4(5)	3/3April - 1/3May
5	*T. gracilis*	+	+		+		+			1-3(4)	May - August
6	*T. gansuensis*[2]	+	+		+		+			2-3(4)	3/3April - 2/3May
7	*T. sachuensis*[3]	+								2-3(5)	June - August
8	*T. taklamakanensis*[3]	+								3-5(7)	July - October
9	*T. hispida*	+	+	+	+		+			1.5-4(6)	July - September
10	*T. karelinii*	+	+		+		+			2-4(5)	August - October
11	*T. arceuthoides*	+			+					2-5(7)	May - September
12	*T. albiflonum*[3]	+								2-4	May - August
13	*T. leptostachys*	+	+	+	+		+			2-3(5)	1/3June - 1/3July
14	*T. jintaensis*[4]				+					1-3(4)	June-1/3September
15	*T. taremensis*[3]	+								2-4(5)	May - September
16	*T. ramosissima*	+	+	+	+		+	+		3-5(7)	May - September
17	*T. hohenackeri*	+	+	+	+		+			3-7(10)	May - August
18	*T. austromongolica*[5]		+	+	+	+	+		+	2-4(6)	May - September
19	*T. chinensis*[5]					+			+	3-5(7)	June - September
20	*T. aphylla*								+	2-4(5)	May - October
21	*T. korolkovii*	+	+	+	+					2-3(5)	May - September

[1] Different on flower structure [2] Unique in Middle Part of Asia [3] Unique in Xinjiang
[4] Unique in Gansu [5] Unique in China

2. THE HABITAT OF *TAMARIX*

Tamarix mainly grows on seashores, sand dune fringes, riverbank sands and gravels, lake banks, saline-alkaline land of lake banks, gravel deserts, salinized meadow lands, moist saline soil, sandy soil, alkaline soil, shifting sand dune, and conglomerates.

In order to endure the extremely dry atmosphere and adapt to the soil salinization and alkalinization, *Tamarix* often forms large desert communities and pure *Tamarix* forest. In the desert vegetation of river banks, *Tamarix* is usually regarded as the important edificator plant, and it makes up the desert forest of river banks together with *Populus diversifolia* and *Elaeagnus angustifolia* etc. The components of the *Tamarix* community are simple, its auxiliary species being *Nitraria sibirica, Etrosimonia sibirica, Unaeda acuminata* and *Atriplex* sp., together with many short-lived and or relatively short-lived plants that it in arid sandy and soil environments.

3. THE PLANT WATER RELATIONS OF *TAMARIX*

3.1 Growth of *Tamarix*

Because of differences in their ability to adapt to the arid environment, each *Tamarix* species registered a different monthly increment in height.

The height increments recorded for the different *Tamarix* species were small in April, increasing rapidly in May and June, falling in July, and reaching a minimum again in August. As temperatures dropped in September, the height increment rate of some species increased somewhat.

3.2 Total Water Content of *Tamarix* and Ratio of Dry/Fresh Weight

Because of the distribution of *Tamarix* in arid/semiarid regions, the total water content of *Tamarix* is not high, but there are differences among species, and the same species have different total water contents in different months (*Fig. 1*).

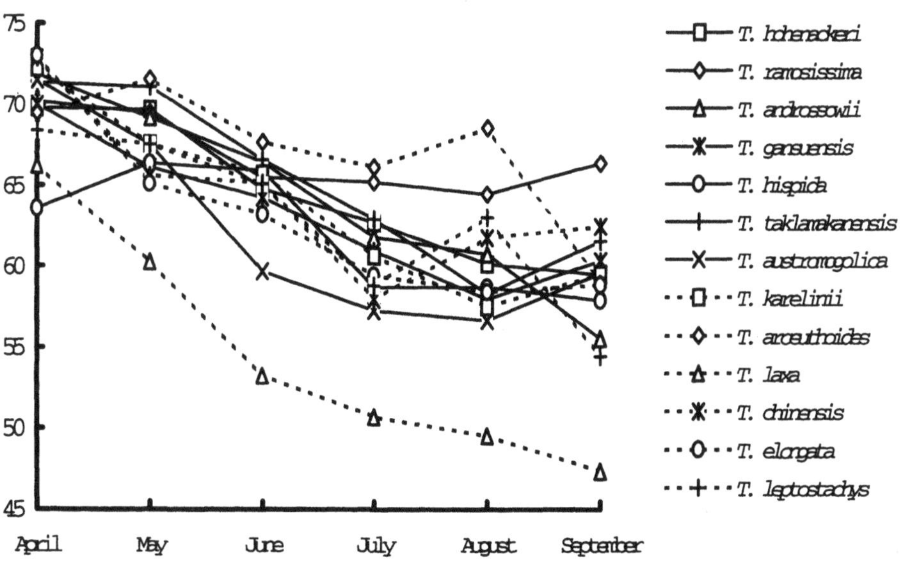

Figure 1. Monthly changes in water content of Tamarix (% of fresh weight)

In April, the total water content of all the Tamarix species is higher because of the younger twigs and lower temperature. As the month goes on, total water content of twigs gradually declines owing to the higher temperatures and lower relative humidity. The degree of decline in total water content varies between Tamarix species because of differences in growth pattern and drought resistance; for example, *T. ramosissina* shows a 4% decline and *T. laxa* a 29% decline.

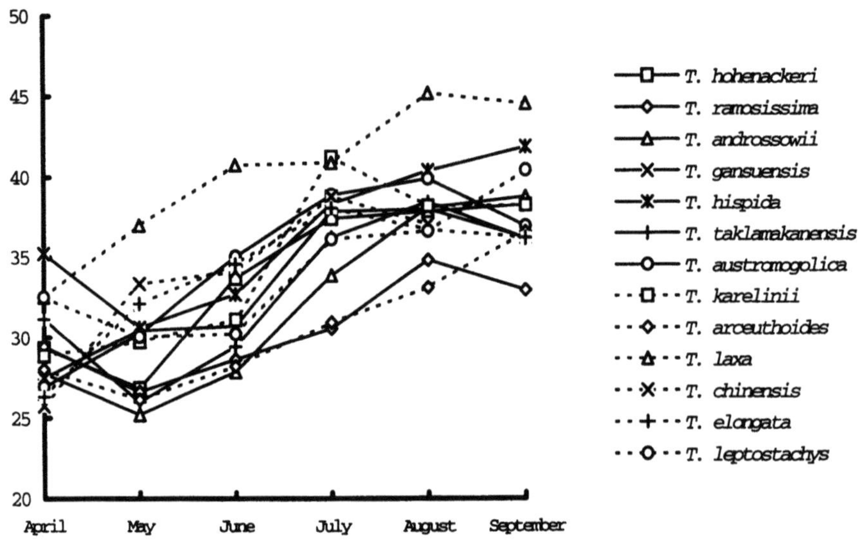

Figure 2. Monthly changes in ratio of dry/fresh weight of Tamarix (gdw/100 gfw)

There is a regular change in the ratio of dry/fresh weight of twigs as the growing stages and seasons unfold, as in other plants (*Fig.2*).

Although differences in the ratio of dry/fresh weight are recorded among Tamarix species in the same month, the general trend over the whole growing period is similar for the 13 species. The twigs of Tamarix are younger and have a higher water content in April, because the growth rate is increasing at this time and the accumulation of dry matter is cannot increase correspondingly. The ratio of dry/fresh weight drops a little in May, because of photosynthesis and an increase in its secondary products, as well as the increase in the degree of lignification. Thus, the ratio of dry/fresh weight gradually rises and reaches its maximum at the final stage of *Tamarix* growth.

3.3 Water Potential of *Tamarix*: Seasonal and Inter-Specific Differences

Water potential is one of the important indices of plant water conditions. Its value indicates the ability of the plant to absorb water from the soil and neighbouring cells to ensure normal physiological activities.

For the same species, water potential is different in every growing period, while remaining within a certain range throughout growth. The range of water potentials varies among the different species *(Fig.3)*.

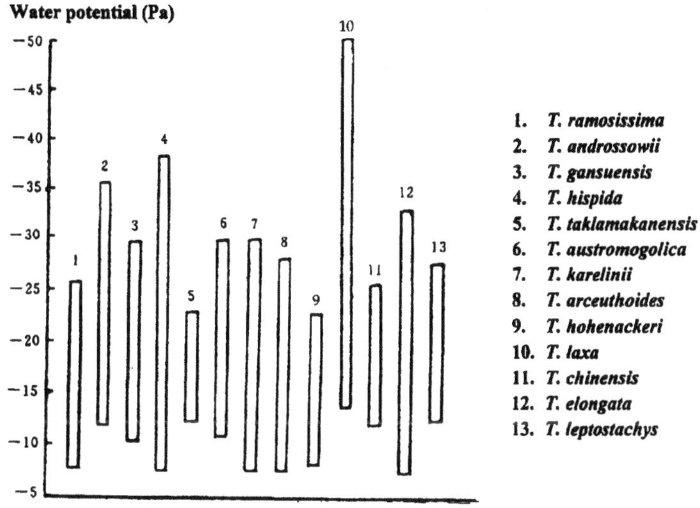

Figure 3. Changes in water potential of Tamarix during growth

From the water potential record of Tamarix, it is seen that the water potentials of different species change as the year progresses. The range of variation can indicate the drought resistance of the plants. Given the same arid soils, plants with lesser water potential can absorb more water than other plants to ensure normal physiological activities and avoid water stress.

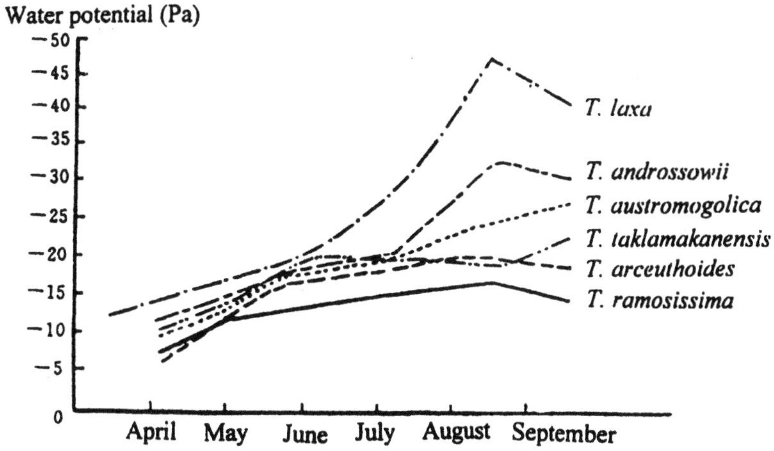

Figure 4. Montly Change in Water Potential of Tamarix spp

The water potentials of a given species in different months are as different as those of different species during the growing period *(Fig.4)*.

The water potentials of all species are greater in April because of the lower temperatures and better soil water conditions. As the months go by, water potentials decline, reaching a minimum in August because plants need to adapt to the hostile environment - extreme drought in July and August.

3.4 Water Saturation Deficit of *Tamarix*

Table 2 shows that, over the growing period, the water saturation deficit of *Tamarix* changes from 5% to 28%. In the majority of cases, the difference between the early and later period of growth exceeds 17%, indicating that some species can endure water stress and grow normally.

Table 2. Water saturation deficit and relative water content of various *Tamarix* species during growth

No.	Item Species	Range during whole growth		Absolute difference (%)
		Saturation & Deficit of natural water (%)	Relative water content (%)	
1	*Tamarix ramosissima*	8.68 - 27.57	72.43 - 91.32	18.89
2	*T. androssowi*	9.35 - 20.14	79.86 - 90.65	10.79
3	*T. gansuensis*	10.17 - 23.39	76.61 - 89.83	13.22
4	*T. hispida*	7.55 - 20.16	79.84 - 92.45	12.61
5	*T. taklamakanensis*	5.60 - 23.69	76.31 - 94.40	18.09
6	*T. austromongolica*	6.40 - 24.82	75.18 - 93.60	18.42
7	*T. karelinii*	10.89 - 28.15	71.85 - 89.11	17.26
8	*T. arceuthoides*	7.80 - 25.11	74.89 - 92.20	17.31
9	*T. hohenackeri*	5.00 - 28.89	71.11 - 95.00	23.89
10	*T. laxa*	4.80 - 23.05	76.95 - 95.20	18.25
11	*T. chinensis*	9.62 - 22.33	77.67 - 90.38	12.71
12	*T. elongata*	10.46 - 23.11	76.89 - 89.54	12.65
13	*T. leptostachys*	4.07 - 21.97	78.03 - 95.93	17.90

3.5 Comparison of Water-Holding Power (Retention Ability)

The retention ability of plants is related to the characteristics of the protoplasm colloids. When the relative content of bound water in the colloids is higher, the content of free water is lower, thereby limiting water loss by plant cells. The plants with greater retention ability have a greater resistance and tolerance to drought. The retention ability of each species is different (*Fig.5*, *Fig.6* and *Fig.7*).

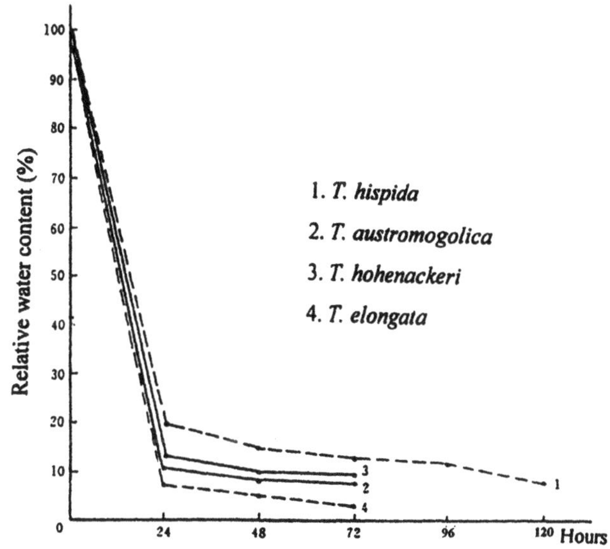

Figure 5. Water-holding capacity of different Tamarix species in July

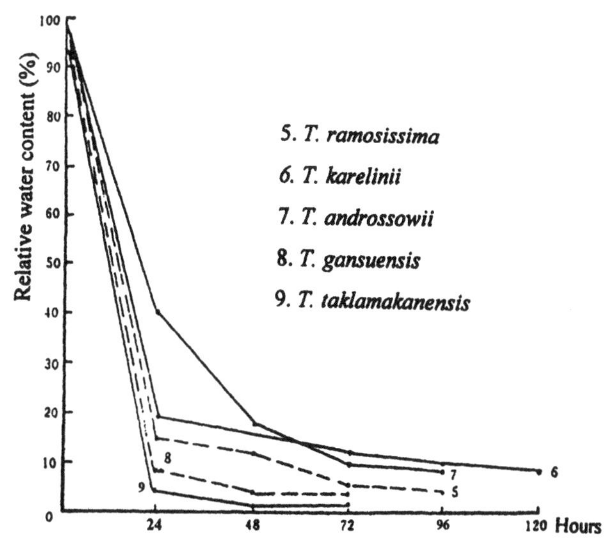

Figure 6. Water-holding capacity of different Tamarix species in July

In the arid environment of July, the total water content of twigs abscised from the plants drops rapidly within 24 hours; over the next 48 hours the rate of water loss becomes slower.

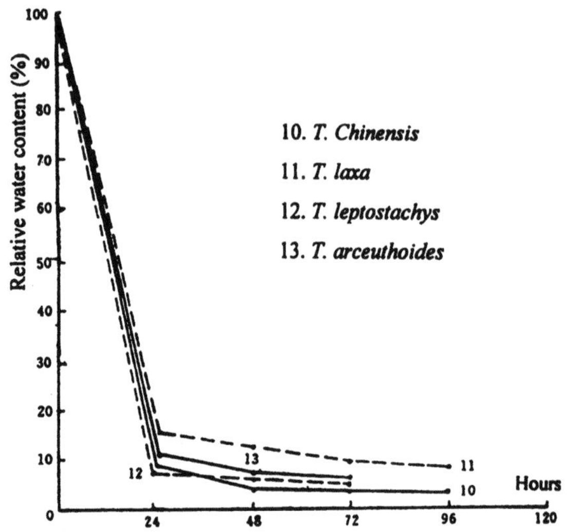

Figure 7. Water-holding capacity of different Tamarix species in July

For each species, the curves of retention ability vary from month to month, as demonstrated by *Fig.8* for the case of *T. arceuthoides.* The April values are the highest because of the lower temperatures and higher relative humidity. As time goes on, the curve gradually dips, reaching a minimum in August; however, values rise again in September as temperatures drop.

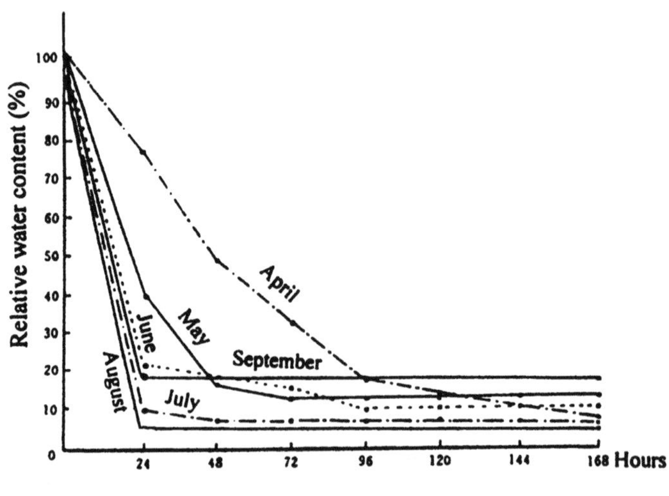

Figure 8. Retention ability of *T.arceuthoides* in different months

Since the retention ability of *Tamarix* is different in each growing period, the drought resistance of plants cannot be determined by examining only the retention ability in a certain month and day.

3.6 Transpiration of *Tamarix*: Intensity and Character

Transpiration is not only one of the components of plant water relations, it is also an important link in the water balance of the plant body. It can reduce the effects of water stress on plant physiological activities by lowering plant temperature, adjusting water loss, and ensuring the efficient use of body water.

As *Fig. 9* shows, the transpiration rates of all the species are low in April, i.e. 300-400 mg H_2O/g fw/h. Transpiration rate rises gradually as atmospheric temperatures and relative humidity change after April. That of *T. ramosissima* drops a little in June, while that of some species rises in August. In September, the transpiration rates of all the species drop because of the lower temperatures and rising relative humidity.

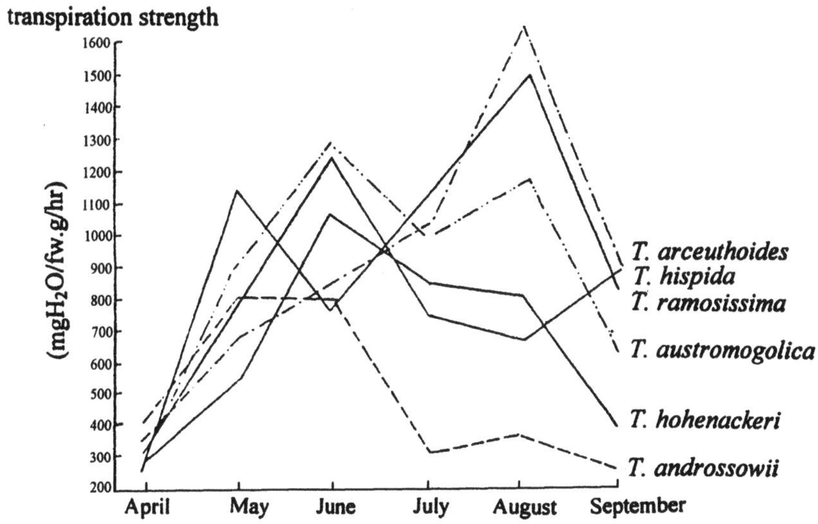

Figure 9. Rate of transpiration of Tamarix in different months

In *T. arceuthoides*, the transpiration rate increases continuously after April, and in the hot month of August it reaches a maximum value of 1520.55 mg H_2O/g fw/h, higher than that of the other species. After the drop in September, its transpiration is still relatively high, namely 906.04 mg H_2O/g fw/h. The transpiration rate of *T. androssowii* stays at a lower level during whole growing period. Two types of *Tamarix* may be distinguished according to the pattern of change of transpiration rate during the growing period: single-peak type (*T. arceuthoides*) and double-peak type (*T. ramosissima*).

4. *TAMARIX* AS WINDBREAK, SAND-FIXING AND SALINE-ALKALINE-RESISTANT PLANTS

The extensive development of *Tamarix* forests has had a considerable impact on the ecological landscape of primary poverty-stricken counties. Greenery now occupies the salt-alkaline lands of the inner oases and large-scale wastelands of the outer oases, in contrast with the original barren situation. A large *Tamarix* forest plays an important role in reducing wind velocity and increasing air humidity. The wind speed at 1.5 m height may be reduced by 40%, that at 1 m height by 60%. At the same time, the humidity in a *Tamarix* Forest may rise by 8-10% compared with a corresponding open space.

Tamarix forests not only effectively control local sand movement, wind erosion and disasters caused by wind-blown sand, but can also stop sand grains brought by the wind. For example, on the 3333.3 hectares of planted *Tamarix* forest at the outer Qira Oasis with 60-70% coverage, the annual sand-blocking amount can reach 700,000 m^3. When the wind passes through a *Tamarix* forest, besides blocking sand, the twigs and leaves of the trees also obstruct the passage of the dust in the air, improving the health situation of humans and livestock. For example, when the wind velocity is 15 m/s, the dust content at 2m height is 1/18 of the content measured at the same height in the outer oasis (observed at Qira County in 1987).

Tamarix not only can be used as a windbreak to reduce wind velocity, raise air humidity, and trap dust floating in the air, but it is also an excellent sand-fixing plant, for the following reasons:

1. The twigs of *Tamarix* readily grow adventitious roots after being buried in sand;
2. The small branches and twigs are soft, thin and elastic, and thus are not easily broken by wind and blown sand;
3. *Tamarix* are shrubs with deep roots, the main root stretching straight down into the ground to the water table;
4. *Tamarix* plants secrete $NaCl$-$NaSO_4$ salts, which can mix with fallen leaves and sand to form a solid litter (*Tamarix* sand dune).

Besides excellent sand-fixing ability, it also has a greater salt-alkali-tolerance, a trait that makes it more adaptable to desert environments than many desert shrubs. The desert areas with more salt-alkali lands are also one of the areas of distribution of *Tamarix*.

Tamarix plants secrete harmful salts from the leaf surface through salt-secretion glands located inside the plant. They owe their ability to tolerate salt to this mechanism, which entitles them to be regarded as the foremost salt-tolerant species in China. According to our experimental results, the salt tolerance of *Tamarix* is better than its alkali-tolerance, and the various species of the *Tamarix* genus differ in their adaptabilities to different salts.

5. OTHER ECONOMIC USES OF *TAMARIX*

5.1 As a Medicine

According to the "Compendium of Materia Medica"(Li Shi-zhen 1518-1593) and "Medicine List of P.R. China", the resin and many parts of the *Tamarix* plant, such as its flowers, leaves, and twigs, can be used to treat fever, carbuncles and inflammation, dropsy and jaundice. In addition, *Cistanche tubulosa (Schrenk.) R. Wight*, a parasite on the root of *Tamarix*, is a well-known Chinese herbal medicine and is called "Desert Gingseng". It has many functions, for example in promoting robustness, as a kidney tonic, and as a cathartic.

5.2 As Fuel

Plants of *Tamarix* have long been a popular fuel owing to their combustibility and high calorific value (*Table 3*).

Table 3. Wood Calorific Value of Different *Tamarix* Species

No.	Species	Calorific Value (cal/g)
1	*Tamarix laxa*	4228
2	*T.elongata*	4129
3	*T.androssowii*	4324
4	*T.gansuensis*	4234
5	*T.arceuthoides*	4271
6	*T.ramosissima*	4206
7	*T.hohenackeri*	4291
8	*T.leptostachys*	4210
9	*T.hispida*	4158
10	*T.karelinii*	3906
11	*T.chinensis*	4114
12	*T.taklamakanensis*	4348
13	*Haloxylon spp.*	4302
14	Standard Coal	6300

(Analysed by the Xinjiang Institute of Coal in February 1985).

From the above table, we can see that the species with the lowest calorific value is *T. karelinii* with 3906 cal/g, or 62% that of coal, and the species with the highest calorific value is *T. taklamakanensis,* with 4348 cal/g, or 69% that of coal. Other *Tamarix* species are characterized by values ranging from 4100 to 4300 cal/g, equal to 63%-68% that of coal.

5.3 As Fertilizer

Tamarix sand dune is composed of sand and *Tamarix* litter, and it can be used as fertilizer because of the relatively high content of N, P and K that develops after a period of fermentation. Moreover, the fresh twigs and leaves of *Tamarix* can also be used as green fertilizer to improve the content of soil organisms. For example, in Jiashi County of Xinjiang, a melon producing area, the local people use *Tamarix* as green fertilizer in the root area to maintain quality and increase the sugar content of the melon fruit.

5.4 As Animal Fodder

The content of protein is an important index of pasture quality.

Table 4. Nutrition Composition of *Tamarix* Species

No.	Species	Nutrition Composition (%)				
		Crude Fibre	Crude Protein	Crude Fat	Ca	P
1	*Tamarix laxa*	11.2	14.1	2.3	2.73	0.24
2	*T. elongata*	11.1	14.1	2.1	2.62	0.23
3	*T. androssowii*	11.6	9.9	1.6	2.31	0.19
4	*T. gracilis*	12.2	12.7	2.1	1.83	0.94
5	*T. gansuensis*	9.2	11.7	2.6	3.69	0.17
6	*T. leptostachys*	9.5	15.7	2.2	3.58	0.30
7	*T. karelinii*	11.8	11.0	1.9	3.58	0.23
8	*T. hispida*	12.4	12.8	2.0	3.49	0.22
9	*T. chinensis*	11.8	11.4	2.5	4.58	0.16
10	*T. arceuthoides*	12.1	15.3	2.1	2.86	0.25
11	*T. albiflonum*	7.0	12.8	1.2	4.30	0.20
12	*T. ramosissima*	10.7	15.1	2.4	2.69	0.23
13	*T. hohenackeri*	11.4	16.8	2.7	2.94	0.20
14	*T. austromongolica*	11.3	16.6	3.2	2.36	0.20
15	*T. sachuensis*	13.6	10.8	1.9	2.75	0.11
16	*T. taklamakanensis*	12.8	11.1	2.3	1.63	0.11

(Analysed by the Xinjiang Academy of Animal Husbandry)

From *Table 4*, we see that the nutritive value of the fresh twigs and leaves of each species are tolerable.

About 20 to 30 herb plants are often associated with *Tamarix* shrubs and make up different communities. They include among others *Glycyrrhiza inflata, Phragmitis communis, Alhagi sparsifolia, Inula ammophila,* and *Karelinia caspica.*

Tamarix shrubs play an important role on easing the fodder shortage in early spring by providing nourishment for sheep, donkeys and camels in arid regions.

Besides the above aspects, the pliable and tough *Tamarix* branches can be used as weaving materials, while the trunk serves to fabricate farm tools, fibreboard, bridges, water gates houses, etc. The tannin content in branches and leaves is about 8%-10%, while that in gall can reach 30%-50% and can be used as leather dye. Its early and long flowering period make *Tamarix* a pollen and honey source plant in early spring, and appropriate varieties can be selected as hedge, potted, landscape and garden plants. *Tamarix* can resist SO_4 pollution because of its S-collecting ability.

REFERENCES

1. LIU Mingting, (1995): *Comprehensive Study on Tamarix and its Large-scale Popularization & Utilization*, Lanzhou University Press.
2. Baum, B. R. (1978): *The Genus Tamarix*, The Israel Academy of Sciences and Humanities.
3. Ying Linke, (1997, 5). *Evaluation on the Resources and Diversity of Tamaricaceae in China*, Chinese Biodiversity, (Suppl.): 85~91.

BIORECLAMATION OF SALINE SOILS OF THE WESTERN COAST OF SENEGAL

Case of introduction of Tamarix aphylla var *erectus, a salt tolerant species*

M.Diatta[1], M. Diack[1], M. Sene[1], and D. Pasternak[2]
[1]*Institut Sénégalais de Recherches Agricoles (ISRA), BP 53, Bambey, Senegal, Tel/Fax: 221-973-6052, email: diatta@ird.sn*
[2]*International Program for Arid Land Crops (IPALAC), The Institutes for Applied Research, Ben-Gurion University of the Negev, P.O. Box 653, Beer-Sheva 84805, Israel, Tel: 972-7-6461972, fax: 972-7-6472984, email: dovp@bgumail.bgu.ac.il*

1. INTRODUCTION

In Senegal, soil degradation is one of the major constraints limiting agricultural development. It negates the efforts of public bodies and partner development agencies to improve food security and increase farmer income as part of their struggle against poverty. Recent studies and diagnostic surveys point to an increase in the area of agricultural land abandoned because of soil salinization. The latter phenomenon has been observed throughout the coastal area, but it is more pronounced in the river and estuary watersheds of Casamance, Gambia, and Sine Saloum Rivers, as well as in the Senegal River delta. It affects approximately 1 million hectares of arable land, located mostly on alluvial terraces, formerly highly valued for rice and sorghum cultivation.

Studies have been undertaken to try and explain the process of soil salinization and to develop technical solutions to overcome this phenomenon. Among these studies, those by Durant (1977), Vieillefont (1977), Marius (1985), Massibot et al. (1946) Pereira Barretot (1985) should be mentioned. These authors have dealt *inter alia* with farmer practices of rehabilitation and improvement of saline soils, gravity irrigation waterworks based on the

Combating Desertification with Plants, edited by Dov Pasternak and Arnold Schlissel.
Kluwer Academic/Plenum Publishers, New York, 2001.

construction of a network of small dams or small salt-containment dams, water and soil analysis, and screening of halophyte species for forestation purposes.

In spite of these various approaches, areas affected by salinization continue to grow leading to a rural exodus. It seems that the lack of effectiveness of these solutions is linked to the fact that the complexity of salinization processes had not been taken into account. These processes, which result in what are called in the Sine Saloum area "tannes" (highly degraded salty plots, literally tanned or burnt out) take place through three main avenues: (i) capillary lifting to the surface of salt from the shallow and highly mineralized aquifer during the dry season; (ii) infiltration of sea water or saline water from rivers; (iii) wind energy, or movement of solid salt particles from the "tannes" towards the cultivated areas which have not yet been affected.

The problem is to find an alternative sustainable solution with a view to reversing the process of salinization of agricultural land.

Within this framework, ISRA (the Institut Senegalais de Recherches Agricoles or Senegalese Institute for Agricultural Research), in cooperation with IPALAC (International Program for Arid Land Crops), has initiated a research project on saline soils of the West Coast of Senegal.

This paper discusses the results obtained in a survey of an area of saline soils to be rehabilitated biologically at Ndifatte, in the southern part of the groundnut growing area about 15 km from the city of Kaolack.

2. MATERIALS AND METHODS

The study area is located between 13°35' and 14°30' lat. N at about 16°45' long. W in the southern part of the groundnut growing area of Senegal.

Administratively speaking, this area is part of the Kaolack and Fatick districts. The population of these two districts is mainly rural and is estimated to be 1,313,098, representing 20% of the country's population.

The climate is of the Sudanese type, characterized by two strongly contrasted seasons, with the region receiving between 400 and 600 mm annual rainfall. From the hydrographic viewpoint, the region is subject to marine influences from the Saloum River and tributaries of the Gambia River. Worsening climatic conditions are associated with an inversion of the hydrographic system and the intrusion of sea water, which causes progressive salinization.

Four main types of soils have been observed in the survey area, all of them halomorphic. These soils can be subdivided into two subgroups

(Maignien, 1965): (i) deep saline sandy-clay textured soils on grassy "tannes" with almost neutral pH and an electrical conductivity (EC_e) of the 1/5 extract amounting to 0.2 and 2 dS/m; (ii) very salty soils on raw "tannes" with salt efflorescence on the surface, with a pH nearing 5 and EC_e of about 20 dS/m.

The main plant species are mangroves and savanna shrubs, with local acacia species dominating. Combretaceae have been destroyed, leaving behind naked or herbaceous spaces. The presence of soluble salts and substances toxic to plants has brought about the disappearance of the trees and consequently a loss in plant bio-diversity.

The main farming system is typically groundnut/millet in rotation. Rice is the domain of women, who grow it in the areas most threatened by salinization processes. (This activity, abandoned in spite of all the efforts expended, should be launched again to ensure an income to rural women).

The scientific hypothesis tested here is based on the use of species very resistant to salt with a high economic value. Planting of these species should lead to a lowering of the water table through transpiration, with a positive effect on the soil. Rain will replace the volume of water displaced from the soil by leaching of the superficial soil layers. Soil aeration would therefore be improved and the process of salinization of the upper layer stopped.

The choice of species is based on a 10-year study carried out by one of the authors (Pasternak, D. and Nerd A., 1995), which resulted in the identification of 150 halophyte species tested for their resistance to salt (using sea water) and economic value (Redouane Choukr-Allah et al., 1985).

The tree species recommended is the straight trunked Tamarix, *Tamarix aphylla* var *erectus*. This is a fast growing tree, very resistant to salt and flooding. It can reach 15 to 20 meters in height and yields an excellent timber. The forage species recommended is the halophyte grass *Distichlis spicata,* a native of the American continent which has remarkable characteristics. It is tolerant to relatively high salt levels, to flooding, to hot and dry climatic conditions, and to high and low pH levels.

This is a C4 grass, with a spreading root system and high production potential. *D. spicata* has been used, among other examples, to rehabilitate the salty swamp of Lake Texcoco near Mexico City. The third species proposed is *Atriplex lentiformis,* also known as "buisson à cailles" or quail bush of the Chenopodiaceae family, an excellent cover plant and resistant to salt. It is used for the protection of earth ridges because of its hardiness, salt tolerance and impermeability to animals. This species is also used as ligneous fuel.

The use of trees to rehabilitate flooded land is an ancient practice. The use of halophyte grass to help in leaching salt is also known. But using a combined tree/grass system to reclaim saline soils and stimulate agricultural crops is a new approach in Senegal. It is also the first attempt to use *T. aphylla* to rehabilitate salty and acid soils.

The trial will be implemented on a 10 hectare plot in the forest experimental station at Ndiafatte, which is characterized by halomorphic degraded soils comprising unleached and alkaline solonchaks on clay alluvia.

3. PLANNING THE TRIAL

The layout chosen was 2-factor trials in a completely randomized pattern of three times five blocks.

The treatment factor has three plant density levels: T1 (3 m x 4 m); T2 (4 m x 4 m) and T3 (4 m x 5 m) and T0 (control without trees). This layout comprises 20 plots of 4,000 sq.m. each. Each of the plots is surrounded by earth ridges 50 cm high and 40 cm wide and is subdivided into two sub-plots: 3,000 sq.m. planted with *T. aphylla,* surrounding a central sub-plot of 1,000 sq.m. of *D. spicata*. The analysis of the effect of various parameters on the planted species will specifically be focused on this central sub-plot.

The factors to be followed up are: (i) height and diameter of tree (every three months), the number of surviving individuals, and the general state of the vegetation, (ii) plant-water relationships, phenology and morphology of the tree roots, (iii) climatic data, (iv) water content in the soil.

4. PRELIMINARY RESULTS AND DISCUSSION

The results in respect to soil water content characterize soil-water relationships and the behaviour of the species introduced into the saline environment.

4.1 Water and Soil Properties

Sixteen piezometers were installed in the field, piezometer No. 1 at the highest spot and piezometer No. 16 at the lowest, beside a river estuary. The piezometers allowed the recording of aquifer depth and water salinity.

4.1.1 Height of the Aquifer

Fig.1. shows the changes in levels of the water table as related to time and relief. The aquifer, relatively shallow in October (first drilling), was subjected to an upward movement followed by a progressive downward trend in the period from November 1997 to June 1998, which represents the dry season.

This regular reduction in level was responsible for drying up at piezometer 1, which is located upstream, and was observed from the month of February. Moreover, an opposite movement was observed at piezometers 14 and 15, which suggests an upheaval of the salt, already perceptible from the month of May onwards at the piezometers located downstream near the waterway.

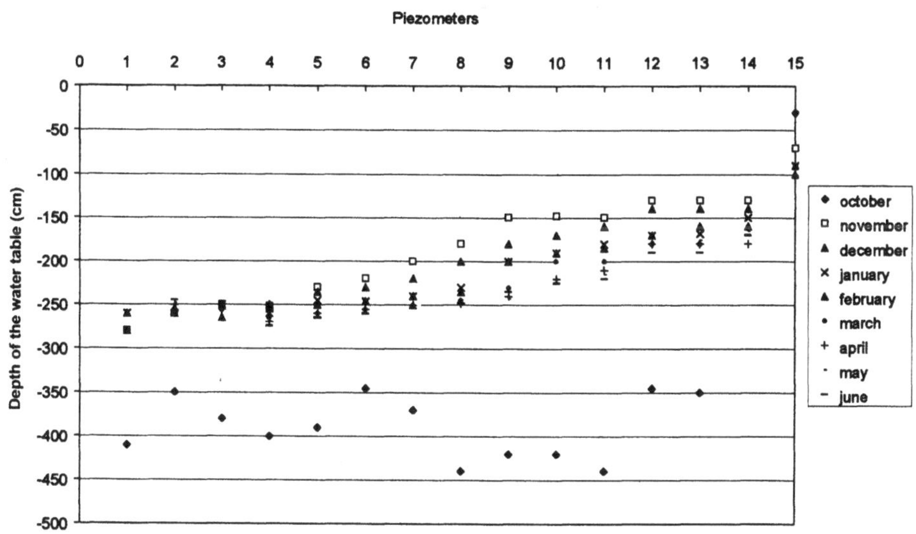

Figure 1. Variations of aquifer levels in function of time (Oct. 97 - June 98) and topography

Fig. 2 presents the changes in salinity of the water table as recorded for the period between December 1997 to June 1998. These results clearly show the variability of the EC, which ranged from 30 to 110 dS/m. Within this variability, piezometers 14, 13, 12 and 11 showed a relatively similar behaviour characterized by a low variation of levels during the year, while at piezometers 10 to 5 another behaviour that does not seem to be dependent on climatic conditions could be observed.

This EC values are very high - 70 dS/m on average - suggesting a tendency to high salt concentration in the aquifer. The pH measured in these same samplings varies between 5 and 7.5, which shows a low acidity level. The absence of data at piezometers 1 and 15 is due to the drying up of the former and to the very high salt concentration at the latter, far beyond the measurement capability of the conductivity meter used.

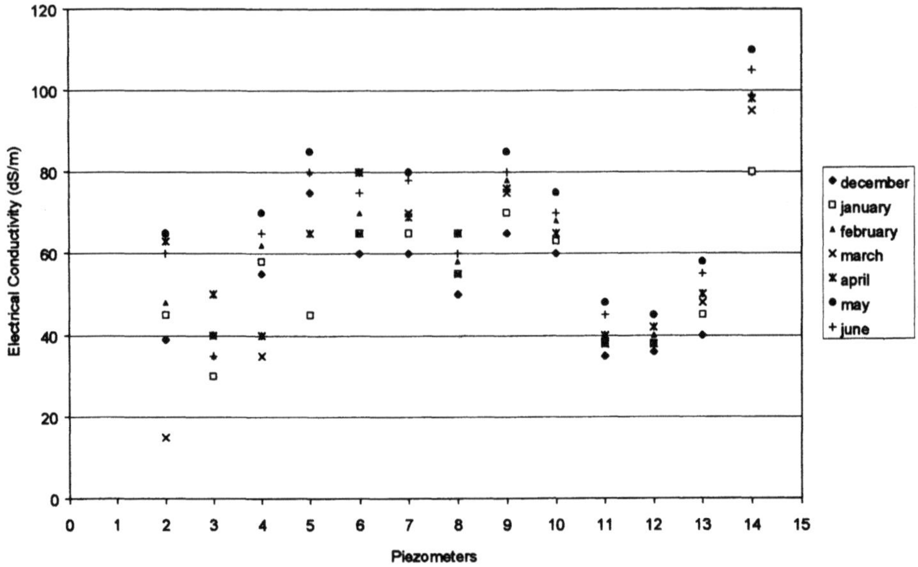

Figure 2. Variations of ground water salinity in function of time (Dec. 97 - June 98) and topography

4.1.2 Monitoring of Soil Salinity

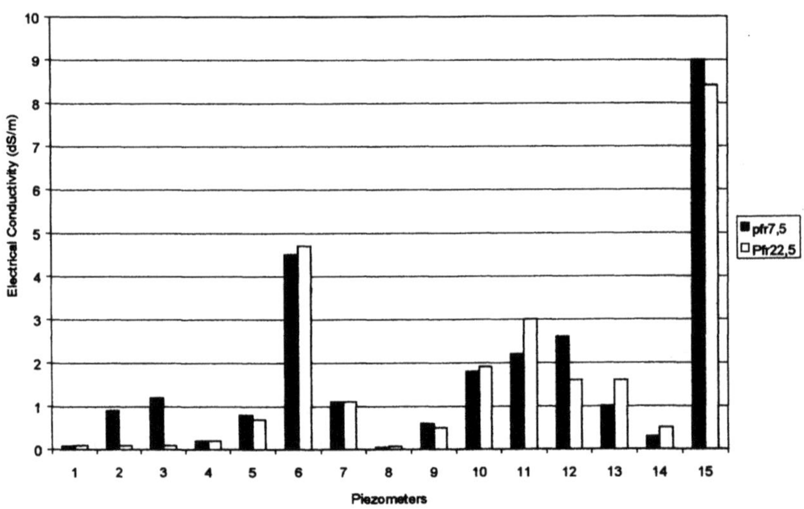

Figure 3. Variation in soil salinity in June 1998, prior to the rainy season.

Figure 3 compares soil salinity as related to depth and topography. Two types of profiles or distinct groups are observed:

- the first group is characterized by a low level of salinity variability between the two horizons;

- the second group distinguishes itself from the first by a higher concentration of salt on the surface on profiles (locations 6, 12, 13 and 15. This seems to be due to the very varied micro-topography, which produces a convex-concave relief that favours raising of salt by capillarity in depressions.

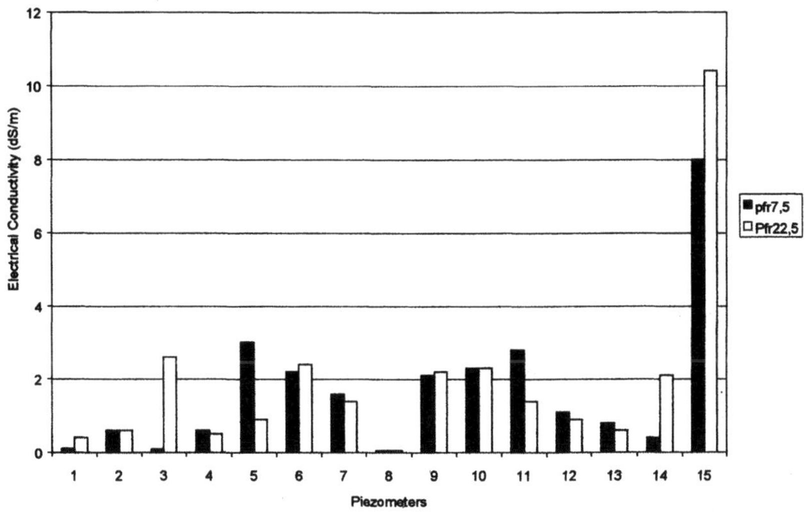

Figure 4. Variation in soil salinity in June 1999, prior to the rainy season.

Figure 4 shows results of measurements of soil salinity taken over 16 months. Here too, two groups can be distinguished:

- the first group, with similar salinity values for both horizons, are in locations 1, 2, 4, 6, 7, 8, 9, 12, and 13;
- the second group is typified by higher salinities at deeper levels in locations 3, 5, 11, 14 and 15.

Comparison of the two figures shows a climatic period effect (rainfall):

- October, corresponding to the end of the rainy season, is associated with lower salinity levels;
- January, i.e. 2 to 3 months after the stoppage of rains, can be considered as marking the beginning of the period of salt accumulation and capillary rise.

4.2 Performance of Planted Species

The performance of *T. aphylla* var *erectus* introduced on salty soils at the experimental station at Ndiaffate six months after planting was evaluated in terms of survival rate and elongation.

4.2.1 Survival Rate

Table 1 shows the survival rate of *T. aphylla* var *erectus* six months after planting on salty soils. It seems that the general average rate of survival is 76% with a standard deviation of 15. The survival rate by treatment varies from 60% ±15 (4 m x 4 m spacing) to 91%± 4 for the larger spacing (4 m x 5 m). Nevertheless, it seems premature to conclude a treatment effect as after six months the trees are still at a young stage.

Table 1: Average survival rate of *Tamarix aphylla* var *erectus* six months after planting as a function of spacing

Spacing	Survival Rate (%)	Plant Height (cm)
E1 (4 x 3 m)	78±12	110
E2 (4 x 4 m)	60±15	98
E3 (4 x 5 m)	91±4	115
Average	76±15	108

Concerning the growth of the species, *Table 1* shows very interesting results in that six months after planting one can observe an average height of 108±9. This result suggests that *Tamarix aphylla* var *erectus* possesses the ability to withstand salinity.

The survival rate of *T. aphylla* is also a function of the level of the saline water, as shown on *Table 2.*

Table 2: Survival rate of *Tamarix aphylla* var *erectus* in relation to depth of aquifer

Depth of Aquifer (cm)	Survival Rate (%)
Level 1 (50-100)	66±16
Level 2 (150-200)	76±19
Level 3 (250-300)	82±9
Level 4 (350-400)	83±29
Level 5 (>400)	83±10

The concentration of salt near the surface explains the higher mortality rate of trees on plots where the aquifer is shallow (50 to 100 cm). It seems therefore that that the major problem to be solved in reclaiming the saline soils is stabilisation of the saline aquifer at a level far from the superficial layers. This seems to confirm our main research hypothesis, namely that a tolerant woody species like *Tamarix aphylla* var *erectus* can be used to lower the saline aquifer by transpiration. Water from rainfall should replace the water volume evacuated from the soil and leach the salt from the surface of the soil, thereby improving aeration

5. CONCLUSIONS

Results are encouraging in spite of late planting during the first week of September, in a situation where rainfall is erratic and concentrated over a short period of 3 to 4 months.

The follow-up data related to soil salinity and to acidity as well as the fluctuation of the aquifer have yielded fairly interesting results concerning the variability of these parameters depending on topography and season. At most sites, a tendency to lowering of the aquifer level and sometimes the drying up of wells during the dry season could be observed. Moreover, it can be noted that at all sites the aquifer is very saline, with a salt concentration often exceeding that of seawater.

Results obtained about the behaviour of *T. aphylla* var *erectus* are important and encouraging. Provided these results are confirmed in the future, this species can be extensively utilized in the reforestation of saline soils of Senegal, because beyond its ecological interest, *T. aphylla* offers various products and uses (construction wood and timber).

It is difficult to draw conclusions regarding the development of the plants at the present stage, i.e. 6 months after planting. The results regarding soil characteristics (presently under elaboration) will allow us to better understand the mechanisms of soil/plant/water/salt relationships.

For the future, these studies should be continued to deepen our understanding of certain aspects related to salt/plant relationship (resistance to salinity through research work on eco-physiology, tilling and soil improvement practices) and the establishment of a herbarium of saline soils.

REFERENCES

1. Anonymous. (1985). Centre Pour Notre Avenir à tous Cap sur terre, édition juillet.
2. Anonymous (1997). Plan stratégique régional de la recherche ISRA/CRA Kaolack.
3. Maignien. (1961). Notice explicative de la carte pédologique au 1/1 000 000 ème du Sénégal, ORSTOM
4. Marius, (1985). *Mangrove du Sénégal et de la Gambie: Ecologie-Pédologie Géochimie-mise en valeur et aménagement* Ed ORSTOM Dakar, 357 pages.
5. Massibot & Charles. (1946). *Mise en valeur des tannes rizicultivables du Sine* (Sénégal) volume 1 n° 9-10 Agronomie Tropicale 451 à 466 pages.
6. Anonymous. (1979). *Les Mangroves du Sénégal Ecologie, Pédologie, Utilisation,* ORSTOM.
7. Pasternak D. and Nerd A. (1995). *Research and utilization of halophytes in Israel.* R. Choukr-Allah, M. Hamdy and C.V.M. Malcolm (Eds). Halophytes and Biosaline Agriculture pp. 325-348. Marcel Decker Inc. New York.
8. Pereira Barreto (1985). *Désertification et Salinisation des Sols et des Eaux* COMIDES II, Dakar/TCP/SEN/4503, 30 pages.

9. Redouane et al. (1995). *Halophytes and Biosaline Agriculture* Copyright by Marcel Dekker, Inc. New York Basel Hong Kong.
10. Sadio. (1991). *Pédogenèse et potentialités forestières de sols sulfates acides salés des tannes du Sine Saloum.* Ed. ORSTOM-Dakar 269 pages.
11. Vieillefon. (1977). *Sols des mangroves et des tannes de basse Casamance (Sénégal: Importance du comportement géochimique du sous sol dans leur pédogenèse.* Mémoire ORSTOM Paris, 263 pages.
12. Glenn E. et al (1998). *L'irrigation à l'eau de mer: La culture de plantes résistant au sel permettrait la production d'aliments pour le bétail dans les zones désertiques,* in Pour la science n°252.
13. Epstein E. et al. (1980). *Saline culture of Crops: A genetic Approach,* in Science, vol.210, pp.399-404.

UTILIZING THE UNDERGROUND FOREST

Farmer Managed Natural Regeneration of Trees

Tony Rinaudo

World Vision, AMEEE team, IIP, 1 Vision drive, East Burwood Vic 3151, Australia, Tel: 613-9287-2309, Fax: 613-9287-2377, email: rinaudot@wva.org.au

1. BACKGROUND

The author had the privilege of living in Niger Republic, West Africa as a Christian missionary and development worker for 18 years. He directed the Maradi Integrated Development Project (MIDP), which has helped farmers to evaluate their problems and work towards solving them using local resources and people since 1975. In his first four years it was particularly frustrating to see the massive scale of deforestation and land degradation, the apparent apathy of the people and the low success rate of both government and non-government forestry projects.

One could travel for hundreds of miles in any direction from the project base and be confronted by the same scene of barren windswept land, stripped of all vegetation. The people seemed to have accepted this situation as inevitable. Women walked for miles to collect materials for cooking with. By the end of the dry season the only fuels available were millet stalks and cow and goat dung. Most home compound fences were either non-existent or falling down. Few people could repair their grain silos or huts. Mud bricks replaced wood as a body-cover in graves in some villages. When MIDP tried to introduce a poultry project, farmers dug up the lateral roots of protected trees at night to use in constructing the coops. Incredibly, if rural people needed fire wood for special occasions, or building poles, they would travel to Maradi, the state capital, up to 50 kilometres away to buy it. Emerging food crops were afforded no protection at all from the strong (up to 70 Km/hr) winds at

Combating Desertification with Plants, edited by Dov Pasternak and Arnold Schlissel.
Kluwer Academic/Plenum Publishers, New York, 2001.

planting time. It was not unusual for farmers to have to replant their crops six to eight times because of losses through sand blasting and burial.

Prior to the early 1980's reforestation in Niger was based on a typical western model, requiring a central tree nursery that was usually committed to producing fast growing exotic species such as Eucalyptus (*E. camaldulensis*) and Neem *(Azadirachta indica)*. These programs were expensive to run, requiring fencing, watering facilities, vehicles, funds or grain to pay labourers and long-term guards or wire netting to protect the trees with. In short, this system of reforestation was not reproducible on a village level because of the cost and the technology involved. Both government forestry services and non-government projects focused on the problem of "deforestation". They ignored the not inconsiderable resentment farmers had towards the forestry service and their perception of trees as "weeds" that competed with their precious food crops.

Since all trees belonged to the state, a farmer needed a permit to clear his land for farming. This was no small feat for someone living in an isolated village. There were heavy fines and even the possibility of a jail sentence for cutting trees down illegally. Because of this farmers rationalised that the less trees on a field the better. Even large, well-meaning projects, whose aim was the "modernisation" of the agricultural sector, required the complete clearing and uprooting of trees from the land. Farmers who did not "modernise" were labelled as backward. This set the scene for the total clearing of all trees on crop land and reinforced the mindset that a "good" farmer is a "clean" (i.e. treeless) farmer.

In addition to all this, people needed to live. They needed wood for cooking, heating, building and income. From the 1970's onward, food shortages were becoming more frequent and more people were affected. Since wood was one of the few marketable natural resources remaining, trees were chopped down in order to supplement income and diet. The forestry policing system was open to corruption and so wasn't effective in stopping the almost total loss of trees from not only farmland but forest reserves and grazing land as well.

In a twelve-year period in Niger it is estimated that 60 million trees were planted. Of these less than 50% survived. Millions of dollars were spent on reforestation projects but there was little to show for it a few years after the end of the projects. Despite the large amounts of money spent, vast denuded areas have not benefited from tree planting schemes. There are very few cases of a community continuing with tree work once a project ends. There was a realisation within the MIDP management that even if the project ran for a decade with an unlimited budget, there would be no significant effect on the deforestation problem in the Maradi area.

It is estimated that 34% of the surface area of Africa is under threat of desertification. The statistics don't measure the degradation of standing trees, the over collection of branches for fuel and foliage for fodder, or the losses of tree seedlings as they sprout to flocks of sheep and goats (Weber & Carol, 1986). Nearly four million hectares of forest are being cleared and degrade every year in Africa (FAO, 1986a). No less than one-and-a-half billion people in developing countries derive at least 90% of their energy requirements from wood and charcoal. Another billion people meet at least 50% of their energy needs this way (Echolm, 1975).

By the year 2010 in Africa, demand for fuelwood is expected to more than double to 953 million cubic meters a year (FAO, 1986). The estimated expenditure from Overseas Development Agencies on forestry in Africa in 1993 was US$ 473 million. (FAO, 1997). Despite this massive effort, in some African countries, the deforestation rates exceed planting rates by a factor of 30:1.

How much money and time and resources will be required to address the problems of desertification and deforestation and meet the growing demand for firewood? On both a local and a world scale, traditional forestry approaches on their own are not solving the problem.

In Niger there were instances of native bush land being cleared to make way for plantations of exotic trees. The bush was seen as useless because it did not fit into any modern-day forestry category of usefulness. It was not uniform, appeared to not grow fast and produced knotted, crooked and forked branches, which could not be used for milling. Little thought was given to its usefulness to the local people.

Closer study of this "useless bush" reveals that in fact, it is extremely productive and useful to local inhabitants, providing many products including timber, firewood, fibre, medicines, foods, fodder and dyes.

In social forestry projects it is critically important to understand what is happening before attempting to change things. Most forestry projects made false assumptions, such as:

1. Desertification and over population were the main cause of deforestation.
2. Most of the trees had been cut down or had died.
3. Native tree species were not worth propagating.

In fact it appeared that:

1. While desertification and over population contributed to loss of trees, inappropriate farming practices and unsuitable tree ownership laws were the main causes.
2. While most of the trees were chopped down, they were not dead.
3. Native tree species are extremely useful and some do grow rapidly.

Simply planting more trees did not solve the problem. Many millions of dollars have been spent unnecessarily in Niger because the wrong questions were asked and the wrong solutions applied.

While on a routine village visit and while feeling particularly frustrated in 1983, the author stopped the car and looked out over the treeless farmland. Then he saw what had been there all along, thousands of what he had thought of as insignificant desert shrubs. In fact, each one was a mass of regrowth from a tree stump. While the trees have been cut down, in many cases they have not died. The trees were already there in a vast "underground forest". This was a literal answer to prayer.

This realisation meant that there was no need to run an expensive nursery to raise more trees. All that was needed was to convince the farmers to change the way they cleared their land. If they protected a certain percentage of the trees, they would have firewood, a source of diverse products and greater protection of their crops at no extra cost. As it turned out, protecting this underground forest through a system of "Farmer Managed Natural Regeneration" (FMNR) proved to be a highly successful means of rapidly and cheaply reforesting large areas of land.

2. THE UNDERGROUND FOREST

On cleared land, where live native tree stumps have not been uprooted, an extensive "underground forest" already exists. Each year, these stumps sprout multiple shoots. It is traditional farming practice to slash and burn all re-growth in preparation for sowing the next seasons' crop. Thus, the "bushes" never grow beyond 1- 1.5 meters tall. Consequently, depending on the time elapsed since the last slashing, the land ranges from being totally barren to being covered with varying degrees of bushy regrowth. It is referred to as an underground forest because most of the time it is not seen or appreciated. With culling and pruning, selected stems from this regrowth grow rapidly.

2.1 Farmer Managed Natural Regeneration (FMNR)

There is no set system or hard and fast rules of FMNR. Farmers are given guidelines but are free to choose the number of shoots per stump and per hectare, the length of the rotation and the method of pruning.

The basic method of FMNR is very simple. The farmer selects the stumps he/she wants to leave and decides how many shoots are wanted per stump. Excess shoots are then cut and side branches trimmed to half way up the stems. A good farmer will return each 2 - 6 months for touch up prunings and thereby stimulate faster growth rates.

FMNR is not new, it is simply a form of coppicing and pollarding. At least 1000 years ago, Europeans learned to run their woods as a self-renewing resource. By coppice farming natural woodlands, "woodsmen" produced poles, small wood for fences and light construction, and firewood without destroying the trees. Stands of Ash still exist that have been cut and allowed to re-sprout on a regular rotation for at least 500 years. (Firewood Crops, 1980). FMNR may be new however in the sense that naturally occurring trees are being managed on cropped land that would normally be totally cleared of tree cover.

As acceptance and understanding by farmers increased, the system of FMNR promoted by MIDP has changed over the lifespan of the project. There have been three main phases:

2.1.1 Sassabin Zamani ("modern – clearing")

Instead of clearing all tree regrowth on cropland, single stems were selected from about 40 stumps per hectare and pruned. Acceptance was very slow at first. It was a big step for farmers to take in light of the ridicule they received. They also feared that the trees would not allow their food crops to grow well.

2.1.2 Sassabin Sabon Tsare (literally "new – modern – clearing")

As people gained confidence with "modern clearing" and realised that there were many benefits to leaving trees, a more intensive form of FMNR was promoted. In this system, five stems were left per stump, with the intent of harvesting one stem each successive year, and allowing a new stem to replace it. For each year that a stem is left, it increases considerably in size and value. Stems on fifty to one hundred stumps per hectare were pruned in this way.

2.1.3 Kowani Gindi Riba Ce. ("Every stump gives a profit")

A significant paradigm shift was occurring. Farmers ceased to see trees as weeds and began appreciating them as a cash crop in their own right. At this stage it was relatively easy to introduce a third modification. In Niger, the dry season lasts eight months. For a significant part of the dry season, native woody vegetation continues to grow and can have a beneficial effect on the land. Farmers were now challenged to leave and prune five stems on every stump growing on their land for the length of the dry season. Just prior to planting their crops they could then cut out all the stems that they did not want.

The stems respond to the culling and pruning by growing rapidly and producing small but useable stakes. Through this method otherwise idle land becomes a productive resource for the length of the dry season.

Acceptance of FMNR was slow at first. A few individuals tried it but at high personal cost in terms of ridicule. Wood was a scarce and valuable commodity and these individuals were in the minority. This led to the theft of their trees and discouragement. A break came in 1984. Radio coverage of an international conference on deforestation had increased peoples' awareness of the link between deforestation and the climate. This was followed by a Niger wide severe drought and subsequent famine. Through a MIDP run "Food for Work" program, people in 95 villages were encouraged to at least give FMNR a try. For the first time ever people in a whole district were leaving trees on their farms. Many were surprised that their crops grew better amongst the trees. All benefited from having extra wood for home use and for sale.

But most farmers had obviously only left the trees in order to receive food. Once the food for work program ceased, over two thirds of the 500,000 trees protected in 1984 - 1985 were chopped out! However, the exposure to FMNR in over 95 villages for a 12-month period was sufficient to introduce the concept and take away some of their fears. After cutting out the regrowth, farmers again experienced the very problems that had plagued them previously: strong winds, high temperatures and infertile soils in addition to a shortage of wood. The large number of trees left during the "Food for Work" program meant that despite stealing, farmers still had plenty of wood to harvest. Over the years, more and more farmers started leaving trees on their land. Today there are over two million trees being left at any one time. By 1989 FMNR in one form or other was fairly well established as a common practice. Today, people who do not practice FMNR are in the minority and they are now the ones being ridiculed! It is reasoned that because everyone needs wood, those not practising FMNR must either beg for it or steal from those who do.

2.2 Species Used in FMNR

The main tree species being regenerated in the Maradi area are *Bauhinia reticulata*, *Guiera senegalensis*, Ziziphus species and *Combretum* spp. The species used in a particular locality will differ depending on factors such as what species occurs naturally, the species' reputation for competing with crops, traditional beliefs about it and its uses and characteristics such as thorniness and growth rate.

2.3 Benefits of FMNR

2.3.1 Firewood and building timber

Today farming communities no longer have to travel to Maradi to buy firewood and building material. Not only has FMNR supplied their domestic needs, they have a surplus for sale, and merchants come weekly to the villages to buy wood for resale in Maradi. Enterprising farmers are turning their wood into value-added products such as hut roofs and tool handles, thus increasing their income. Financial security from the sale of wood has increased to the extent that during recent famines, food was no longer given free as it had been previously, but sold at a subsidised rate.

2.3.2 FMNR contributes to land reclamation

Farmers discovered by accident that they could restore hard pan soils. When they harvested their trees, they did not want to cover good farmland with piles of wood so they stacked it on hardpan sites. Several things happened

1. Termites burrowed through the hard crust breaking it up as they did so and began eating the leaves and lighter twigs.
2. When it rained, instead of running off the hardpan, water penetrated it and softened it.
3. As wind passed over the woodpile it deposited fine silt and organic matter, building up topsoil to a depth of 30cm in some cases.

Within one season, and at no cost, land, which had been unproductive for decades, was restored. Hundreds of hectares of formerly useless land have been restored by this method around Maradi.

2.3.3 FMNR positively affects crop yields and animal productivity

When land was cleared of trees the effect it would have on livestock production was barely understood. The long dry season, high stocking rates and harvest of grass for sale results in severe fodder shortages for a large part of the year. Low food value millet stalks made up the bulk of livestock diets, resulting in malnutrition during the dry season. Suckling young sometimes died and working oxen were in poor condition for ploughing at the start of the rains.

As the trees resulting from the practice of FMNR matured it was realised how important they were. They produce edible pods eagerly sought out by livestock. Farmers were also able to increase incomes by selling surplus pods.

Livestock were attracted by the shade and the tree pods and thus began to spend more time on treed farms, adding to soil fertility through their droppings. Trees provide shading for crops, protect them from winds, which may exceed 70 kph, reduce evaporation and enrich the soil through recycling of nutrients from foliage, flowers, fruits and twigs. Dust depositions on the leaves and droppings of birds attracted to the trees can also be significant (Buerkert & Hiernaux, 1998) Where there is no shading soil temperatures can easily exceed 60 degrees C. These high temperatures enhance evaporation and make it difficult for emerging crop seedlings to thrive.

In the past, every stalk of straw was taken off the fields for use in cooking fires or for animal fodder. Large increases in crop yields have been recorded with surface application of crop residues on acid Sahelian soils. In the past farmers did not broadcast crop residues and other organic material as surface mulch. Reasons included the many competitive uses of residues and difficulties of weeding (Buerkert & Hiernaux, 1998). Since animals benefit from the tree pods, for the first time farmers could afford to leave the crop residue on the fields for mulching.

2.3.4 FMNR increases biodiversity and reduces dependence on pesticides

Indigenous wildlife is returning to the area. One site saw monkeys return for the fruiting season of certain trees after a 15-year absence. Ten years ago a sighting of anything larger than a lizard was extremely rare. Because of FMNR, today there are areas where rabbits and wild guinea fowl have returned. Insect predators such as birds, lizards and certain insects (preying mantis, wasps etc) find shelter and places to breed in trees. These contribute directly to crop yields by reducing pest populations. Pre FMNR there were virtually no natural predators to crop pests as there was no habitat for them.

2.3.5 FMNR contributes significantly to the local economy

From the first year, light firewood is collected from trimmings. From the second year on, cut branches are thick enough to sell. Many thousands of dollars have been earned in the district through sale of firewood and building poles.

2.3.6 FMNR contributes to the quality of life

It could be argued that quality of life has improved aesthetically and tangibly. In areas where FMNR is practiced the velocity of strong, dust-laden winds has been reduced. Shade and respite from the heat has returned and the once barren, highly reflective countryside is much more relaxing and pleasing to the eye.

2.3.7 FMNR provides food for people

A number of the species being left provide edible leaves and fruits for humans. Some of these are considered as famine foods only while others are sought after and eaten at anytime. In recent times, fruits and leaves from stands of protected trees have been the only thing keeping some communities from starvation.

2.4 Constraints to Transferability

2.4.1 Drought

Generally, drought does not affect FMNR negatively. Whereas shallow rooted annuals such as millet are completely dependent on rains, deep-rooted perennial species used in FMNR survive and even continue growing despite poor rainy seasons or even absence of rains.

2.4.2 Insect pests

Locusts and termites have not caused major problems on the native species used in FMNR. Native species which are susceptible to locust attack usually recover after the locusts have passed by.

2.4.3 Distance to markets

Most areas in rural Niger are wood deficient. Even without good proximity to markets, the benefits far outweigh any possible disadvantages. Farmers nearer large centres stand a better chance of making more income from wood and other tree products though. Villages even 50 kilometres from Maradi sell their wood in the city. Wood is cheaper in remote areas, but due to severe shortages, it is always marketable.

2.4.4 Presence of live tree stumps

FMNR depends on the presence of suckering tree stumps of species which can be regularly coppiced or pollarded and which provide useful wood. However, even land where the original trees have been removed has been restored by farmers who have broadcast seeds of desirable species in their fields. Seeds that germinate and grow to maturity are eventually managed in the same manner as regrowth from tree stumps used in FMNR.

2.4.5 Climate

In areas of low rainfall e.g. below 200 mm, tree growth rates will be slower and the cutting rotation should be reduced accordingly. In these areas tree establishment by direct sowing of seed in the field will be less successful and take longer than in higher rainfall areas.

2.4.6 Culture

Various cultural factors will affect the adoption of FMNR.

There needs to be respect for other peoples' property. In Niger there is free access to farmland during the dry season and in the past farmers not only had no recourse if someone else cut their trees, it was considered anti social to complain about it. Consequently a "what's the use of even trying" attitude prevailed.

In many cultures innovation and individuality are frowned upon. Resulting ostracism and ridicule will discourage most from attempting anything new. Language may reflect deeply ingrained attitudes. e.g. in Hausa the word for tree (itce) is the same as the word for fire wood. i.e. Trees are given no intrinsic value of their own, apart from their utility for firewood. Obviously, if a community as a whole can be persuaded to practice FMNR, adoption should be more rapid and less traumatic.

Historically, when soil fertility diminished, rural communities survived by moving to virgin bushland and clearing it. There is rightful pride in the pioneering spirit that accompanied land clearing and enabled farmers to feed their families. If not presented properly, FMNR may appear to attack a community's pride in their forefather's "taming of the land" and thus not be readily accepted.

Each ethnic groups' world view will affect adoption. Fulani cattle herders, who see their lifestyle as the best in the world find it humiliating to consider chopping wood and selling it the way Hausa tribesmen, their traditional enemies, do.

Inappropriate government laws can have an enormous effect. If the farmer does not have the right to harvest the trees he has protected, there will be little incentive for him to do so.

2.5 Reasons for successful adoption of FMNR in Maradi

- There was a severe wood shortage.

- District wide exposure to and teaching of FMNR through a Food for Work Program and continued project activities. People perhaps for the first time linked drought and famine with the disappearance of trees and saw that they were not only partly responsible, but it was within their means to do something about it.
- Farmers trusted MIDP staff.
- Farmers were effectively given ownership of the trees on their land. This point is very significant. If farmers had no guarantee that they would benefit from their labours it is very unlikely that they would bother practicing FMNR.
- A community based regulatory system developed. Cases of stealing were dealt with by the local chief and not a distant forestry agent. Farmers gained confidence that their rights would be respected and defended so they began to take the normally culturally unacceptable step of turning in offenders – even if related to them.
- FMNR is easy to implement and requires no investment to start.
- FMNR is profitable.
- The presence of trees in fields did not lower millet yields as feared and actually helped increase them. This factor is very important to poverty stricken farmers who regularly face hunger.
- FMNR increased available fodder for livestock.
- FMNR did not depend on a project to spread to other areas, but spread from farmer to farmer by word of mouth.

3. CONCLUSION

Traditional forestry approaches to solving problems of deforestation and desertification are inadequate. Deforestation is proceeding at a rate at least 30 times greater than reforestation in some countries. Traditional approaches are expensive and often cannot be continued by the local population, who have little sense of ownership once the project ends.

Where practicable, Farmer Managed Natural Regeneration (FMNR), which relies on the management of existing indigenous species, should be considered as a viable approach to reforestation. It is easy to adopt, low cost, and has the potential to quickly increase tree cover on a large scale. FMNR provides benefits to soils, crops, the environment and local communities. Because of its low cost, good economic return, and ease of implementation it has the potential to become a people movement that spreads by word of mouth from farmer to farmer.

In short FMNR's potential to impact the environment and the welfare of whole communities is enormous and deserves greater attention by foresters, agriculturalists, project planners and farmers world-wide where conditions allow it.

In describing the benefits of FMNR, this paper is not discounting the value of tree planting schemes. In some regions, there are no live tree stumps in the fields which can be regenerated. A number of tree planting projects have been very successful (e.g. CARE's Windbreak project in the Maaja valley of Niger). For certain tree species such as fruit trees and valuable rare or exotic species, a nursery may be the only way of propagation. However, for rapid, cheap and ongoing reforestation (beyond the life of the project) FMNR should be given high consideration.

ACKNOWLEDGEMENTS

Canadian International Development Agency which funds the ongoing work of SIMs' Maradi Integrated Development Program.

ECHO (Educational Concerns for Hunger Organisation) – for permission to use material from ECHO Notes article "Working with Nature: Farmer Managed Natural Regeneration. Issue 58 Nov 1997.

IPALAC (International Program for Arid Land Crops) for sponsoring the author's participation.

SIM (Society for International Ministries) International, SIM Australia and SIM Canada and the many individuals who support this work through orld Vision Australia for encouraging the author attend the IPALAC conference (Combating Desertification with Plants) and allowing time for preparation of this document.

REFERENCES

1. Buerkert A., & Hiernaux P. (1998). *Nutrients in the West African Sudano-Sahelian zone: losses, transfers and role of external inputs in Z. Planzenernahr.* Bodenk., 161, 365-383. WILEY-VCH Verlag GmbH, D-69451 Weinheim.
2. Echolm, E. (1975). *The Other Energy Crisis: Firewood.* Worldwatch Paper 1. Worldwatch Institute, Washington, D.C., USA.
3. FAO (1986). Atlas of African Agriculture. *African Agriculture – the next 25 years.*
4. FAO (1997). *State of the World's Forests.* P110, 108.
5. Firewood Crops (1980). *Shrub and Tree Species for Energy Production.* National Academy of Sciences, Washington, D.C.
6. Weber, R. & Stoney. C. (1986), *Reforestation In Arid Lands.* Volunteers in Technical Assistance. USA.

SAND DUNE STABILIZATION WITH PLANTS IN CHINA

Weishou Shen
Desertification Research Center, the State Environmental Protection Administration of China, Nanjing, 210042, Tel: 86-25-5472856, Fax: 86-25-5411611, email: shenws@public1.ptt.js.cn

1. BRIEF INTRODUCTION OF SANDY DESERTS AND SANDY LANDS IN CHINA

Sand deserts and sandy lands in northern China are situated in the temperate region between 35-40^{0}N and 75-125^{0}E. The area to the west of Helan Mountain Ranges belongs to the desert zone with annual precipitation below 150mm and sandy deserts makes up about 80% of the total area of sand deserts and sandy lands in northern China. In this highly desiccated desert zone there exist Taklimakan Desert, Gurban Tonggut Desert, Kumtag Desert, the Deserts in Qaidam Basin, Badain Jaran Desert, Tengger Desert and Ulan Buh Desert. There are many sand mountains in the vast expense of deserts, some of them are more than 200m in height.

The middle sand area between Helan Mountain Ranges and Wenduermiao-Dinbian, a narrow belt, belongs to semi-desert zone with annual rainfall 150 to 250 mm and sand area occupies only 4% of the total area of sand deserts and sandy lands in northern China. The west part of Qubqi Desert and Otindag Sandy Land, the Northwest part of Mu Us Sandy Land are situated in the belt. In this area, the bottoms among the dunes are rich in ground water.

The area to the East of semi-desert zone belongs to semiarid steppe zone with annual rainfall 250 to 400 mm, the sandy area accounts for about 16% of the total area of Chinese northern sandy area. There exist Horqin and Hulun Buir Sandy Lands, the Southeast parts of Otindag and Mu Us Sandy

Combating Desertification with Plants, edited by Dov Pasternak and Arnold Schlissel.
Kluwer Academic/Plenum Publishers, New York, 2001.

Lands. In this area, more than 80% of annual rainfall is concentrated in the period from July to September, winds are strong and frequent in winter and spring. This area was covered with typical steppe and forest steppe, destroyed by man activities, vegetation on the sandy lands has been seriously damaged. As a result, shifting dunes and fixed sandy lands co-exist at present. In bottoms among dunes, ground water level is 0.5 to 2 m, generally fresh water, can be used by plants.

Most of sand deserts in China are located in the inland. To the west of Wushaoling and Helan Mountain Ranges, except Gurban Tunggut Desert in Junggar Basin, the overwhelming majority is mobile dunes, which occupy about 75% of the desert area. To the East of Wushaoling and Helan Mountain Ranges, only Mu Us Sandy Land in southern Ordos and Horqin Sandy Land in the lower reaches of Xiliao River where exists a part of mobile sands, the overwhelming majority is fixed and semi-fixed dunes, occupying 80% of the sand area in this region (*fig.1*).

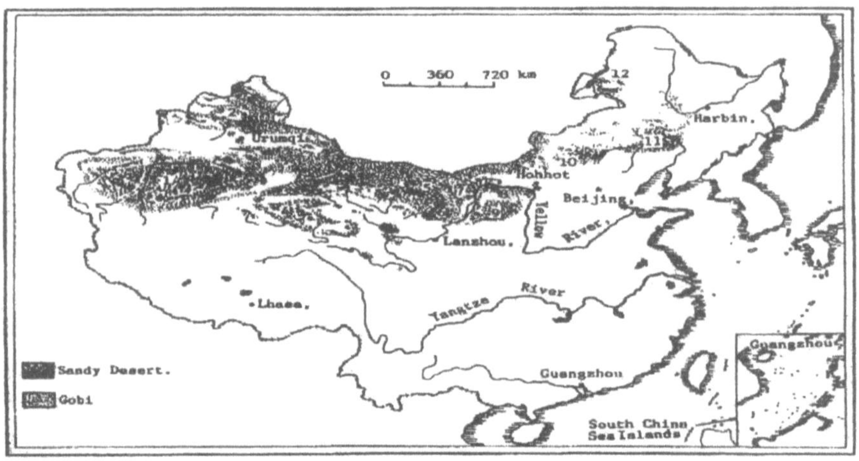

Figure 1. Distribution of the Deserts in China

Limited precipitation, strong and frequent winds, sparse vegetation and loose sandy ground surface, makes the natural conditions of the sand areas adversely. It is recorded that the dust storm and sand devils occurred every year is about 50-100 days, especially in shifting sand areas, dust storm is more serious. The strong blown sands and shifting dunes incursion bring about serious endangers to farmlands and grasslands, irrigation works, roads and railroads and residential areas etc. Therefore, how to control shifting sands and improve the living environments and conditions of industry, agriculture and animal husbandry has been an arduous task over the past 40 years.

Sand dune stabilization with plants conducted in China began after the founding of the People's Republic of China in 1949. Through many years of research and practice, the dune stabilization in China has achieved remarkable results. So far at interior of desert oases, the areas bordering sand deserts and Gobi, the interlocked areas of agriculture and animal husbandry, the coverage of artificial trees and shrubs vegetation has been greatly increased, natural psammophytic vegetation has also recovered to a certain extent. Windy and dusty devils are much weakened.

Based on our 50 years of sand stabilizing experience, following is a brief introduction of fundamental principle and measures of sand stabilization with plants in China.

2. POSSIBILITY OF SAND DUNE STABILIZATION WITH PLANTS IN CHINA

At the area where plants can grow, stabilizing sand by plants is a cardinal measure. What is called biotic sand stabilization is directed against artificial sand barriers and other abiotic measures, is actually the vegetation reconstruction in sand areas. Which mainly include establishing artificial vegetation, recovering natural vegetation and forming an artificial and natural united vegetation system.

Artificial sand barriers, chemical sand-holding preparations and other mulches, as the temporary measures, are aimed at stabilizing sand surface, protecting plants from wind erosion and buried by sand, and creating a stable living environment for plants. However, limited by materials, transportation, labours and funds etc, these assistant measures were used only when it is indispensable.

2.1 Semiarid Zone of China

Mu Us Sandy Land, north part of Qubgi Desert, Otindag Sandy Land, Horqin Sandy Land and other adjacent scattered sand areas, are all located in semiarid steppe zone where annual precipitation is more than 250mm, generally 300-400mm. In these sand areas, the dune is composed of fine sands with the particle size 0.05-0.25 mm. in diameter, water regime of the shifting sand dunes is relatively better although it is infertile. The thickness of surface dried sand is generally no more than 10 cm, downwards, the stable water content in wet sand layer is normally 3-4 %, thus the available water that can be used by plants is 2-3%. At the same time, shifting sands has a strong permeability and can be wet through with rainwater for several times each year. In addition, a great quantity of permeated rainwater supplies the ground water, which results in the ground water level in bottoms rising, generally within 1-2 m.

Shifting and semi-fixed sand dunes and wind-eroded lands in semiarid zone of China belong to desertified lands. They had been formed from the destruction of primary steppe vegetation, and there remains some buried steppe soil. That is to say, in semiarid area, there is seldom single-phase sandy land. Moreover, most of the wandering sand dunes are in shapes of crescent dunes and the dune chains. The height of the dunes is lower, normally 3-8 m. The bottoms among the dunes are wide with some plants.

To sum up, in semiarid area of China, the wandering dunes and semi-fixed dunes are in a relatively advantageous position. They can not only be stabilized by plants, but also be planted proper ratio of coniferous and broadleaf trees. In other words, it is possible to establish a efficient artificial and natural united vegetation system on the sandy area.

2.2 Semi-Desert Zone of China

As regards the wandering dunes in semi-desert zone of China, water content in wet sand layer is normally 2-3%. The dune area with deep ground water generally can be only planted with shrubs, while the dune area with ground water level in 2-3 m can still be planted with proper trees in bottoms.

2.3 Desert Zone in Northwest Part of China

As to the desert area in Northwest part of China, precipitation is very limited. In east part of Alashan Desert, precipitation is only 150mm, east part of Hexi Corridor 100 mm, to the south edge of Taklimakan Desert only 10 mm. In these areas, therefore, it is impossible to stabilize sands by plants. As for the oases, although situated in desert areas, they are not deserts. This is because the wandering dunes at the contiguous zone of oases and sand desert and scattered dunes at interior of oases had been covered with vegetation. As a result of man activities, especially gathering firewood and overgrazing, natural vegetation gradually vanished, and the dunes revive and spread. Therefore, this kind of wandering dunes belongs to desertified land to a great extent, there is a latent possibility for this kind of desertified land to be reversed.

Wandering dunes on the borders and interiors of oases are generally lower, with the dune height 2-10 m. and most of them are crescent dunes and the dune chains. With small dune density, the ratio of the bottoms is more than 30%, even 50-60%. Beneath the dunes and in bottoms, there exists thicker soil with ground water level 1-3 m. That is to say, the water and soil conditions of the wandering dune areas on the borders and in oases is favourable to revegetation.

In brief, in the areas where the sand control measures has been taken or is being taken, the natural conditions are relatively favourable to biotic sand stabilization.

3. MAJOR MEASURES TO STABILIZE SANDS IN CHINA

Wandering dune area can be divided into three topographic positions, i.e. windward slope, leeward slope and bottom. Among which the lower part, a third of the windward slope, is wind erosion section, the middle part is wind erosion and sand crossing section, and the upper part, another a third of windward slope belongs to dune top. The leeward slope is steep with thicker dry sand. There is no wind erosion or slight erosion in bottoms. The different measures and plant species should be selected according to these topographic positions in biotic sand stabilization. The practice and experience of China has proved that the following measures are very efficient to control and stabilize wandering dunes.

3.1 Afforestation in the Bottoms

Regardless of the sand areas in desert oases or steppe zone it is possible to make fully utilizing the bottoms and flat sections in dune area to build lumpish trees and shrubs forests so as to break up and surround dunes. 1-2 years after the first time afforestation, with the dune moving forward, the position of formerly foot of windward slope gradually become flat section where the second and third times afforestation can be made. In this way, the artificial forests continuously expand and the dune area gradually reduced. With the shifting sand moving forward and artificial forest developing, dune height cuts down and the coverage of lumpish forests reaches 30-50% and even more. So far the shifting sand is basically controlled and stabilized. The large-scale of sand -break forests in Ulan Buh Desert, Tengger Desert, Hexi Corridor and Mu Us Sandy Land, were all constructed by utilizing bottoms of dune areas.

3.2 Planting Sand Binders Directly on Wandering Dunes

Sand stabilization with plants in semiarid zone of China generally needs not artificial sand barrier, but closely plant sand binders in lines, serving a double purposes of sand barrier and biotic sand stabilization. In Horqin

Sandy Land, what had been done is, on the lower and middle parts of windward slope and even on the whole small dunes, perpendicular to main wind direction, to plant *Artemisia halodendron* and *Salix gordegevii* closely in lines in spring. In Mu Us Sandy Land and east part of Qubqi Desert, The same planting method are used for *Artemisia ordosica*, *Salix psammophylla* and *Amorpha fruticosa*. Once the lower and middle parts of dunes are stabilized by plants, after 1-2 years of wind erosion,, the upper part of dunes tend to become gentle. At that time, the another planting can be done on it. Usually, by 2-3 times of planting, the shifting sand can be finally stabilized.

At some places when the dunes are stabilized to a certain extent, assisting proper artificial sand barrier, *Pinus sylvestris* var. *Mongolica*, *P. tabulaeformis* and other coniferous trees are planted. This is the basic model of sand stabilization with plants in steppe zone of China.

3.3 Planting Sand Binders Under the Protection of Artificial Sand Barrier

Sand stabilization with plants in semiarid zone of China, except the places which need urgent protection, can be successfully done without artificial sand barrier. While in the desert oases and semi-desert areas, only under the protection of sand barrier could sand binders survive and grow on wandering dunes.

In desert oasis areas, there exists clay in the bottoms; it can be used to make clay sand barrier. In north part of Ulan Buh Desert and Hexi Corridor, using clay to make sand barrier in check or lines has been proved very successful. Such as Minqing and Linze oases, when planting sand binders, the clay sand barriers in lines on the windward slope of dune was made 1 year before. The barriers was 20-25 cm in height with row spacing 2-4 m. After several strong winds, the sand surface between barriers was slightly eroded but the depth no more than 10 % of the row spacing. Under this circumstances, *Haloxylon ammodendron*, *Tamarix ramosissima*, *Hedysarum scoparium*, *Caragana korshinskiii* and *Calligonum* spp. were planted between the barriers. If the dune height is more than 7-8 m. the measures were only taken on lower and middle parts of windward slope at first. when the dune top was cut down by wind force, the second and third times sand binding were put into practice. In addition, after 2-3 years of rainwater erosion, the sand surface in the barriers is solidified by clay, at this time the height of sand binders is more than 1 m and dunes are wholly covered with artificial vegetation.

In Jinta area of Hexi Corridor, annual precipitation is only about 50 mm, the thickness of dried sand on the surface of windward slope is 30-40 cm, and the water content in wet sand layer is only 1-2%. In case of deep ground water and no irrigation condition, it is impossible to stabilize sand only by plants. What had been done is to collect crushed stones on the spot and cover the whole dunes with a thin sheet of the stones, generally 3-4 cm in thickness. After that some xeric shrubs are planted on it. These measures have played a very important role in preventing Jinta Reservoir from harms of wandering dunes and blown sand.

In the 1950s, China built Baotou-Lanzhou, Lanzhou-Xinjiang, Gantang-Wuwei, Wuda-Jilantai, Jining-Erlian and other railways in the north sand areas of China. Among them the sections which were endangered by wandering dunes and blown sand is totally about 750 km in length. Therefore, researches on the principle and measures of shifting sand stabilization along the railway line have important practical significance. According to the research and practice in Shapotou area, setting up 1×1 m of wheat straw check barriers along the railway line had played a very important role in protecting trains going smoothly and creating a good condition for sand binders in initial stages. Under no irrigation condition, the wheat straw check barriers normally maintains about 5 years.

The dune top is susceptible to wind erosion, and the straw barriers needs repairing in early years. *Hedysarum scoparium*, *Caragana korshinskii*, *Artemisia ordosica* and other sand binders were planted among the barriers, 10-15 years later, the coverage of artificial vegetation reached about 30%. In this case the sand surface was basically not eroded by wind, artificial vegetation replaced the wheat straw barrier and play a permanent protective role.

3.4 Enclosing Sand to Recover Natural Vegetation

Both in desert and semi-desert areas of China, enclosure is usually used as a basic measure of revegetation. Generally speaking, combining with planting sand binders, proper enclosure is taken, forbidding firewood collection and grazing in early years, natural vegetation will gradually recover, especially when the scattered shifting sand is inter-locked with fixed and semi-fixed dunes. Artificial vegetation and natural vegetation together can form a stable protective system.

Up to now there is a large area of desert vegetation characterized by *Haloxylon ammodendron* and desert riverbank forest featured by *Populus euphratica* in Northwest part of China. They are normally used as grazing land and have been destructed in varying degrees. Practice has proved that proper enclosing and limited grazing have a good effect on the natural regeneration of the desert vegetation.

Natural vegetation has the ability of natural regeneration and is more stable than artificial vegetation is. Therefore, when the sand dune stabilization with plants or the vegetation reconstruction is put into practice, it is necessary to evaluate the possibility and the speed of remained plants recovering. Meanwhile, complementary sowing, enclosing and other protective measures can promote the recovery of natural vegetation. The experience in China can be summarised as follows:

3.4.1 Semiarid Steppe Zone of China

Abandoned farmlands, overgrazed grasslands and seriously damaged fixed and semi-fixed dunes, once enclosed, *Agriophyllum squarrosum, Artemisia* spp. and other plant species increase quickly in number and quantity. After 3-4 years of enclosure, wind erosion on the ground surface is basically controlled and scattered small wandering dunes tend to be fixed. Although *A. squarrosum* and other annual plants can not steadily stabilize shifting sand, they can create a stable environment for the developing of perennial plant species.

Under enclosure, grass and willow scrubs in bottoms and flat sand area grow and spread. Among them, the representative species are: *Pennisetum centrasiaticum* , *Calamagrostis epigeios, Agropyron cristatum, Leymus Chinesis, Salix gordejevii, S. microstchya* var. *bordensis* in Horqin Sandy Land; *Agropyron mongolicum, Leymus secalinus, Salix psammophila, S. cheilophila* etc. in Mu Us Sandy Land.

In order to improve grassland quality and resistance, for the desertified lands in semiarid area of China, combined with enclosure, the following species are usually sown: *Caragance microphylla, C.intermedia, C. Korshinskii.* Under the condition of water irrigation, sowing still more some species of *Artemisia, Nitraria, Glycyrrhiza* and *Caragana* etc. have remarkable results for the vegetation recovering.

After years of enclosure, natural vegetation recovers to a certain extent, and artificial vegetation develops too. At this time, limited grazing is usually underway. Cutting sprouting shrubs from ground surface can not only provide firewood, but also promote the scrub's rejuvenation.

3.4.2 Semi-Desert Zone

After sand area is enclosed, natural vegetation can also recover to a certain extent. Taking Yinchuan-Gantang section of Baolan railway line as an example, both in artificial sand stabilization area and in semi-sand stabilization area, natural vegetation grows and propagates very fast in rainy years. The main species are *Agriophyllum squarrosum, Coripermum patelliforme, Bassia dasyphylla, Artemisia sphaerocephalla* and *A. Ordosica* etc. and the coverage can reach 20-50% in rainy season.

3.4.3 Desert Zone

For the semi-fixed sand dunes at interior and edge of oases, enclosure combines with planting sand binders, natural vegetation can gradually recover. The representative species are: *Artemisia* spp. *Nitraria* spp. *Glycyrrhiza* spp. *Sophora alopecuroides*, *Alhagi sparsifolia*, *Tamarix* spp. *Inula salsoloides*. *Phragmites communis*, *Achnatherum spendens*, *karelinia caspicaa*, *Psammochola villosa*, *Limonium bicolor* etc.

4. SAND REVEGETATION BY AIR-SEEDING IN CHINA

Air-seeding is a fast and effective sand control measure. The air-seeding on sand areas in China began in the end of 1950s. At first, it started from Southeast part of Mu Us Sandy Land, and gradually expanded into the whole Mu Us Area, Horqin Sandy Land, Southeast part of Tengger Desert and Gurban Tonggut Desert. The systematic researches on the selection of plant species, best seeding time, prevention of seed displacement, techniques of seeding operation and management after seeding etc. have been made and wealth of experience has been accumulated through over 40 years of experiments and practice.

Among the plant species used for air-seeding on the moving sands in Mu Us Sandy Land, *Hedysarum laeve* comes first on the list both in its sand binding ability and its economic value. With fast root growth during sapling period and sprouting feature it can rapidly expand the colonizing range and make sand surface fixed. *Artemisia sphaerocephala* is another species used for air-seeding. Although the wind erosion hardiness is inferior to *H. laeve*, it has higher sapling density and can stabilize sand surface in certain period, thus creating a stable growth site for other species. Therefore, on the dune area with higher mobility, as the protective species of other purpose ones, *A. sphaerocephala* is still a sound air-seeding plant. *Hedysarum scoparium*, a pioneer plant on psammosere in desert zone and introduced into steppe region used for air-seeding, can adapt the ecological environment of moving sands. But because of its sparse growth habit, it is always scattered in seeding area with lower density. Other plant species, such as *Artemisia wudanica*, *Astragalus adsurgens*, *A. mellilotoides*, *Caragana korshinskii* and *C. intermedia* etc. sometimes are also used for air-seeding.

Based on the research in Mu Us area (Shen Weishou, 1998), among the geomorphic factors of dunes, the key one which has the most important influence on the sand stabilization after air-seeding is the windward slope of dunes, the second one is dune height. With the windward slope below 5 degree and dune height lower than 5m, the proportion of the fixed state on the windward slope of dunes can reach 80% after seeding. When the slope is 10 degree and dune height is 10m the proportion accounts for 50%. When the slope is more than 15 degree and dune height is over 15m, the proportion can only reach about 30%, that is to say, only the lower part of dunes can be fixed.

REFERENCES

1. Zhu Zhenda, Liu Shu et al. (1980). *An Introduction to Sandy Desert in China.* Beijing: Science Press.
2. Zhu Zhenda, Zhao Xingliang et al. (1998). *Sandy Land Rehabilitation Engineering.* Beijing: China Environmental Science Press.
3. Shen Weishou. (1998). *Process of Moving Sand Dune Stabilization after Air-seeding in Maowusu Sandy Land, China.* Chinese Journal of Arid Land Research (Allerton Press), 11(1), pp. 27-31.
4. Shen Weishou. (1997). Population Dynamics of Main Plants with which Maowusu Sandyland was Air-seeded. Acta Phytoecologica Sinica, 21(4), pp, 328-334.
5. Shen Weishou. (1999). Successional Stage and Rate of the Air-seeding Vegetation in Maowusu Sandyland, China. Scientia Silvae Sinicae, 35(3), pp, 103-108.
6. Zhu Zhenda, Wu Huanzhong et al. (1998). Study on Combating Desertification/Land Degradation in China. *Beijing: China Environmental Science Press.*

DEVELOPMENT OF A TECHNOLOGY FOR SAND DUNE STABILIZATION IN NORTHERN NIGERIA

Olaolu Babalola
Department of Agronomy, University of Ibadan, Ibadan, Nigeria, Tel: 234-2-8102461-4 or 8102070-5, Fax: 234-2-8103118, email: babalola@ibadan.skannet.com

1. INTRODUCTION

When a fragile ecosystem is exploited beyond its carrying capacity, the system breaks down. This is the problem of the arid and semiarid region of Nigeria, which constitutes about one-third of the (924,000 km^2) entire land area. Indeed, nature has imposed physical and biological constraints on the area such as drought, low and variable rainfall, sparse vegetation, thin, droughty and infertile soils mainly of aeolian origin which are highly erodible and compactible, high atmospheric evaporativity induced by high wind velocity and low relative humidity.

High carrying capacity of man and animals coupled with the use and overuse of the fragile resource base has led to serious land degradation. Deforestation, overgrazing, soil compaction by human and animal traffic and repeatedly running over of the land with heavy machinery have exposed large tracts of land to the climatic elements resulting in desertification or the expansion of desert-like conditions. FAO, (1977) reported that about 55% of the world's area of existing desert is contained in Africa. Desertification is one of the greatest threats to increased food production and food security in Nigeria. The boundaries of the affected areas appear diffuse and not clearly defined. However, in the 80's, desertification problems were confined to areas north of latitude 12° N. Today, this boundary has moved southwards more than 1° to a latitude below 11° N. (FORMECU, Personal

Combating Desertification with Plants, edited by Dov Pasternak and Arnold Schlissel.
Kluwer Academic/Plenum Publishers, New York, 2001.

Communication). Unless drastic measures are taken, the southward progression would continue unabated.

One aspect of desertification which until recently received little or no attention in Nigeria is the potential danger which sand dunes constitute to the rapid spread of the phenomenon. Extensive active dunes and migrating sands occur in different parts of northern Nigeria upwards of latitude 11° N. In 1976, sand dunes covered only 0.1% of the entire land area or 812 km^2. In 1995, this has increased to 0.5% or 4,829 km^2 (FORMECU, 1998). The rate of expansion of sand dunes is thus about 200 km^2 per year. Unstabilised dunes become mobile due to wind action resulting in further desertification of the area and posing further threat to life and property (Babalola, 1985). In Sokoto state of Northern Nigeria, whole communities have had to relocate as sands buried their homes and farmlands. Highways have been blocked by migrating sands. Extensive crop failure has occurred in the field because seedlings are either buried by sands or desiccated as their roots are exposed by deflation or wind action. Thus the urgent need to stabilize the dunes cannot be overemphasized.

2. STABILIZATION METHODS

There are three main methods used for dune stabilization, namely, the chemical, mechanical and biological methods. In the chemical method, water, oil or chemicals are used as surface stabilising sprays to bind the soil particles together or to cover the sand thus preventing it from eroding (Hagedorn et al, 1977).

The mechanical methods are designed to alter the wind speed and cause either sand deposition or deflect the wind and its freight of sand before they get to the object to be buried. The biological method makes use of vegetation in terms of afforestation or reforestation.

Permanent stabilisation is best accomplished by the maintenance of a vegetal cover. Vegetation, however, will not grow as long as there is still a strong movement of sand. Therefore, successful stabilisation of sand dune requires a combination of the mechanical and biological methods.

Most of the sand dune fixation efforts in Nigeria have achieved little success. This is because the problem has been looked at as routine afforestation. The construction of shelterbelts has been very useful in arresting wind erosion on tracts of long and exposed land areas and has helped to stabilize farmers on their farms. It has not, however, per se, solved the problem of sand dune fixation. Besides, only a very small percentage of the desertified area requiring shelterbelt construction has been treated till today.

3. METHODOLOGY

In an attempt to develop a technology for sand dune fixation, a number of pertinent questions need to be asked.

1 - What is the chemical and physical nature of sand dune?

2 - How do you ensure that seedlings raised on sand dunes survive in view of the sandy nature and the very high atmospheric evaporativity which creates a high demand for soil moisture in the sand dune area?

3 - In order to survive and avoid burial, seedlings planted must be drought tolerant, have good root system development and must grow fast. Which plants meet these requirements?

4 - How can the chemical and physical conditions of the sand dune be improved to ensure survival, fast growth and establishment of selected plants?

5 - What is the best pot-size for raising seedlings?

6 - What kind of mechanical defences should be provided for the seedlings?

7 - How should the mechanical defences be arranged in the field?

In order to answer these and other questions, studies were carried out in the laboratory, nursery and the field.

3.1 Laboratory Studies

Laboratory studies were carried out to (i) determine the particle size distribution of the dunes and adjacent flood plain, (ii) determine the moisture characteristics of some potting mixtures to be used for raising seedlings and (iii) determine the effect of a soil conditioner (Alcosorb) on soil physical properties of three contrasting soil textures found in Northern Nigeria. Alcosorb, a polyacrynamide is capable of absorbing 400 times its own volume of water (Anonymous, 1987).

Table 1. Potting mixtures used in the experiments.

Treatment	Ratio	Mixtures
1	3 : 1	Sand : Manure
2	3 : 2	Sand : Manure
3	3 : 1 : 2	Sand : Manure + Fadama "clay"
4	3 : 1	Sand : Manure + 0.5 kg /m^3 alcosorb
5	3 : 1	Sand : Manure + 1.0 kg /m^3 alcosorb
6	3 : 1	Sand : Manure + 1.5 kg /m^3 alcosorb
7	3 : 1	Sand : Manure + 2.0 kg /m^3 alcosorb
8	3 : 1	Sand : Manure + 2.5 kg /m^3 alcosorb
9	3 : 1	Sand : Manure + 3.0 kg /m^3 alcosorb

Table 1 shows the potting mixtures tested in the laboratory and in the nursery. An attempt was made to improve the physical and chemical condition of the traditional potting mixture (3:1 sand/manure mixture) by adding different levels of manure, fadama clay and alcosorb.

3.2 Nursery Studies

In the nursery, studies were carried out to (i) determine the effect of the potting mixtures on the growth and development of test plants, namely neem (*Azadirachta indica*) and *Eucalyptus camaldulensis*), (ii) determine the most suitable pot-size for raising seedlings, (iii) determine the effect and mode of application of alcosorb on the growth of seedlings and (iv) determine the frequency of watering and hence water-use efficiency of neem and *Eucalyptus* on each potting medium. Neem and Eucalyptus were selected as test plants based on previous studies. In the nursery, there were four experiments.

Experiment 1 was a potting medium trial using the nine potting mixtures indicated in *Table1*. Experiment 2 examined the mode of application of alcosorb, either as a gel or as granule. Alcosorb normally available as granules may be "activated " before application by applying 300 ml of water per gram of alcosorb or can be applied in a granular form. Experiment 3 was a pot-size trial (normal pot, 20 x 5 cm, 20 x 10 cm and 30 x 10 cm). Experiment 4 was another potting medium experiment designed to confirm Experiment 1. Here, tensiometers were used to regulate irrigation water which was applied whenever soil water potential reached -40 cb. *Neem* and *Eucalyptus* were used as test plants in all experiments. Treatments, which were replicated 22 times, allowed for destructive samplings of plants at various stages of growth. Seedlings were raised in black plastic bags for 4 to 5 months. Plant height, leaf number, shoot and root weights and depths of root penetration were determined weekly or at sampling.

3.3 Field Studies

In the field at Gidan Kaura in Sokoto State, studies were carried out to (i) determine the effectiveness of various mechanical defences for seedlings, (ii) determine the appropriate spacing, heights and arrangement of sand fences and (iii) test the effect of identified potting mixtures on the survival, growth and development of neem and *Eucalyptus* on sand dunes. Twigs of shrubs and trees normally used as sand fence (mechanical defence) materials were abandoned for fear of denuding the environment. Instead, guinea corn stalks commonly available in the field after crop harvest were woven into mats of about 3m lengths using the fibre rope from *Borasus ethiopium.* The weaving was done in such a way as to attain about 50% porosity. The prefabricated mats were treated with a mixture of kerosene, engine oil and dieldrin against termite attack because preliminary studies had shown that untreated mats were highly susceptible to insect attack. There were 2 sand fencing treatments in the first experiment. Sand fence heights of 1 m and 0.5m with or without a lifehedge of *Euphorbia balsamifera* were installed at spacings of 5, 7 and 10 m, to give 4 treatments (T_2 to T_5). Treatment T_1 was life hedge only. Wind erosion pins installed at the middle and edges of each plot were used to monitor sand movement.

In the second field experiment, life hedges were eliminated as treatments because of their very slow growths. Also a shorter spacing of 3 m was added so that there were four spacings: 3, 5, 7, and 10 m. The boundaries between plots were also fenced with guinea corn stalks giving a kind of checker board fencing so that whatever the direction of the wind, there was protection for the seedlings planted in each plot.

In yet another experiment, sand dune was ameliorated with each of the nine different potting mixtures used in the nursery. Amelioration was done by side dressing and by subsoiling *(Fig 1)*. Seedlings which were raised in the traditional way of using 3:1 sand/manure mixtures, were "seated" on enriched subsoil or were side-dressed. Plant height, growth and survival of neem and *Eucalyptus* were determined periodically over a year.

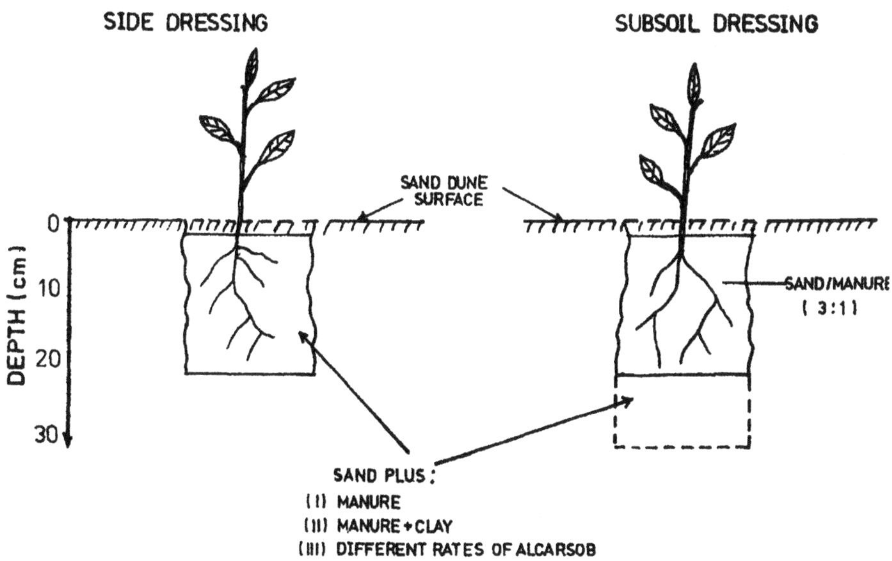

Figure 1. Amelioration methods of sand dunes planted to neem and eucalyptus in the field

4. RESULTS

4.1 Laboratory

Sand dune was predominantly coarse sand with 86% falling into a diameter range of 125-250 microns. Whereas the average silt plus clay content of the sand dune was 4.5%, the adjoining fadama (flood plain) soil had a range of 8-37%(Tables 2a and 2b). Similarly, whereas mean contents of organic carbon, total N, and CEC were 0.06%, 0.012% and 0.17cmols/kg on sand dunes, corresponding values were o.58%, 0.06% and19.4 cmols/kg in fadama soils. Manure and fadama top soils, when added to the traditional potting mixture (3:1 sand/manure), could significantly improve the soil chemical and physical properties of the growth medium.

Table 2a: Soil physical and chemical properties of sand dunes (0-30cm) at the experimental site in Gidan Kaura village of Sokoto State.

Descriptive Statistics	Sand %	Silt %	Clay %	pH-H 2O	Org .C %	Total N(%)	Ca	CEC (meq/ 100g)
Mean	95.5	-	4.5	6.5	0.018	0.06	2.37	0.18
Minimum	94.5	-	4.0	6.2	4x 10(-3)	0.04	2.22	0.13
Maximum	96.5	-	4.8	6.8	0.01	0.09	2.58	0.19
Coefficient of variation (%)	0.66	-	8.1	4.7	33.5	35	6.75	18

Table 2b. Some physical and chemical properties of surface (0-15cm) fadama soils in Sokoto State.

Descriptive Statistics	Sand %	Silt %	Clay %	pH-H2 O	Org .C %	Total N(%)	CEC (meq/ 100g)
Mean	80	8	12	7.7	0.58	0.06	19.4
Minimum	63	2	6	6.9	0.20	0.02	10.0
Maximum	92	19	29	8.1	0.90	0.10	29.3
Coefficient of variation (%)	12	71	53	3.6	41	45	18

The most beneficial effect of the soil conditioner, alcosorb, was found on sands in terms of increased moisture retention (*Fig.2*), reduction in soil water evaporation and increased percentage of stable aggregates. In sand, the percent stable aggregates increased with the rate of application and ranged from zero in the control to 28.7% at the highest rate applied of 3.0kg/m3. Rate of application of Alcosorb, for effective use, depended on the soil texture. Coarse textured soils, like sand dunes, tolerated higher rates with added advantages than clay soils in terms of moisture retention bulk density, infiltration rates, soil evaporation, soil strength and aggregate stability.

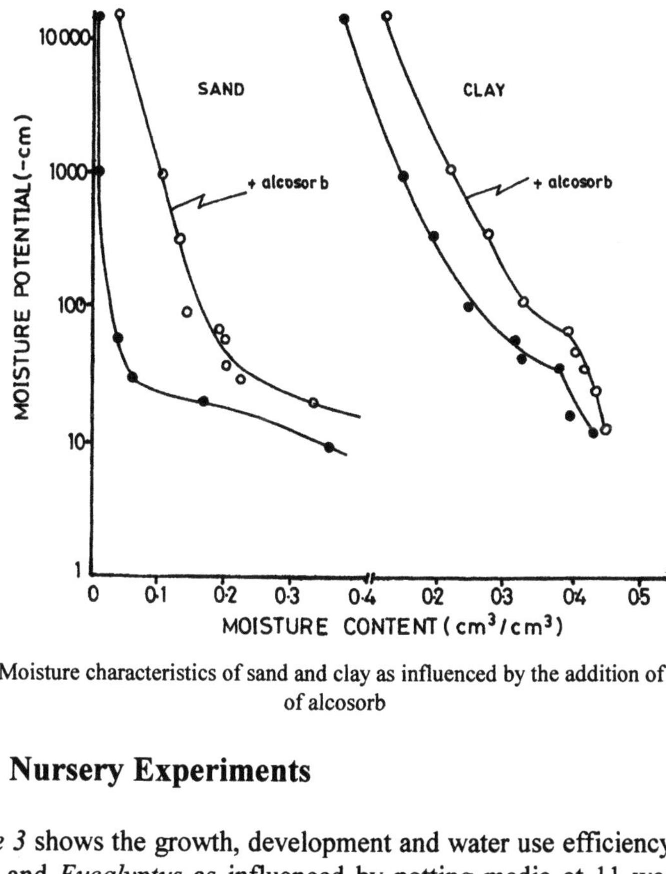

Figure 2. Moisture characteristics of sand and clay as influenced by the addition of 2.0 kg/m3 of alcosorb

4.2 Nursery Experiments

Table 3 shows the growth, development and water use efficiency (WUE) of neem and *Eucalyptus* as influenced by potting media at 11 weeks after germination. The germination percentage was close to 100% on all treatments. At the 11th week, the plants had received 156, 156, 156, 106, 99, 89, 74, 56 and 54 cm of water under treatments (potting mixtures)1, 2, 3, 4, 5, 6, 7, 8, and 9 respectively. Generally 4 conclusions were drawn from *Table3*:

(i) Treatment 3 (sand/manure /fadama clay mixtures) gave the best growth performance among soils without conditioners especially for *Eucalyptus*.

(ii) Among treatments with soil conditioners, Treatments 6 and 7 (i.e. the addition of 1.5 and 2.0 kg/m^3 of alcosorb) performed significantly better than other treatments.

(iii) Water use efficiency (WUE) increased generally with increasing rates of alcosorb to up to 2 to 2.5 kg/m^3.

(iv) WUE was much higher for neem than for *Eucalyptus*.

Gel application of alcosorb is not as effective as granular application for both test plants. All growth parameters were higher in granular application than gel application. For instance, the mean plant heights for neem were 26.33 and 22.7 cm for granular and gel application respectively and correspondingly for *Eucalyptus*, 14.7 and 14.0 cm. These differences were however not significant.

Growth of plants in normal pot size (20 x 5 cm) was significantly different from that in larger pot sizes (either 20 x 10 cm or 30 x 10 cm). For instance, mean plant heights at 15 weeks after planting were 15.8 and 24.5 cm for 30 x 10 and 20 x 5 cm, respectively for *Eucalyptus* . Corresponding figures for neem were 26.2 and 28.0 cm, respectively.

Table 3. Growth, development and water use efficiency (WUE) of *Azadiracta indica* and *Eucalyptus camaldulensis* as influenced by potting media at 11 weeks after germination.

Azadirachta indica

	TREATMENTS								
Growth parameter	1	2	3	4	5	6	7	8	9
Wt of fresh	3.55	4.37	3.48	4.16	5.36	4.50	6.04	5.26	4.59
Plant HT (cm)	18.0	18.8	19.0	27.0	25.7	27.3	26.5	23.7	21.7
Root Wt (g)	0.33	0.38	0.38	0.56	0.94	0.61	1.45	1.26	1.3
No of leaves	29.7	29.3	25.0	33.3	38.3	29.0	30.7	26.0	25.7
Depth of root (cm)	11.3	9.7	9.5	10.0	14.0	13.6	17.3	12.2	4.0
WUE (mg/cm)	0.227	0.280	0.223	0.392	0.541	0.506	0.816	0.68	0.607

Eucalyptus camaldulensis

	TREATMENTS								
Growth parameter	1	2	3	4	5	6	7	8	9
Wt of fresh plants	0.05	0.59	0.83	0.65	0.48	0.67	0.37	0.67	0.25
Plant HT (cm)	5.1	10.0	11.3	10.7	9.0	12.5	8.8	10.7	9.0
Root Wt (g)	0.01	0.04	0.06	0.04	0.05	0.08	0.09	0.06	0.03
No of leaves	8.0	8.0	14.5	15.0	13.0	13.3	10.0	12.3	11.3
Depth of root (cm)	1.47	4.6	6.5	7.0	7.8	3.8	3.3	5.0	4.2
WUE (mg/cm)	0.003	0.038	0.053	0.042	0.061	0.075	0.05	0.120	0.046
Irrigation water (cm)	156	156	156	106	99	89	74	56	54

The results of the second potting mixture experiment confirm the conclusions drawn from Experiment 1 that (I) Sand/manure/fadama clay (3:1:2) or treatment 3 was significantly the best potting mixture for raising seedlings among the soils not treated with soil conditioners.

(II) The effect of treatment 3 on height development of neem and Eucalyptus compared favourably with mixtures containing soil conditioner.

Treatment 7 which received 2.0 kg/m^3 of alcosorb produced the best crop performance.

4.3 Field Studies

Data collected in April 1997 shows that the least deflation and accretion occurred at 5m spacing and highest values at 10 m spacing. There were 2 plots, which recorded no sand movement at 5 m spacing in Treatment 5. The minimum deflation of 3.1 cm at 5m spacing can expose plant roots and facilitate mortality. The accretion of 9.7 and 6.1 mm recorded at the same spacing can bury slow growing seedlings. These seem to suggest that perhaps a shorter spacing than 5 m would be more suitable. Results obtained for several other dates were variable. This may be attributed to the fact that only two sides of each plot were fenced whereas there was a need to put fences on the four sides because of the variable wind directions.

Table 4 shows the mean monthly wind speed and direction between 1990 and 1995 at Sokoto airport in the sand dune area.

Table 4 Mean monthly wind speed (10^{-1}km/h^{-1})and direction between 1990 and 1995 at Sokoto airport. (Source: Nigerian Meteorological Station, Sokoto)

Month	Wind speed	Minimum	Maximum	Predominant direction
January	341	187	443	NE - E
February	351	16	541	NE - E
March	301	200	537	NE - SE
April	263	176	311	NE - SW
May	350	267	497	S - W, S - NW
June	385	295	480	E - W
July	279	157	393	S -W
August	192	113	250	S - W
September	152	45	224	SE - W
October	116	30	169	NE -SW, NE - W
November	193	70	280	NE -SE
December	168	176	393	NE-SE

The highest wind speeds, 26.3-38.5 km/hr were recorded in the months of December to July. This period coincided with the dry season or early rain period (May, June and July). It is interesting to note that the period immediately after the rains in October recorded the lowest wind speed throughout the year. Sand fencing and repair of sand fences are best carried out during this period.

Table 4 shows that the wind direction is multidirectional and this corroborates the data on accretion and deflation obtained in the field. It was on this basis that another experiment was carried out using checker board arrangement (i.e. fencing a plot on all sides) and a shorter spacing of 3 m. Monitoring of sand movement had to be stopped abruptly for lack of funds.

Table 5: Effect of sub-soil amelioration on the mean height development (cm) of (a) neem (*Azadirachta indica*) and (b) Eucalyptus grown on sand dune soils at Gidan Kaura, Sokoto State

Treatment	Date		
	4/9/1997	3/12/1997	16/8/1998
(a)			
0 kg/m^3 Alcosorb	36.0	39.3	-
0.5 kg/m^3 Alcosorb	37.3	37.3	25
1.0 kg/m^3 Alcosorb	35.7	41.0	-
1.5. kg/m^3 Alcosorb	42.3	44.3	40.0
2.0. kg/m^3 Alcosorb	35.3	39.7	-
2.5. kg/m^3 Alcosorb	28.0	33.3	-
3.0. kg/m^3 Alcosorb	35.0	38.7	-
3:2 Sand/Manure (3:1)	36.7	43.3	-
3:1:2 Sand/Manure/Fadama	31.3	35.7	-
(b)			
0 kg/m^3 Alcosorb	59.0	65.5	65.0
0.5 kg/m^3 Alcosorb	62.7	72.5	90.0
1.0 kg/m^3 Alcosorb	58.3	68.5	65.0
1.5. kg/m^3 Alcosorb	66.0	72.0	100.0
2.0. kg/m^3 Alcosorb	69.0	84.0	100.0
2.5. kg/m^3 Alcosorb	63.0	70.0	73.3
3.0. kg/m^3 Alcosorb	75.0	82.7	95.3
3:2 Sand/Manure (3:1)	61.7	72.3	86.0
3:1:2 Sand/Manure/Fadama	55.3	76.7	105.0

Without any irrigation during the dry season, *Eucalyptus* had 100% survival over the one year period. Neem suffered 50% mortality while corrasion damaged another 37% (*Table 5*). Soil amelioration by side dressing was more effective than subsoil amelioration treatment in the short term. Percent increases in plant height due to side dressing of *Eucalyptus* and neem over a 3 month period were on the average, 27% and 13%, respectively. Corresponding values for subsoil treatments were 17% and 11%, respectively. However the placement of sand/manure/fadama soils in the subsoil produced the highest increase of 90% in the height of *Eucalyptus* within a year. Alcosorb at the rate of 2.0 kg/ m^3 placed in the subsoil also gave a fairly high percent increase of 45% after a year.

5. SUMMARY OF THE DEVELOPED TECHNOLOGY

(i) Normal potting mixture of 3:1 sand/manure is unsuitable for raising seedlings for effective sand dune fixation.
(ii) Enrichment of this mixture with fadama clay soil is recommended at the rate of 3:1:2of sand/manure/clay for raising seedlings in the nursery.
(iii) Normal pot size (20 x 5 cm) produced better plant growth than bigger pot sizes.
(iv) Soil conditioner, alcosorb, has great potentials for afforestation purposes. It produced higher water use efficiency than control. Thus, plant growth can be sustained for much longer time without rainfall or irrigation. Granular application is preferable to gel application. The most beneficial effect was on sands in terms of increased soil moisture retention, reduced evaporation and increased aggregate stability.
(v) Neem and *Eucalyptus* are recommended as economic trres for sand dune fixation because they are drought resistant, have good root system development and grow faster than most indigenous trees. Neem exhibited much higher mortality (50%) than Eucalyptus (0%) on sand dunes over a year period without irrigation. Corrosion damaged another 37% of the neem stands. Grass seeds should be broadcast at the time of planting.
(vi) Side dressing of plant root with alcosorb was better than subsoil dressing in the short run. The subsoil dressing with 3:1:2 sand/manure/fadama clay soil or using 2.0 kg/$m^3$of alcosorb was better in the long run than side dressing.
(vii) Seedlings raised on potting mixtures enriched with manure and fadama clay or 2.0 kg/m^3 of alcosorb should be "seated " on equally enriched subsoil on the sand dune for optimum crop performance.
(viii) Guinea corn stalks fabricated into mat forms are effective as sand fence materials. Spacing ranging between 3 and 5m may be more effective in reducing wind speed and curtailing accretion and deflation. Sand fences

should be arranged in a checkerboard fashion because of the multidirectional nature of the wind. Height of sand fence material (1m or 0.5 m) is not of much consequence where spacing of sand fences are appropriate.

6. CONCLUSION

There is an urgent need to embark on sand dune stabilization in northern Nigeria using an appropriate technology. The technology developed here needs to be tested on selected sand dunes in different parts of the north. There will be a need to introduce as minor crops¬ (in terms of numbers) some indigenous, economic trees such as *Isoberlina doka* and *Khaya senegalensis* normally adaptable to the environment. The allelopathic nature of Eucalyptus will call for a special arrangement of plants on dunes. Thus the technology will be fine-tuned as attempts are made to stabilize dunes in the very near future, if research grants become available. Large scale use of the soil conditioner, alcosorb, will also be tested and this may well be the panacea for afforestation and shelterbelt construction problems in northern Nigeria.

ACKNOWLEDGEMENT

The author acknowledges the collaboration with the Forestry Research Institute of Nigeria and the contributions of its staff: Dr. M.A Odeinde, Mr. A.B. Momodu, Dr. Frank Adekiya, Mr. Ogigirigi, Mr. A. Idowu and Mr. A. Kolawole. Also the funding of this project which was a research grant from the World Bank assisted Nigerian Agricultural Research Project (NARP) is hereby acknowledged.

REFERENCES

1. Anonymous (1987) *"Grow-soak 400: Polymeric soil conditioner"* 10 pp.
2. Babalola O. (1985) *"Technology for sand dune fixation in the semi-arid zone of Northern Nigeria"* pp 131- 156 In V.O.
3. Sagua et al (eds.) *"Ecological disasters in Nigeria: Drought and Desertification"*. Federal Ministry of Science and Technology, Lagos, Nigeria.
4. F.A.O.(1977)*United Nations Conference on Desertification*, Rome, Italy.
5. FORMECU (1998) *Forestry Monitoring and Evaluation Co-ordinating Unit*, Personal communication. Abuja, Nigeria
6. Hagedorn H: K. Giebner; O. Wiese; D. Busche and G. Grunert (1977). *"Dune stabilisation"* German Agency for Technical Cooperation Ltd Eschborn 194 pp

SAND DUNE STABILIZATION PRACTICES IMPLEMENTED IN TURKEY

Bahri Çevik
Çukurova University Faculty of Agriculture, Adana, Turkey, Tel/Fax 90-322-3386386/6827, email: unlu69@pamuk.cc.cu.edu.tr

1. INTRODUCTION

Wind erosion is a detrimental problem in Turkey specially in the Southern Central Anatolia which is one of the driest zones of the country. Nearly 466 000 ha of land is subject to wind erosion varying from slight to excessive, and approximately 70 % (322 474 ha) of this area is within the boundaries of the Konya province, in Karapınar which is endangered by a rapid desertification due to improper cultivation.

Removal of vegetative cover, in addition to bad cultivation practices, has triggered extensive erosion. Shifting sand dune areas, expanding large areas each year, by wind effect is wide spread. Resulting decrease in soil fertility has disturbed daily life and well beings of unfortune dwellers of Karapınar town.

Wind erosion control and sand dune stabilization practices were implemented by the Soil Water Conservation and Irrigation Service of Turkey (TOPRAKSU) during the period from 1962 to 1972. Project area of about 16 000 ha, was near Karapınar in the drainless depression area between Konya-Ereğli in Central Anatolia. The elevation of the area is 1000-1020 meters from sea level (*Figure 1*).

The aim of this paper is to present and discuss important aspect of the project which was implemented succesfully by a small group of agricultural engineers.

Combating Desertification with Plants, edited by Dov Pasternak and Arnold Schlissel.
Kluwer Academic/Plenum Publishers, New York, 2001.

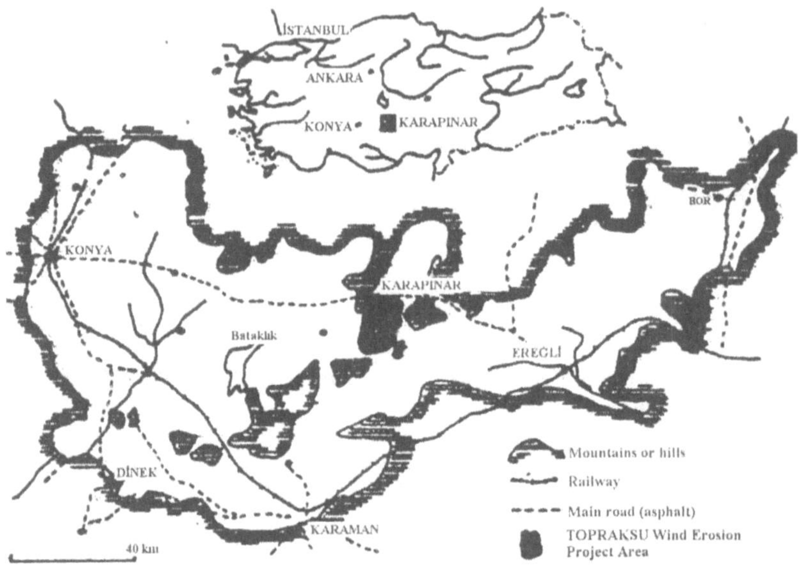

Figure 1. Situation map of the Great Konya Basin and project area

2. CAUSES OF WIND EROSION IN KARAPINAR

The reasons of wind erosion depend on four major factors:

2.1 Climate

Climate is the first factor causing wind erosion. Central Anatolia is in semiarid belt, surrounded by high mountains, causing long spells of periods.

Wind velocity, direction, frequency and duration effect soil erosion. Spring winds in Central Anatolia is very strong. At this time there is not enough vegetation on soil surface. Specially in April and May the weather is hot and dry. The dominant winds are commonly from the South-Western, mainly from South-Southwest directions. In the area, there is always 2 or 3 days of wind storms. Southerly winds take moister from the soil and increase evapotranspiration. Average velocity of wind is 20-25 m/sec, maximum velocity is about 30-35 m/sec, at this time wind carries particles of soil and covers other areas in the region. Average total precipitation is 276 mm yearly, 207.4 mm in driest years and 457.2 mm in wettest years.

Average temperature is 20.9 °C in Summer, 10.5 °C in Autumn, -2.2 °C in Winter and 9.9 °C in Spring. Evaporation rate is of an average value of 1190 mm/year. Relative humidity ranges between 40 % in summer to 80 % in winter.

2.2 Soil

Texture, structure and moisture of soil are important factors on wind erosion.

2.3 Vegetative conditions

Also unsufficient vegetation and organic matter in the soil increase erodable area and increase wind erosion risks.

2.4 Man

(a) Improper land use, (b) improper use of tillage machines, (c) intensive grazing, (d) cutting of the grasses and bushes with their roots for fuel.

3. LANDSCAPE FORMATION IN PROJECT AREA

Landscape formation is divided into four sections (Figure 2).

3.1 The Aelion Sandplain

This plain has been the bottom of the lake in Pleistocene times, occupying largest part of the project area. During the process of drying, the wind removes and shifts finer particles and sands. The vegetation covering gradually drying shores at a certain distance, traps the sand. Ancient dunes must form through this process.

But in recent times, overgrazing and human activities, specially improper land use, destroyed the vegetation and so wind eroded the old ancient dunes. The shifting sand accumulated and had built up new dunes. These "cresentic dunes or barchans" are 2 to 12 meters high, light gray or pale brown and carbonatic sand material.

3.2 The Lacustrin Plain

This plain occurs mainly in the North-Western part of the project area. The topography is level to nearly level, and the parent material is lacustrin sediment. Soils of the lacustrin plain are very deep, well drained, carbonatic, gray or light gray to white, commonly loamy or clayey. But the plain was covered with sand layer (15-25 cm thick) and small sand dunes, which were shifted from active sand dunes section.

3.3 The Volcanic Hills

They occur in the North-Eastern and South-Eastern part of the area. The largest one is in the North-Eastern part of the project area, with 1120 meters of altitute. This hill protects Karapınar town against the drifting sand.

3.4 The Limestone Platform

It occurs in the Western part of the project area, along the border line which is fenced.

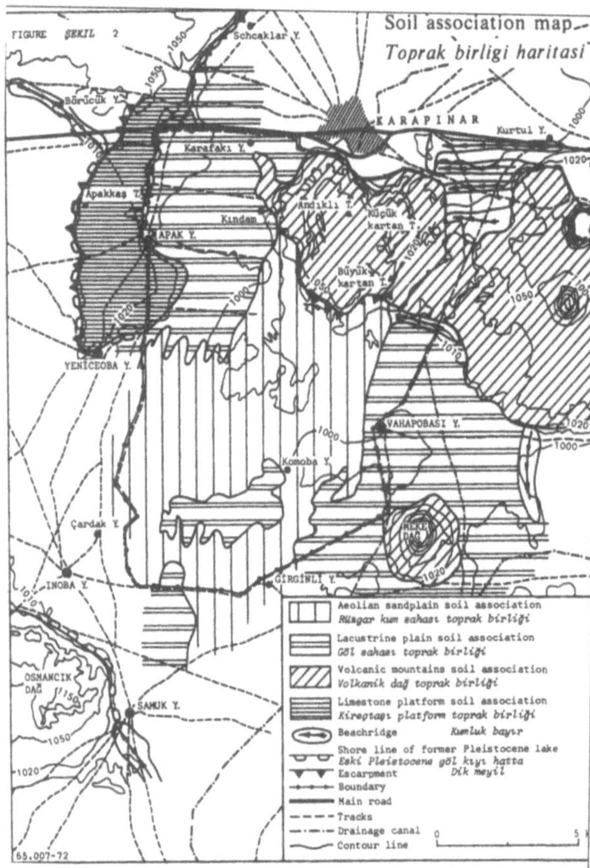

Figure 2. Soil association map

4. LAND USE AND VEGETATION

The largest part of the project area has been idle land for the last years and was in native pasture formerly. Sheeps and goats have been grazed in herds, probably for centuries. Livestock graze selectively and they remove more leaves from some plants than from others. Their selection of plants for grazing varies with the season and the degree of use. For this reason, pasture responds to grazing in different ways. Some kinds of plants decrease "decreaser", some increase "increaser", and others originally not present, may invade "invaders".

In this overgrazed area, the composition of the vegetation is determined mainly by intensive grazing sheeps and goats.

The plants are mainly increasers and invaders. *Perganum harmala* (Üzerlik) and *Marrubium parviflorum* (Tapir) which are not liked by live stock occur mainly on clay and sandy soils respectively.

To this effect, the list of scientific and local plant names are given below (Groneman, 1968):

<u>Plants living mainly on calcareous sandy soils</u>
Artemisia fragrans (Piren)
Artemisia scoparia (Piren)
Bromus tectorum (İpelek)
Ajuga chia (Şaplak)
Marrubium parviflorum (Tapir)
Salvia cryptantha (Adaçayı)
Alhagi camelorum (Yandak)
Verbascum cheiranthifolium

<u>Plants living mainly on calcareous loam and clay soils</u>
Halimione portulacoides
Achillea santolina (Bin yaprak otu)
Centaurea triumettii (Gökbaş)
Cirsium acarna
Inula
Tragopogon pratensis (Yemlik)
Convolvulus lineatus (Kahkaha çiçeği)
Isatis tinctoria (Deli sarı)
Sinapis arvensis (Hardal)
Euphorbia tinctoria (Sütleğen)
Phlomis ameniaca
Teucrium orientale (Alanotu or Hazeran)
Alhagi cameloru[1](Yandak)
Onobrychis sativa
Glaucium corniculatum (Gelincik)
Melilotus officinalis

[1] Occurs on sandy, loamy and clay soils.

Adonis flammea (Kan damlası çiçeği)
Reseda lutea
Verbascum ballsianum
Verbascum lasianthum (Sığır kuyruğu)
Zozimia absinthifolia (Çörtük)
Perganum harmala (Üzerlik)

5. PRELIMINARY STUDIES AND IMPLEMENTATION OF THE PROJECT

5.1 Preliminary Studies

- Determination of climate conditions: Collecting data about temperature, precipitation, evaporation, wind direction and wind speed.
- Soil survey and mapping: Detailed taxonomic soil maps and land use capability maps of the project and surrounding area were prepared.
- Hydrogeolical surveys of the area: Collecting data for the ground water resources to utilize for irrigation.
- Planning for the implementations in the problem areas.

5.2 Implementation

The application of the project, and the actual field works were implemented in the following order:

(1) The project area was fenced in by barbed wire. Cultivation of the land, and collection of plant roots for fuel by villagers were prohibited, and the domestic animals were kept out of the area.

(2) According to the hydro-geological studies, 32 deep wells were bored, and water for irrigation was pumped from 60 to 70 meter depths.

(3) Project area was divided into four sections, and different sand stabilization and erosion control practices carried out on these sections.

6. WIND EROSION CONTROL PRACTICES OF THE PROJECT AREA

Four different sections were idendified according to topography and the prevailing erosion hazard type, and different erosion control practices were implemented in these sections.

6.1 Wind erosion control practices on sand dunes

This section was about 4300 ha large, and lied in an area 7 km from Karapınar and extended in South-West direction. The parent soil material consisted of pleistocene lake sediments. After a long time of extensive grazing and digging the roots of the plants for fuel, the area became unprotected against the winds and the area was covered by sand dunes (barchans) some of which were as large as 150 by 240 meters wide, and 40 meters high.

The most important task of the project was to stabilize these dunes. To achieve this, mechanical and vegetative control practices were used together. In order to initiate any kind of crop cover on sand dunes, first, semi-permeable "reed windbreak curtains" were constructed. Locally available and quite abundant reeds were cut to the size and woven together with wires.

Two types of reed curtains were used. One type, woven 2 m long and 2 m high, tied together to the 0.05 x 0.10 x 2.00 m wooden poles which were erected on every two meters. The bottom half meter were buried into the sand, and the curtains were about 1.50 m high from the surface. The other type of curtain used was 3.0 meter wide and 1.5 m high, and tied to the 0.05 x 0.04 x 1.50 m wooden poles, seperated by 3 meters. Both type of reed barriers gave about 50 % porosity. The second type was found to be more economic and easier to construct. The windbreak curtains run perpendicular to the prevailing wind direction at intervals of 6-10-12-20 meters with distance being determined according to the downhill or uphill slope of the sand dunes.

After the construction of windbreak curtains, rows between the barriers were seeded. The crops used for vegetative cover were given in *Table 1*. The most promising of them was found to be rye. Rye adapted to the harsh conditions easily, its vegetative development was fast and gave good strong, tall stands.

Table 1. Annual and perennial grasses used vegetative cover on sand dunes

In English	Scientific name
Rye	*Secale cereale and Secale montanum*
Sorghum	*Sorghum Halepense*
Sudan Grass	*Sorghum Sudanense*
Wheat Grass	*Agropyron Cristatum and Agr. Elongatum*
White flowered sweet clover	*Melilotus Alba*
Yellow Flowered Sweet Clover	*Melilotus Officinalis*
Sainfoin	*Onobrychis Sativa*
Common vetch	*Vicia Sativa*
Hairy Vetch	*Vicia Villose*

In the same year, seedlings of some trees (*Table 2*) were planted into initially vegetated rows. They were known to be growing in the climate conditions of Central Anatolia. The seedling were irrigated in the first year of establisment with the water pumped from the deep wells by water tanks.

Table 2. Trees and bushes used tree windbreaks on sand dunes

In English	Scientific name
Wild Olive or Russian Olive	*Elaegnus*
Accacia	*Accacia*
Honey Locust	*Gleditchia*
Alder	*Acer*
White Ash	*Fraxinus*
Tamarix	*Tamarix*
Poplar	*Populus*
Villow	*Salix*
Pine	*Pinus Nigra and P.Silvestris*

In about 3 to 5 years the grass and trees have developed to such extend that the reed windbreaks, which detoriated and collapsed were no longer needed.

6.2 Small dunes or miniature barchans area

These miniature crescentic sand dunes or barchans, which measure 0.5 to 2 m across and 50 to 60 cm high, covered 9 700 ha in the project area. They were formed when the windblown sand was immobilized by a clump of native plants or small bushes.

List of activities implemented in area of small dunes:

(a) First of all land was ploughed by tractors for cutting into the dunes.

(b) Establishment of vineyards and orchards, and improvement of any existing pastures on some of the unirrigated parts of the area.

(c) Preparation of the land near deep wells for irrigation by subdividing according to water supply and by land leveling.

(d) Practice of stripe cropping under dry farming conditions was started and extended each year to cover 1 000 ha planted to *gramineae* (rye or wheat) and fallowed or 1 000 ha planted to *gramineae* and annual or perennial *leguminosae*. Stripes were tried in 40-50 meters width to the direction of prevailing winds.

On this part of the project area, it was possible to grow any crop which was adapted to the Central Anatolian conditions; and soon this area became the seed and seedling producing centre for the whole project area.

6.3 Highly Erodable Flat Areas

These areas which were covered by 15-20 cm thick fine material blown from sand dune areas, extended over some 500 ha of land along the Konya-Adana highway. Due to the absence of plant cover, and their suitability to wind erosion, these areas were constituted a source of hazard for the traffic on the nearby highway. The wind erosion on these areas were controlled fully by planting perennial grasses on strips perpendicular to the direction of prevailing winds.

6.4 Volcanic Hill Area

These areas which were underlaid by basaltic formations, covered 1 500 ha of area extending from the vicinity of Karapınar town to the Northern edge of the sand dune area. The soils of this area were partly covered with sand transported from dune plains by wind. The project activities on this area, which was covered by forest in the recent past, consisted of re-establishment of oak woods (*Quercus*) and almond (*Prunus amygdalus*) orchards, in dry conditions.

The implementation of the project in this area, especially the construction of deep weels, were set on good example for the local farmers. Afterwards more than 3 000-3 500 deep weels were opened by the farmers and irrigation of some 30 000-40 000 ha irrigated land was realised in agricultural lands of Karapınar.

7. CONCLUSIONS

Today the Karapınar province displays and signifies the results of joint efforts of a small group of dedicated agricultural engineers and stands as a living monument of victory against wind erosion. The success of the over all project proves not only that the control of wind erosion is possible, but also it can be done inexpensively if man puts his mind into it.

It is obvious that, the total investment in monetary terms is not big, and in no way comparable to the efforts of the man who put their minds and muscles into work. And, greatest them all is the joy of bringing the life back onto a piece of land which you call home.

The project staff, agricultural engineers only had desire and the will to combat the wind erosion which was turning their mother land into desert. Unbearable working conditions, limited funds, and lack of scientific information caused a great deal of hardship; but at the end they succeeded, and turned the desert into a garden of orchards and green fields; utilizing their imagination, and try and error techniques.

The benefits obtained with the project:

- Prevented to move people of Karapınar to some other location.
- Increased value of land.
- Agricultural structure developed by means of pumping irrigation and increased net income of the farmers.
- Project area, presently, is a national park; it is a recreation area for the amusement, holiday and hunting pleasure of the people from nearby provinces.
- And turned the desert into a garden.

REFERENCES

1. Çevik, B. and Berkman, A., 1987. "*Wind Erosion Control and Sand Dune Stabilization. Practice Implemented in the Great Konya Basin of Turkey*", Eremology (Desert Sciences), Proceedings of a Post-Graduate Course Held in Ghent, Belgium, (31 August-25 September, 1987).
2. Groneman, A.F., 1968. "*The Soils of the Wind Erosion Control Camp Area Karapınar, Turkey*". Agr.University Wageningen the Nederlands.
3. Topraksu, 1975. "*Activities of Wind Erosion Control in Karapınar*", Ministry of Rural Affairs, Soil Water Conservation and Irrigation Area Office, Number 44/1, Konya.

DESERTIFICATION IN THE WEST BANK AND GAZA STRIP

Present Status and Future Aspirations

Issam Nofal and Tahseen Barakat
Ministry of Agriculture, Ramallah, Palestinian National Authority, Tel: 972-22961081/8, Fax: 972-22961212

1. BACKGROUND

Soil deterioration, which gradually causes desertification, is one of the greatest disasters still threatening human communities. Scientifically, desertification is expressed as full or partial deterioration of soil occurring as a result of the unsustainable utilization by man of his own natural resources, which in turn causes imbalance of the elements of the earth and of the environmental system. This process results in damage to those properties of the soil that bear a direct relation to its productivity. Decreasing soil productivity entrains conversion to desert-like uncultivated land. In this situation, natural productivity does not meet the minimum requirements for the survival of man and his livestock, and imports from other regions become a necessity.

Soil deterioration that causes desertification is considered to be one of the dynamic phenomena which occurred throughout history as a direct result of human mismanagement of his natural resources. Other factors, especially drought and demographic development, also contribute to deterioration. The salient point is that desertification has spread very quickly in the last decades. In this respect, the situation in the West Bank and Gaza does not differ much from the situation in the neighboring countries. However, there are important elements in the Palestinian case which differ from other countries. First, there is the inability of the Palestinians to govern their natural resources (water and land) over the last three decades because of the Israeli occupation. The second is that, even after the peace agreements signed

Combating Desertification with Plants, edited by Dov Pasternak and Arnold Schlissel.
Kluwer Academic/Plenum Publishers, New York, 2001.

beween Israelis and Palestinians, considerable confusion and contradictions as regards shared responsibilities impede implementation of the remedial measures proposed by the Palestinians to alleviate soil deterioration phenomena.

In the Palestinian case, we see that we have to confront desertification at two levels. The first is the national level, which implies optimal utilization of our natural resources (land and water) through managing them freely to take the needed actions, namely land and water reclamation or water harvesting projects, as well as reforestation and rangeland and natural reserves development. Second is the regional level, which implies regional cooperation with neighboring countries in alleviating desertification, a problem that is threatening all the region regardless of borders.

2. CAUSES AND ASPECTS OF DESERTIFICATION IN THE WEST BANK AND GAZA

2.1 Water Resources Deterioration

Water is considered one of the limiting factors for production in arid and semiarid areas. The unsustainable utilization of water resources has caused a noticeable decline of water in many regions in this world. In Palestine, different factors negatively affect water resources, causing water deterioration, which can take many aspects, as discussed below.

2.1.1 Exhaustion of Ground Water Aquifers

Mismanagement of ground water aquifers is one of the factors that negatively affects agricultural production and sometimes results in land degradation phenomena. Some of the ground water aquifers that have a shallow water table (near the ground surface), like the coastal aquifer in Gaza, have been heavily invested since the early fifties, in a way that does not suit the storage capacity and the replenishment of that aquifer.

2.1.2 Salinization of Water

Fresh and saline water co-exist in a natural balance in many aquifer systems, with saline water occurring in the lower part of these aquifers, and a hydrostatic balance prevailing between fresh and saline water. According to Ghyben-Herzberg relation, however, this balance may occur with seawater. Over-pumping of the aquifer causes discordance of the hydrostatic balance between fresh and saline water, with saline water rushing towards the fresh water and increasing its salinity. Recent studies show that salinization and seawater intrusion phenomena are explicitly present in the coastal aquifer in Gaza. They also show that salinization is increasing in the eastern aquifer in the Jordan Rift Valley in the West Bank.

2.1.3 Water Scarcity

The deficiency in water quantity and quality is considered one of the main reasons behind soil deterioration and desertification. Water scarcity and inaccessibility to water resources are very clear in the West Bank and Gaza. In spite of the severe water problem in the coastal aquifer in Gaza, settlements are abstracting a significant amount of fresh water for use in irrigation, which means banning Palestinians from using a part of their fresh water to the benefit of people in settlements.

In the West Bank, since the early days of the occupation in 1967, 140 pumping stations that were abstracting water from the Jordan River and irrigating thousands of dunams in the Jordan Valley were damaged. Thousands of irrigated dunams in the Jordan Valley were confiscated under what is called "closed military zones" to be converted to desert land. Indeed, restrictions imposed on abstraction of ground water in the West Bank prevent the development of new water projects and make it difficult to develop the old ones. It is very clear that in the absence of sovereignty by the Palestinians over their natural resources, a sound and integrated development plan for land and water cannot be implemented. This is of great importance in combating desertification.

2.2 Soil Deterioration and Desertification

Direct deterioration is that produced by irrational human practices in agricultural land, which minimizes the productivity of land and has a direct bearing on human life and stability in the region. This definition of deterioration does not single out overgrazing, deforestation, or lack of water availability. However, all of these processes are related to each other. In Palestine, soil and land deterioration related to human activities could be attributed to two principal classes of factors, as detailed below.

The first class comprises four factors:

- Deterioration as a result of transforming or transferring of soil from one place to another, with negative effects on the original and the new site.
- In-situ deterioration of the soil profile as a result of either mismanagement or misutilization of the soil, e.g. salinization, and deterioration of physical, chemical and fertility properties of the soil.
- Soil and land degradation as a result of plant cover removal, deterioration of rangeland, forests and water resources.

- Quarries: In the last few years, stone crushers and quarries are spreaded rapidly in many localities of the West Bank, causing soil deterioration and removal of the plant cover and negatively affecting human life and plants.

The first of these four factors has a minor effect on the Palestinians compared with the second and the third. The latter prevail in large parts of the West Bank and Gaza, especially salinization problems, degradation of soil fertility, deterioration of rangelands, forests and water resources.

The second class of factors is related to Israeli practices and activities in the West Bank and Gaza since 1967. Restrictions have been imposed on land use, which are strongly interrelated with plant cover removal and consequently desertification. Those activities are summarized below.

2.2.1 Land Confiscation

For different reasons, Israelis have confiscated large areas in the West Bank and Gaza Strip, imposing restrictions on land use in these areas. The total area of confiscated land constitutes 35.8% of the West Bank and Gaza Strip area, and it is distributed over all the agricultural zones.

2.2.2 Settlements

The built-up area of the Israeli settlements constitutes about 3.9% of the West Bank and Gaza Strip. They occupy strategic sites that, for several reasons, lead to the exhaustion of the nearby lands. They affect negatively the continuity and integrity of the land, and some of them are of agricultural nature, consuming large amounts of water and negatively affecting the availability of water for the Palestinians.

2.2.3 Bypass Roads

These roads were established at the expense of about 2.3% of the area of the West Bank and Gaza Strip. These roads are splitting Palestinian lands into isolated entities, and imposing restrictions on land use in the adjacent areas.

2.2.4 Closed Military Areas

Thousands of irrigated dunams were converted to desert land in the Jordan Valley. Closed military areas constitute about 20.94% of the West Bank and Gaza Strip. Closed areas are concentrated in the Jordan Valley and in the Eastern Heights.

2.3 Forest Degradation

Prior to the establishment of the Palestinian National Authority (PNA) and Palestinian Ministry of Agriculture in 1994, no significant interest in forest preservation was witnessed. Forest area in the West Bank and Gaza strip decreased from about 343,560 dunam in 1970 to 125,100 dunam in 1995 (64% of the original area) as a result of cutting trees, fires, plantation discontinuity, and erection of settlements in forest areas. The above factors have led to decrease of the plant cover and to soil erosion and transportation.

2.4 Rangeland Degradation

Prior to 1967, natural rangeland formed more than 25% of the West Bank and Gaza area. Most of the natural rangeland is concentrated on the eastern slopes of the West Bank, with annual rainfall ranging from 100 up to 300 mm/annum. Large parts of the rangeland (about 1.5 million dunam) are not accessible for grazing as a result of land confiscation and closure. The decrease in area has caused over-grazing of the available rangeland, which in time caused plant cover removal and soil erosion. Other factors that have contributed to rangeland deterioration in the last thirty years are:

- Over-grazing, cutting trees and fires.
- Agricultural expansion in the marginal areas and urbanization.
- Mismanagement and mis-development of rangeland.
- Constraints and difficulties facing execution of development projects, as a large part of the West Bank land is not under Palestinian sovereignty.

3. PROGRAMS OF THE MINISTRY OF AGRICULTURE (MOA) TO COMBAT DESERTIFICATION

After the establishment of the Palestinian National Authority (PNA) in1994, and the establishment of the MoA, a general natural resources directorate was formed to include water, soil, forest and rangeland departments. Seventy-two employees from different fields and backgrounds are working under the natural resources directorate. They established new nurseries to meet the need for forest and range seedlings. Six public nurseries were constructed in the last five years, which resulted in maximizing tens of old forest seedling production units. Water, which is a valuable resource, is a central consideration in the development plan of MoA, PNA and NGO's in Palestine. Rehabilitation of water resources - including springs and wells -

has priority in the development program. Water harvesting projects, construction of wastewater treatment plants and networks, pilot projects and researches are found in the development plan, either in an on-going phase or planned.

For the time being, MoA is executing different projects aiming at combating desertification. Among them are:

- Greening Palestine Project
- One Million Tree Project
- Land Reclamation Project
- Reclamation of the Eastern Slopes of the West Bank (Water Harvesting).
- Researches and pilot projects.

REFERENCES

1. Ministry of Agriculture
2. Agricultural Policy (1999).
3. Water and agriculture (Arabic copy), ACSAD, Volume 17, (1997).

REVERSING DESERTIFICATION THROUGH LARGE, HERDING UNGULATES

Shannon A. Horst
Center for Holistic management, 1010 Tijeras, NW Albuquerque, NM 87102, USA, Tel: 01-505-8425252, Fax: 01-505-8437900, email: shannonh@holisticmanagement.org

1. INTRODUCTION

The Allan Savory Center for Holistic Management helps people learn to make decisions that are simultaneously financially, environmentally, and socially sound. This decision-making process helps people - from all walks of life - establish first what is important to them and then figure out how they will produce it without depleting their own resource base. It has worked with people throughout the world to help them restore the loss of biological diversity that leads to desertification.

This paper is based on the work of Allan Savory, a wildlife biologist from Zimbabwe (formerly Rhodesia). It is based on the work of hundreds of people, most of whom have limited formal education - at the most bachelors' degrees. Essentially, it is the work of lay people.

Underlying the work that the Center does with people who earn their income directly from the primary resources (farmers, mining companies, and conservationists) is the Holistic Management decision-making framework. This work includes trying to understand, and to simulate, how the natural whole might have functioned before humans entered the picture. How did animals, plants, weather, water and other things function and interact.

Combating Desertification with Plants, edited by Dov Pasternak and Arnold Schlissel.
Kluwer Academic/Plenum Publishers, New York, 2001.

2. THE TECHNOLOGY

The seasonal rainfall environments of the world, which represent 2/3 of the earth's surface, once supported a mass of large, herding and grazing animals. On the North American continent, for example, at least 80% of the "genera" of large mammal life were wiped out thousands of years ago by early hunters. On the Australian continent, aboriginal peoples hunted to decimation, among others, at least eight species of "hoofed" kangaroo.

The perennial grasslands of the world, which are found in the seasonal rainfall environments, co-evolved over millions of years with large (and small) hoofed graziers. These animals played the role of cycling the mass of vegetation that grew each year back into the ground - through grazing, trampling, dunging and urinating.

In the Center's work with families who work livestock in these environments, we begin with helping them determine what they want as a quality of life in the long run, what they will have to produce to have that and what their resource base will have to look and function like to sustain that over a very long time. These families, generally, describe a landscape that is healthy - a variety of species, abundant water, covered soil, and clean air.

To produce a landscape such as this, we help these families learn to manage their animals in a different way. We use the animals to actually produce healthy grasslands, covered soil, and effective water cycles, mineral cycles and energy flow. To do this, we change the way the animals are managed. This requires keeping the animals tightly bunched and at high densities for short periods of time on any given area.

During the last thirty years, wherever we have been able to help people focus on *what they want* and how the land will have to function, these people have achieved some pretty great results in:

- Improved soil cover and water cycle;
- Improved mineral cycle and energy flow (the photosynthesis process);
- Improved productivity per acre;
- Increased profitability.

The series of slides are offered to highlight some of our successes. These include:

- A fenceline contract slide from Lebowa showing one side where animal behavior, density and recovery periods for the plants have been changed and grasses are beginning to return;
- A site in Montana where the animals were gathered once a year while the surrounding area were grazed as they had been for many years; the gathering area shows dense grass; the surrounding range is denuded;

- A site in Texas where the family was able to produce tall, lush grasslands even through a serious drought; this family doubled the stocking rate;

On some advanced sites, animals are being used to reclaim mine sites; the tailings are a substance somewhat like powder or sand and can be biologically dead or have some mix of acidity, phosphorous, or other minerals and chemicals; research plots have shown that feeding the animals on the slopes at high densities (unique to each site) for short periods have produced stable slopes - able to withstand flood-stage waters of 14 " per minute without soil/tailings movement. After the animals have fed and moved on, the slopes are covered in mulch, dung and urine left by the animals. The animals have been known to incorporate the biomass (leftover from feeding) as deep as 18". The slopes begin to cover, sometimes drawing on moisture in the tailings along with the moisture from the urine, with grass within 3-6 months; earlier if rains come.

When considering this technology, one may use any hoofed animal - cattle, sheep, goats, bison, etc.

About 10 years ago, farmers in the midwestern US began putting animals out on the stubble that was left after they had harvested their crops. But, this work is not similar. In such settings, farmers are not putting the animals at high density, so stubble is generally not ground in, and the animals are left over long periods such that any naturally occurring grasses will potentially be grazed and re-grazed.

In some of the most advanced work tied to the Center for Holistic Management, farmers are now growing annual crops in the midst of perennial grasslands. The crop produces good income, the grass keeps the soil covered and weeds down, and the animals produce additional income. By using a robust grazing planning process, developed by Allan Savory, farmers (or pastoralists) can plan for:

- The possibility of drought;
- Areas needing specific treatments, such as gullies or weed outbreaks;
- Areas needing to remain ungrazed and trampled because of the presence of nesting birds or other wildlife, tree seedlings, etc;
- Specific conditions for the cattle and offspring;
- Problem areas where noxious weeds or other dangers might result in harm or death.

The theories behind this use of animals to improve deserts and rangeland conditions is based on the work of Allan Savory. A Rhodesian who was exiled in the latter years of his country's civil war, Mr. Savory is a biologist. Many years spent in the "bush" as Zimbabwe's youngest civil servant managing major game areas, gave Savory ample time to observe the behaviour of big game. It also gave him time to watch the continued deterioration of its habitat – despite the best scientific effort's to understand what was causing the deterioration and changes made, including shooting elephants, to save the game. Later, drawing on the work of Jan Smuts in "Holism and Evolution" and Andre Voisin, in "Rational Grazing", Mr. Savory began to re-think the scientific theories of the day concerning grazing and grasslands.

By the 1960s, Mr. Savory had concluded that perennial grasslands cannot survive without large, herding and trampling ungulates. And, he was employed as a consultant throughout the world to help private land owners use his emerging theories concerning the need to simulate, with domestic stock, the bunching and trampling behavior of wild graziers.

Desertification is a symptom of the loss of biological diversity. And, it is occuring all over the world even in areas that are considered quite wet and lush. Long before wind and rain begin to move, soil, plants, animals and micro-organisms have already disappeared. To reverse desertification, then, the solutions lie in those technologies, activities and efforts that produce or re-build biological diversity. The most important aspect to building up biological diversity and reversing desertification is getting the soil covered, so that water will cycle, minerals will cycle, and plants can maximize solar energy flow.

In seasonal rainfall environments, the large herding, grazing animals once played a crucial role in: cycling the minerals of the once per year build of vegetation, breaking the earth's surface so that seedlings can germinate, providing fertilizer in the form of dung and urine, and grazing perennial grasslands so that sunlight can get to their growth points, which are at the base of the plant. Humans, long ago, messed up the functioning relationship between graziers and grasslands by removing the pack-hunting predator and bringing in domestic livestock. Domestic stock do not behave in the same way that these wild graziers once did.

When perennial grasslands are either completely rested or lightly continuously grazed by wildlife and/or domestic animals the biological whole breaks down. Some grasses get repeatedly grazed, thus, being "overgrazed." Other grasses are not grazed at all and slowly "oxidize" because sunlight cannot get to their growth points. When this occurs, grass plants begin to die and ground spacing opens up. Once spacing opens up, weedy and woody vegetation begins to move in but does not cover the ground, grasses continue to die, the surface crusts and seedlings cannot get established, water sheds and soil begins to move.

A series of slides shown at this point, indicated that:

- Areas that had been rested (stock removed entirely) for as many as 60 years, have not recovered their grasslands; and, there is no perceptible difference between those areas and areas just across the fenceline where livestock has continued to graze lightly throughout the year;

- In season rainfall environments where there is a high amount of precipitation (30-60"), there is a mass build up of vegetation. Without the animals, there is no way to cycle the minerals and bio-mass back into the ground.
- Lands that have reached a stage of desert-like conditions can be returned to perennial grasslands using animals. Removing animals will not restore these lands, and managing the animals in the way they currently are managed will further destroy them.

3. SOCIAL AND CULTURAL RAMIFICATIONS AND CHALLENGES

In instances where the biological deterioration has been reversed and financial data was being kept, we have consistently seen that productivity and earnings go up. We have also seen that costs go down as animals become healthier and less vet expenses are incurred and as supplements are no longer needed because animals are getting all of their forage needs off the range. In one case, a farm that was able to support only one family is now supporting three - and at a higher level of prosperity than the one family was experiencing.

No data has been consistently collected from communal lands. However, long-term work with the Navajo Nation communal lands in northwestern New Mexico has made it possible for Navajo families to make it through severe drought better than any other families in the tribal area. Additionally, communal government lands in western Colorado that were once a battle ground between ranchers and environmentalists, are now being actively managed, through Holistic Management, by both groups (together) and have begun to recover.

Like many alternative approaches and technologies, access to training and adoption of the methodology are the largest challenges faced by those who wish to use this technology. And, the grazing technology alone will not effect long lasting change.

Our experience, like that of many other organizations, shows that, until a farm family, a lone pastoralist, or a tribal group have decided that their very existence is tied to the health of their land and that their own actions may be responsible for the deteriorating health, they will generally not consider new approaches. And, until they have created a vision of what they want that is beyond their current desperate conditions, they are virtually unable to effect the changes needed to move out of their current conditions. People, it has been proven, need to believe that it is in their own best self-interest to make the changes. Then, they need to have the knowledge and skills to implement what needs to be done.

In the Holistic Management process, individuals (in both the developed and developing worlds) who adopt some simplified version of bunching and moving their animals - without first identifying **what they are doing and why** - have experienced little success with the change in animal management. This is not surprising. Technologies, external inputs and new approaches will not, alone, bring about the changes needed to reverse desertification around the world.

REFERENCES

Allan Savory and Jody Butterfield, *Holistic Management: A New Framework for Decision Making,* Island Press, Washington, DC 1999

Part IV

MECHANISMS OF PLANT TRANSFER

PASSIVE TRANSFER
The Eden Experience in Niger

Arne Victor Garvi
Eden Foundation, BP 174, Zinder, Niger Republic, Tel: 227-510601, Fax: 227-510263, email: josef@eden-foundation.org

1. INTRODUCTION

Aid is like business; the only difference is that the customer doesn't need to be satisfied. Projects are often designed in order to obtain donor money without regards to what the recipient really needs. Reports are written after a project period of a couple of years, and if everything looks good until then, the project is ok even if everything collapses just after it has pulled out.

Sustainability is found in the hearts of the local population, their motivation being our biggest asset. We need to tap into their motivation, offer a product that meets their need, and have mechanisms like with a customer. He must show initiative, be free to choose and be satisfied with the product. Goals of a donor need to be in harmony with goals of a project that in turn needs to be in harmony with goals of a recipient. In order for this to work one has to start with the goals of the recipient.

2. EDEN PROJECT

2.1 Background

Eden Foundation was founded in 1985 on the basis of the belief that there were effective means available for fulfilling the goals of the people living in the arid areas of North West Africa. The means were drought tolerant edible

Combating Desertification with Plants, edited by Dov Pasternak and Arnold Schlissel.
Kluwer Academic/Plenum Publishers, New York, 2001.

perennial plants that could potentially grow in such a harsh environment without artificial support.

A project was designed based on farmers' goals with plants being the effective means for fulfilling these goals. Eden's purpose was to support the farmers in fulfilling their goals.

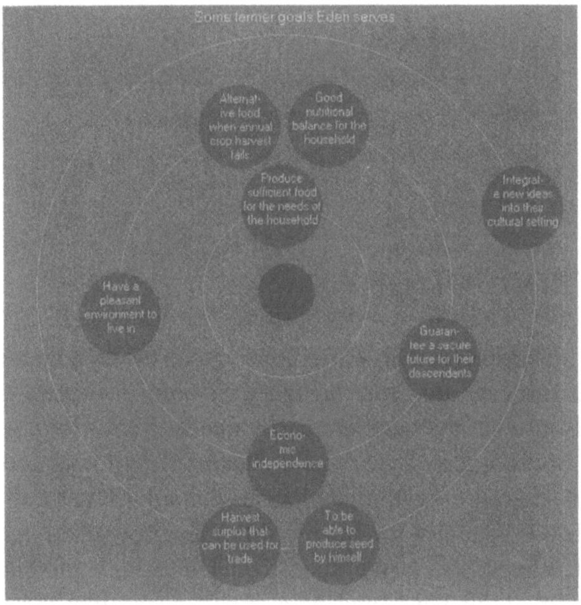

This image was created with Goalscreen and has been used with the kind permission of Unilogo Invest AAB, Sweden, www.unilogo.se

It was decided to set up a project in Niger, which Eden considered the most needy country in the target area. It was seen as important to work for farmers' food self-sufficiency, to allow them to live within their means without dependency on cash input or loans, and to use resources which could thrive without 'crutches'.

2.2 The plants

Drought tolerant edible perennial plants were chosen, as they would fulfil several goals these farmers have. There were reports saying that a lot of useful drought tolerant species existed, and that many of them were little researched and underexploited. The species should be natural, not hybrids nor superspecies, but with great genetic variety for their resistance and the farmers should be able to reproduce them from seeds themselves.

A database was set up to register information about any promising drought tolerant species from around the globe. From this information it was decided which species could be established in such an arid area. By 1991 data on 282 species considered to be sufficiently drought tolerant was compiled.

2.3 Direct seeding

The technique chosen for the establishment of the plants was direct seeding which was seen as the natural choice for such a dry environment. It meant that the farmers could direct seed them at the exact location where they wanted them, and was based on the assumption that they would not be watered.

2.4 The field station

In 1988, a 20 hectare field station, almost totally denuded of vegetation was set up in a 200 mm rainfall area, 13 km south of Tanout, Niger, in order to conduct direct seeding experiments, and hopefully arouse the interest of surrounding farmers if successful.

The field station in 1987

The area was chosen to resemble the difficult conditions for the farmers, and had formerly been used as a millet field. A fence was completed around the field station in 1989. 341 direct seeding experiments were carried out on 34 species from 1988 – 1990, with 360 seeds for each experiment. The main variables were non-toxic seed treatments and sowing depths. Neither irrigation nor fertilisers were applied. A completely randomised design was used. Seeds of the most promising species were available for distribution, at no cost, to interested farmers.

2.5 The seeds

Most of the seeds were collected by Eden, but some local and exotic species were purchased. Genetic variation was sought when collecting seeds so collection was made from several mother trees in different areas. Seeds where given lot numbers, so the origin could be traced. Deseeding techniques were developed for each species. Seeds vulnerable to insect attacks were treated with non-toxic Neem oil. They were stored in room temperature to simulate the storage capacity of a farmer, and lab germination tests were carried out to check their viability.

3. DIRECT SEEDING METHOD

A method for direct seeding for the farmers was developed. Seed envelopes with advice as symbols (icons) were produced as most of the farmers were illiterate. Seed quantities in each envelope were calculated so that one plant would reach maturity from each envelope. The tools that would be recommended for use were tools the farmers already had.

3.1 Extension method

From the very beginning of the project the intention was not to approach the farmers, but have them contact Eden because of their own curiosity and initiative. Eden would let the farmers observe the field station, which acted as the shop window to the farmers.

It was estimated that it might take up to 10 years for the first farmer to approach the project. If they did not, then the project would be considered a failure and be closed down. It was not envisaged to contact the farmers to convince them. They would have to see for themselves that the approach might be appropriate for them, and implant it in their culture. The approach should be contagious in that once adopted other farmers would do likewise. The farmers were to direct seed on their own farms, so that someone would be responsible for the plants.

Eden would never approach a village without first being invited by a farmer. The first farmer in a new village who had either seen the field station or the direct seeded plants of another farmer would be told after he had contacted Eden that he could ask others in his village if they were interested in direct seeding. When a farmer placed his order for seeds he would be free to order any species, but then Eden would explain what was available. The reason for this was to register what species the farmers really preferred irrespective of what the project offered. Farmers who had adopted direct seeding would be paid an annual visit as long as they wanted to continue, where their results would be monitored and new seed orders taken.

No expatriates would be involved with the farmers in extension work. Local extension workers were chosen on the criteria that they would respect the farmers in a manner that they saw themselves as serving the farmer. They were talented, but not highly educated and were trained by the project. They would often travel by local transport or foot when visiting farmers so as not to distance themselves from the farmers too much. Eden's logo was used extensively for 'brand' recognition on the field station sign, project cars, extension workers' shirts and seed envelopes. The farmers would be trained by an extension worker in direct seeding, but they themselves would choose the locations and sowing patterns (i.e. intercropped, forming hedges, etc.) in their own fields.

4. REGISTRATION OF RESULTS

On-farm research was done by monitoring all the farmers' results yearly, as well as other pertinent information. The farm was seen as the outer field station and the farmers as receiving free seeds as a compensation for participating in this research. In this way survivability under actual circumstances could be calculated. Their results were interrelated with the results from the field station and adjustments based on this information were made for the next season.

4.1 The farmers

Before Eden arrived the farmers in the area did not know how to direct seed. Their staple crop was millet. Their area was lacking vegetation, but they did not know of an effective way to establish new perennials. Erosion was a great problem, but many welcomed it as a way to get soft sand on top of areas of harder red soil so their millet would grow better. Many were hostile to trees as birds that ate their crops nested in them. Some did not like bushes, as the snakes would hide there. Neither did they know which species would be best to establish. Most of them would slash and burn their fields yearly, leaving virtually no vegetative cover for long periods during the dry season, when strong winds eroded their fields.

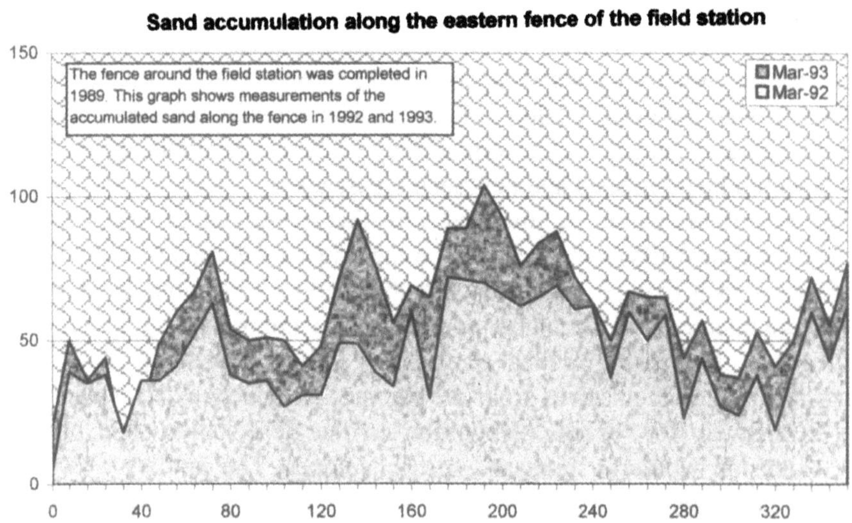

4.2 Results

The farmers did not understand at all what the project was about in the beginning. Their comments were that this was not a real project as there were no lorries and equipment. Rumours started to spread, but already in 1991 the first farmer asked for seeds. He saw an exotic plant growing at the field station, and got it confirmed that we didn't water it and that it produced something edible. A total of 9 farmers from the 2 neighbouring villages ordered seeds that year.

One thing that made direct seeding catch on was that in 1992, Moussa, the farmer to the east of the field station, got the biggest harvest in a 20 km radius. The field station had by that time already quite a lot of revegetation from natural revegetation that had occurred from rootstocks that had formerly been cut down yearly and annuals that were not burned down in winter anymore. Moussa's field was therefore protected from the worst winds by the field station, but the farmer on the opposite unprotected side did not get a good harvest. He had to resow his millet plants several times because of the strong winds, while Moussa only had to resow those millet plants further away from the field station.

Farmers who heard about this were convinced that Moussa got his bumper crop because the vegetation of the field station protected his field.

From then on many farmers became friends of trees instead of enemies, as one farmer put it.

Moussa's field that was protected by the field station

Further away, that was not protected from the wind

The first farmers had viewed the plants as something exotic and had sowed them in their gardens, but now they began to see the plants as beneficial for reducing the strong winds in the fields that adversely affected the millet plants. It became clear to them that the benefits of the perennials (i.e.windbreaks) outweighed the disadvantages (i.e.nesting place) and many started to direct seed and in turn convinced others, who then contacted the project. Since then the number of farmers and villages increased yearly, reaching 1,514 farmers from 96 villages in 1999.

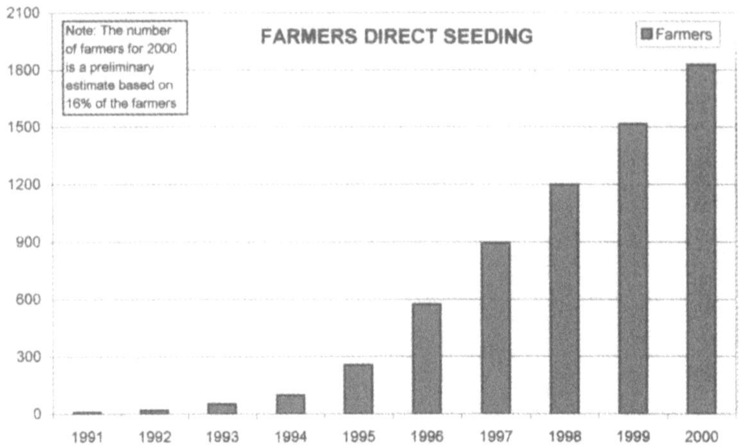

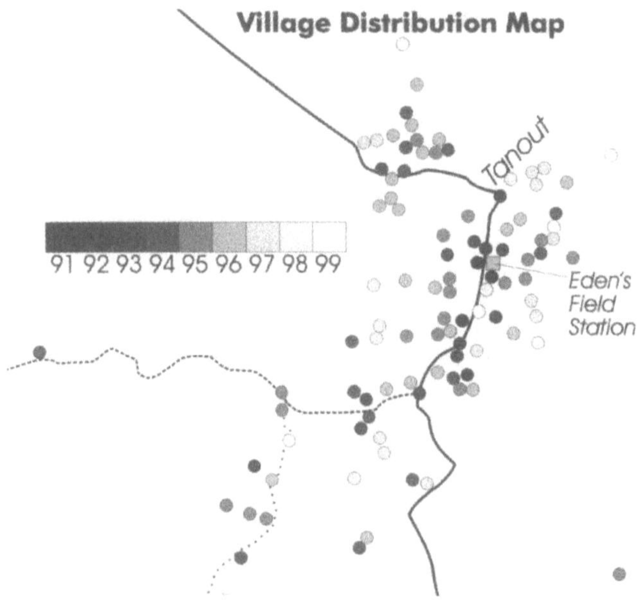

The results from their direct seeding has also increased yearly to a total of 244,621 plants in 1998 (this figure refers to seedlings and young plants, not mature perennials). They also had 10,810 perennials from 32 species growing from natural revegetation. This proved to be a 'side effect' of direct seeding as participating farmers stopped slashing and burning perennials in their fields. The natural revegetation came mostly from old rootstocks that previously had been cut down annually, and many of these became large in a short time, as they already had a developed root system.

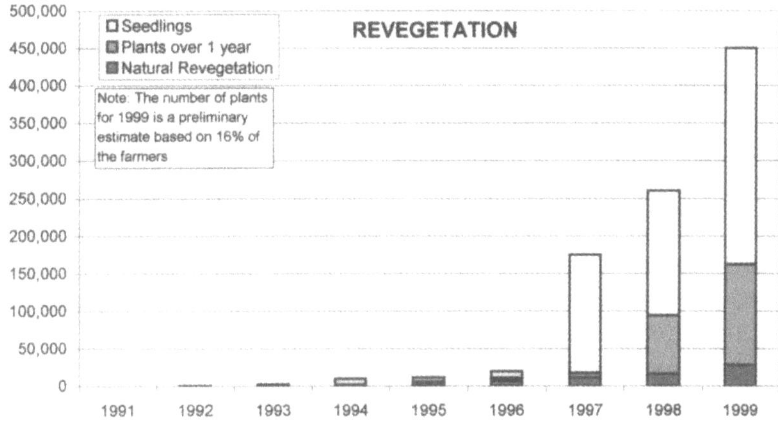

Some of these plants have started to give food and in 1998 the farmers harvested 8,680 litres of food and sold plant produce for 235 FFR. The women and the children are the main beneficiaries of this as the men often leave the villages in the dry season to go off for temporary work somewhere else. Another expression of attitude change towards trees that has been noted is that some villages have introduced tree laws. After starting with direct seeding they have taken the initiative to ban the cutting down of trees, allowing themselves only to take branches without destroying the trees.

The farmers who are dissatisfied with direct seeding are registered yearly, and only 0.3% of those that have participated in the project have expressed that they did not want to continue.

5. DISCUSSION

In business, the customer and the one who pays for the product are usually the same. This is not the case with aid where the donors are actually sponsoring the 'products' that a project delivers to the recipients. The project is 'selling' its solution to the recipients, but the project receives its money from the recipients' sponsor – the donor. The project's loyalty is often where the money comes from and it tends to be more concerned with getting 'paid' for the products it delivers, than the real needs of the recipients. Therefore a project often uses incentives in order to get the recipients to accept the product, even though the recipient often does not really want it. Many products are fashionable ideas imported from the developed world that are not really appropriate to the recipients. In the worst cases the product is even forced upon them. This is why so many use an active transfer of their "solution" to the recipient. However, people have always tended to adopt new practises in their culture if they have wanted it, like with television. It is important to let people choose what they want, and not force things on them in a neo-colonial manner.

Passive transfer is a mechanism for registering customers' interest, where such a mechanism does not exist. They need to see what you have to offer and then choose what they want. Like with a paying customer there must be a cost, even when they are receiving the aid for free in order to confirm that they really want it. With Eden they show what they want by having to contact the project themselves on their own initiative. They pay a cost by investing time and room in their fields by direct seeding even if they are receiving the seeds for free. If they are dissatisfied with the product they are free to stop, and their reason for that is registered. For passive transfer to work you must have a solution the customer can see and a way for him to contact you when he has decided it is something he really wants. Otherwise you can close for lack of customers, so it becomes the recipient that evaluates you. The donor and the project should evaluate how satisfied the recipient is with the product, instead of following a plan to the letter. The project should serve the recipient, not the donor. It should be the link between donor and recipient. The donor and project should serve the fulfilment of the recipient's goals with effective means.

The constraints are that the donors often have their own agenda and try to transform the projects into their puppets, making it difficult for the projects to serve the recipients even if they really want to. Passive transfer is often a long-term process, while the donors often think in short term projects. A long-term project can find itself without donors as donor fashions change, and it can then become a temptation to change with fashion instead of continuing to serve the recipients' goals. It is difficult to present a plan in facts and numbers, as it is a process outside of the project's control, and something donors are not used to. If and when donor money is secured, the project still has to develop a product that the recipients would want themselves, and it must be presented in a way so that the recipients can adapt it into their culture.

6. CONCLUSION

Farmers are not stupid or ignorant and should not be treated as children who do not know what is best for them. They know what they want once they see it, and they will adopt it and integrate it into their culture if it is affordable and appropriate. We need to treat them with respect and serve the fulfilment of their goals, and they need to be free to choose which approaches to adopt.

The Eden experience has shown a way to spread a solution without infringing on the farmers' dignity and right to decide for themselves, namely passive transfer. The results have so far been satisfactory, and many farmers have adopted the solution. However, it is difficult to control the spread, penetration and time frame. Whether such control really ought to be exerted would be the subject of another discussion.

RESPONSE TO AGROFORESTRY INTERVENTION

The Case of Subsistence Farmers in Semiarid Homa Bay District, Kenya

Alice A. Kaudia

Kenya Forestry Research Institute, PO Box 20412, Nairobi, Kenya, Tel: 254-154-32541/32891-3, Fax: 254-154-32844, email: akaudia@arcc.or.ke, kefri@arcc.or.ke

1. INTRODUCTION

In Kenya, the promotion of agroforestry practices can be traced back to 1971 when the Forest Department started the rural afforestation extension scheme. Through this scheme, tree-growing practices were promoted to rural farmers as a strategy to reduce their dependence on planted and natural forests as sources of tree products. Controlling environmental degradation was another major reason for promoting agroforestry. Intensification of agroforestry promotion started early 1980s.

Research on agricultural land use in semiarid areas of Kenya have shown that intensification of tree-growing by farmers is associated with human settlement (Tiffen, Mortimore and Gichuki,) policy intervention to encourage farmers to grow trees and reduce land degradation (1994; Scherr, 1994, Rocheleau, *et al.*, 1995)), socio-economic factors like profits from grown trees as well as endowment with resources (Scherr, 1995). Apart from knowing the socio-economic factors that determine response to agroforestry interventions by small-scale farmers, knowing the extent to which trees grown by households meet household needs is necessary. This can indicate success or failure of agroforestry promotion strategies and trigger appropriate strategies for improving agroforestry promotion

This paper describes the introduction of agroforestry to semiarid Homa Bay District, discusses observed response to promoted agroforestry tech-

Combating Desertification with Plants, edited by Dov Pasternak and Arnold Schlissel.
Kluwer Academic/Plenum Publishers, New York, 2001.

nologies as explained by socio-economic differences between the survey households, and the implications of the findings to promotion of agroforestry in semi arid areas.

2. METHODOLOGY

Formal sample survey was the main method used for obtaining primary data on agroforestry practices. Recording of time spent in different household reproduction and production activities by respondents in pre-designed data sheets was used to obtain information on labour-time distribution between various activities by households. Respondents were asked to recall activities done from the time of waking -up to the time of going to sleep. The data sheets were completed at the end of each day. The recording of information by respondents was used to eliminate biases associated with enumerator s' perspectives in interviews and also biases by respondents in situation where researchers collect information by observation or participant observation methods (Wollenberg, 1985). Data was also obtained through review of project reports, District Development Plans, observation during field interviews, enumeration and informal interviews with key informants. All administrative divisions of the district climatically categorized as semiarid were selected for the study. These are Karachuonyo, Asego, Rangwe (Eastern part), Nyarongi and Dhiwa divisions.

A sampling frame was contracted by first listing all projects that have promoted agroforestry technologies in the district since the early 1980s. Records and reports of the projects - where available- were studied and names of farmers/members of tree planting and agroforestry groups listed to construct a sampling frame. A representative sample of 322 households was selected. Pre-tested questionnaires were used to record responses from interviewees.

Five socio-economic attributes of survey households were considered relevant to this study. These were: gender of head of household, level of education of head of household, age of head of household, area of land owned by households and wealth status of a household.

Based on an ascending scale of: 1 = Very small scale (only a few trees) to 5 = Large scale (many trees and the household can meet most of its needs from on-farm sources), respondents were asked to record the scale of application of each of the agroforestry technologies observed in their farm and the types of trees grown for each of the agroforestry practices. Most of the respondents could not record the activities in English. Responses including names of grown agroforestry trees and shrubs were reported in the local Luo language.

Data extracted from the questionnaires were summarised and analysed using SPSS software. Frequency distribution of the responses was done and where necessary, in the case of continuos data, outliners were removed. Continuos data was grouped to enable analysis through cross-tabulation.

Assessment of farmers' response to introduced agroforestry technologies could not be done for the different types of agroforestry practices by farmers in relation to the five independent variables selected for this study. Alley cropping being a new method for growing trees together with food crops was selected as a test technology to determine farmers' response to new farming technologies as explained by the socio-economic attributes of their households.

3. RESULTS AND DISCUSSION

3.1 A Profile of Homa Bay District

The district is comprised of a total area of 1,155.5 square kilometers. It has two physiographic regions: (a) Lake Shore lowland in the western part and (b) the Upland Plateau in the eastern part. The altitude of the district ranges from 1,143 meters above sea level in the western part to 1,560 meters above sea level in the eastern part.

Temperatures vary with altitude and average 17.1° C and 34.8° C in the eastern and western parts respectively.

Two main rainy seasons are experience in the district. The main rains occur during February to March period. This is also the main crop-growing season. Minor rains fall between August and November. Rainfall average 500 to 1000 mm per annum in the relatively high potential eastern parts and 250 to 700 mm per annum in the drier western part.

Relief is the main determinant of types of soils. The Lake Shore Lowlands are typically covered with sandy loam alluvial soils. In the high parts, well-drained brown clayey soils predominate. Generally, productivity of soils in the district is low because of continuos cropping associated with small land holding units. Average land holdings per household in the district is 2.04 Hectares (GOK, 1997).

Homa Bay District has no gazzetted forest. Both urban and rural communities of this district depend on bush and scrubland wood resources for tree products such as woodfuel, timber for construction and poles. Local communities have open-access to wooded hilltops scabland under the control of local authorities. But, the open-access wood resources are limited. Trees grown on farms are therefore an important livelihood supporting resource.

There are no major industries in the district and off-farm employment is limited. Mainly fish processing industries provided opportunities for off-farm employment. However, since early 1997, because of extensive colonization of the lake surface by the water hyacinth plant, fishing at industrial and artisanal scale has declined to a halting level. Subsistence farming is therefore important in supporting local livelihoods. The main food crops grown are sorghum, maize cassava, beans and millet.

3.2 The Survey Population

Characteristics of a survey population can provide useful explanation of observed agroforestry practices. The population was characterized in relation to the five socio-economic attributes assumed to influence decisions on adoption of agroforestry; namely age of head of household, level of education, gender of head of household, wealth status and area of land owned.

Most of the households were reported to be headed by men (85% of the responses). Low level of education and hence poor literacy level was observed. Most (84% of male heads of households) attained primary level of education. In addition, most of the heads of households were aged. Fifty eight percent of the population was comprised of heads of households over 52 years old. In terms endowment with resources as indicated by type of housing units and ownership of other on-farm production resources which were used as indicators of wealth status, the survey population was mainly comprised of poor households (94.4%) owning small areas of land. On average, households owned 1.8 ha of land. Up to 53 % of the survey population reported owning land less that 1.4.ha.

An analysis of information gathered on labour-time use on various activities for 197 members of the surveyed households over a period of 98 days spread over seven months indicated that most of the labour-time is spent in agricultural production and household reproduction (*Table 1*). This would be expected given the lack of industries and limited non-farm employment opportunity in the district. Farming is not an economically important activity as reported by Scherr (1995). Indeed, households in this region spend a lot of time in activities that support household reproduction instead of being involved in income generating activities. There is a need for creating non-farm employment opportunities to address the continued mining of natural resources and hence embedding of poverty on the already impoverished living conditions in the region. Being a warm and relatively dry area, promotion of high value fruit trees as part of an agroforestry program can be a remedial intervention.

Table 1. Proportion of time spent per day by households on livelihood supporting activities

Activity category	% time in hours per day
Community work	2.0
Agricultural production	31.2
Agroforestry/Forestry	4.2
Household reproduction	48.2
Income generation	6.2
Livestock	8.2
Total	100

3.3 Promoted Technologies

Lack of extensive woodland resources in the district necessitated promotion of tree-based farming technologies like alley cropping and general growing of trees. Since the early 1980s, 20 different projects - of 3 to 6 year duration - have promoted agroforestry and related land use and farming technologies in the district.

Methods for growing trees and shrubs together with food crops either simultaneously or sequentially and the different types of trees and shrubs are the agroforestry technologies that have been promoted in the district. Multipurpose trees and shrubs with soil ameliorating characteristics like *Calliandra callothyrsus, Leucaena leucocephala, Sesbania sesban and Grevillea robusta* were introduced for intercropping. Other species like Euaclyptus and *Cupressus lusitanica* were promoted for growing in woodlots.

It was found that farmers grow or deliberately conserve different types of trees and shrubs in their farms. The study documented 54 different types of trees and shrubs grown or conserved by farmers. *Table2* lists three main species grown based on six different agroforestry practices.

It is notable from *Table 2* that in the case of indigenous practices such as management of grazing land; households deliberately conserve indigenous trees and undertake limited planting of trees in such niches.

An assessment of the scale of application of the cited agroforestry practices and the scope for the households to meet their tree products general revealed that intensification of growing of trees either through deliberate planting or conservation is mainly in the crop fields and in woodlots (*Table3*).

It was noted that overall, few households practiced alley cropping. Yet, this was a champion technology promoted to address land degradation, low food crop yields and deficiencies in fuelwood among other product and services. A X^2 test of independence on the practicing of alley cropping and the five socio-economic attributes of the survey households at 95%

probability level revealed that the practicing of alley cropping depends on age of head of household and area of land owned.

Table 2. Agroforestry practices and species

Type of agroforestry practice	Tree/shrub species commonly cited in descending order
Alley cropping	*Calliandara calothysurs* (24)[1], *Leucaena leucocephala* (16) and *Grevillea robust* (10)
Trees on boundaries of farms and farm plots	*Eucalyptus saligna* (55), *Grevillea robusta* (44), *Cupressus lusitanica* (17)
Trees conserved in grazing land	*Combretum molle* (26), *Croton macrostachys* (7), *Makharmia lutea* (6) *Eucalyptus saligna* (6)
Trees grown on perimeter of homes	*Eucaluptus saligna* (34), *Cupressus lusitanica* (23) and *Grevillea robusta* (17)
Scattered trees in cropland	*Eucalyptus spp* (35*), Rhus natalensis* (25) and *Markhamia lutea* (19)
Woodlot	*Eucalyptus saligna* (107*), Eucalyptus camaldulensis* (22) and *Grevillea robusta* (15)

Further analysis using the Kendall correlation two-tailed test procedure, showed that the association between practicing of alley cropping and the five independent variables was very weak. However, an important observation was the negative correlation between gender of head of household, wealth status and area of land owned. This can be so because the survey population was mainly comprised of poor households, mostly headed by men and owning small pieces of land.

However, results of this research suggest that households headed by women, those owing small plots of land (0.1- 0.4 ha) are likely to respond positively to new tree growing interventions. That most households headed by women may respond positively to new agroforestry technologies involving integration of trees and crops is not unusual because most women provide agricultural labour in subsistence agriculture. But, there is need for research do determine reasons for differences in response to agroforestry between households headed by men and those headed by women. This could not be determined by this study.

[1] Numbers in parenthesis indicate the frequency of reported cultivation of a species in a particular tree growing practice

Table 3. Scale of application of agroforestry technologies

	Scale of application of agroforestry practice				
Agroforestry practice	1	2	3	4	5[2]
Trees in crop fields n = 111	65	11	3	2	30
Woodlot n = 135	79	14	8	5	29
Trees on perimeter of homes n = 115	76	5	8	4	22
Alley cropped trees/ shrubs n = 52	30	3	1	2	16
Trees on boundaries n = 111	81	5	6	5	14
Conserved trees on grazing land n = 63	47	6	4	1	5

A positive correlation was observed between age of head of household and level of education. Households headed by older people (over 52 years) with higher levels of education tend to accept new land use innovations. That most of households were reported to be headed by older people can be associated with the fast rate of retrenchment from formal employment. This has a positive implication for agroforestry adoption. Often, retired people in Kenya today, are well-informed individuals with scope for understanding the value of growing trees. Because of their level of education, agroforestry extension materials that can be used for promoting agroforestry to them can easily be produced as printed material.

4. CONCLUSION

This research has demonstrated that in semiarid areas where subsistence farming is predominant, economic theories which explain rational allocation of factors of production including labour-time, may not adequately explain response to introduced agroforestry technologies. In this case, for instance, it would be expected that wealth status of a household would have a positive correlation with adoption of new farming technologies because such households would have the capacity to absorb risks associated with trying new

[2] Scale at which respondents said they are able to meet their tree products needs from on-farm sourses

ideas. On the contrary, it was found that poor households owning small land units and who have limited alternatives are more responsive. A framework of livelihood strategies provides a better explanation of empirical actions by subsistence farming communities.

Results from this research suggest that agroforestry extension programs should target poor households, those headed by mature individuals and households owning small pieces of land.

Because of the low level of schooling of most members of the community in the survey area, it will be necessary to package information in non-print formats or if printed material is used, it should be adapted to levels of communication and language suited to the people. Result and technology demonstrations as well as audio-visual material could be useful tools for disseminating agroforestry in this region.

ACKNOWLEDGEMENT

This research was done with financial support from the International Foundation for Science (IFS). The author is very grateful to IFS for the support.

The directorate of the Kenya Forestry Research Institute provided an enabling environment for the fieldwork and research. This support is acknowledged. I thank farmers who willingly provided information for this research.

Extracting data from questionnaires and generating summary sheets was done by Charles Nyogot and Judith Nyala of KEFRI. I thank them for their inputs.

REFERENCES

1. Rocheleau, D. E., Steinberg, P. E. and Benjamin, P. A. (1995). "*Environment, development, crisis and crusade: Ukambani, Kenya*", 1890-1990. World Development, 23:6, pp. 1037-1051.
2. Scherr, S. J. (1994). "*Agroforestry in the East African highlands: A sustainable strategy for land use intensification*". National Geographic Research and Exploration 10: 2, pp. 144-157.
3. Scherr, S. J. (1995). "*Economic factors in farmer adoption of agroforestry: Patterns observed in Western Kenya*". World Development, 23:5.
4. Tiffen, M.M., Mortimore, M. M. and Gichuki, F. (1994). *More People less Erosion: Environmental Recovery in Kenya*. John Wiley and Sons, UK.
5. Wollenberg, E. (1988). *An Evaluation of Methodologies used in Time Allocation Research.*: In Poats, S. V.; Schmink, M. and Spring, A. (eds).

6. *Gender Issues in Farming systems Research and Extension.* Westview special Studies in Agriculture Science and Policy. Westview press. Boulder and London.

FAO, PLANTS AND SUSTAINABLE DEVELOPMENT IN DRYLANDS

M. C. M. Porto[1]
Crop and Grassland Service, Plant Production and Protection Division, FAO, Rome, Tel: 39-0657054129, Fax: 39-0657053152/6347, email: Marcio.Porto@fao.org

1. INTRODUCTION

Approximately 30% of the world's terrestrial surface is covered by drylands, namely the arid, semiarid and dry sub-humid areas (FAO, 1997). The extent of the area and the fact that 900 million people live in these environments are the main reasons for the priority given by the Food and Agriculture Organisation of the United Nations (FAO) to the sustainable development of drylands.

Activities related to the control and/or prevention of desertification have been part of FAO programme of work since its creation, as the Organisation has always been concerned with issues related to agriculture and food production in marginal environments. Activities were emphasized in a keynote paper prepared for the meeting of the Committee on Agriculture in 1987, where the basis for FAO's strategy for the sustainable development of drylands was formed. This emphasis was reinforced during the Earth Summit in 1992 and, in 1993 FAO presented an important contribution to the resolution of dryland problems by defining the principal elements of a strategy for sustainable dryland development (FAO, 1993). The strategy has as its main objective the increase in the productivity of croplands and

[1]The opinions expressed in this paper are those of the author alone and do not imply any opinion whatsoever on the part of the FAO.

Combating Desertification with Plants, edited by Dov Pasternak and Arnold Schlissel.
Kluwer Academic/Plenum Publishers, New York, 2001.

rangelands, while alleviating poverty and reversing the process of land degradation in dry areas.

An important step in the process of combating desertification was achieved in 1993, with the creation of FAO's Interdepartmental Working Group on Desertification, with the main objective to provide specific support to programmes launched through the United Nations Convention to Combat Desertification (CCD). The Working Group, which comprises representatives of 15 technical units of FAO, also facilitates, in a participatory way, FAO's support to the implementation of the CCD (FAO, 1999b).

2. THE ROLE OF FAO IN COMBATING DESERTIFICATION

FAO activities related to desertification are summarized in *Table 1.* The Organization undertakes both normative and field activities in areas suffering from desertification (and/or threatened by this process) in all continents and in collaboration with countries, regional and international organizations. The number and scope of the activities listed in *Table 1* are considerable. In 1998-99, in addition to the large number of normative activities being coordinated by FAO, a total of 240 country projects and more than 70 regional field projects were in operation. In 1999, approximately 100 FAO on-going field projects are directly related to the assessment and control of desertification (FAO, 1999b).

Although recognizing the several technical areas relevant for addressing the problem of desertification, the importance of a complementary approach and the potential benefits that can be obtained by the use of interdisciplinarity in a strategy to address FAO's work on drylands, this paper focuses on activities related to crop and grassland production and management, the mandate areas of the Crop and Grassland Service, a technical unit of the Plant Production and Protection Division of FAO.

Before presenting information on the activities related to agricultural production in drylands it is necessary to provide some information on the classification of drylands as viewed by FAO. The major climatic characteristics of drylands are the availability of water, concentrated in a short rainy season and the high temperatures during the rainy season that are responsible for the loss of much of the rainfall by evaporation and the intensity of tropical storms that ensure that most of it runs off in floods (FAO, 1999a). For these reasons, aridity and variability are the two dominant characteristics of dryland climates (FAO, 1999a). Detailed information on FAO's work climatic characterization can be obtained from FAO's WebPage (www.fao.org).

Table 1. A summary of FAO activities related to desertification

Type	Activities
Normative Activities	**Support to CCD: FAO's Interdepartmental Working Group**
	Technical advise to Member Countries on: - technical assessment of development projects - formulation, technical backstopping and monitoring of projects
Field Activities	**Formulation of Investment Projects** **Inter-regional projects for Participatory Upland Conservation and Development in** Bolivia, Burundi, Nepal, Pakistan and Tunisia **Activities in the Near East and North Africa** - land and water management and sand dune stabilisation - integrated management of rangelands and other natural resources - management of wood resources and reforestation **Activities in Sub-Saharan Africa** - regional initiative on introduction and implementation of an integrated approach to drought management - special programme for food security (Senegal and Ethiopia) - forests and food security in Sahelian Africa - land cover mapping of Eastern Africa **Activities in Asia** - management of natural resources and the environment - farming systems, soil and water conservation -support to regional network on agroforestry and to soil conservation **Activities in Latin America** - sustainable rural development in arid zones - prevention/rehabilitation of land degradation - drought alleviation - prevention of irrigated soils salinization - sustainable forest production and farming systems development

Source: FAO, 1999. Report of the Food and Agriculture Organization of the United Nations (FAO) for the Third Conference of Parties of the Convention to Combat Desertification (CCD). November, 1999. 14p.

The typology used by FAO to assess crop suitability to different agro-ecologies considers the Length of the Growing Period (LGP), or the period of the year when both moisture and temperature conditions are favorable for crop growth (FAO, 1999a). Arid, semiarid and dry sub-humid areas have, respectively, LGP of 75, 120 and 180 days.

2.1 SELECTED INTERVENTIONS

As shown in *Table 2*, a number of normative and field activities related to crop and grassland production in dry areas is being implemented by FAO, or with the support of the Organisation in different countries and regions. Two of these experiences described below demonstrate our efforts to increase agricultural production in drylands as well as contributing to combating desertification.

Table 2. **FAO activities related to crop/grassland production in dryland areas in 1998-99.**

GROUP	ACTIVITY
Field Food Crops	**Information Management** ▪ Development of maize and wheat databases ▪ Book on wheat, biotic and abiotic stresses in arid/semi-arid areas ▪ Book on "Breeding for Marginal Lands **Training, Education and Awareness** ▪ Development of grassland and cereal systems
Horticultural Crops	1.1. **Support to the International Technical Cooperation Network on Cactus Pear** 2. **Promotion and improvement of date palm cultivation** 2.1. **Other Technical Cooperation Projects** ▪ Rainfed orchards of arid zone fruits ▪ Cactus pear 2.2. **Other Initiatives** ▪ International Technical Cooperation on temperate zone fruit trees in the tropics and sub-tropics ▪ Cashew nut rehabilitation in African countries
Industrial Crops	**Technical Cooperation Projects and Initiatives** Irrigated sugarcane Drought tolerant sweet sorghum and safflower **Information Management** Plant Info of GPPIS and Production of a CD-ROM on New Crops
Grassland and Pastures	**Technical Cooperation Projects and Initiatives** Prosopis Range Rehabilitation/Establishment of Wildlife Reserve Projet de Gestion des Resources Naturelles Winter fodder crop production Breeding cold tolerant alfalfa for dry areas **Information Management** Books on Prosopis species from Latin America, Pastoral Mobility, Grassland Resource Assessment and Utilisation of Cactus Pear Publication of Newsletters and Databases Development of Website to include dryland material

2.2 Range Rehabilitation, Wildlife Re-Introduction and a Participatory Approach to Range Management in the Syrian Steppe[2]

2.2.1 Background

In Syria, as in a number of Near East countries where the land is State-owned, there has been a breakdown of traditional systems of management. Rising population and livestock numbers, and a high degree of mechanisa tion, have increased pressure on the range and resulted in a severe deterioration of large areas of the Al-Badia rangelands.

The Syrian Al-Badia Steppe, climatically characterised by low and erratic rainfall (less than 200 mm per year), cold winters (often below 0° C) and hot summers (exceeding 45 ° C), comprises more than 55% of the land area of Syria, or some 10.2 M ha. In spite of the harsh conditions, the Badia remains an important area for livestock raising (Mirreh, 1999). Current range sheep production is highly mechanised through the extensive use of water tankers, trucks and tractors.

Programmes for better veterinary care, more watering points and feed subsidies contribute to keeping livestock raising in the Badia an economically attractive business for herders as well as for merchants who have purchased large flocks and put them onto the steppe with sheep herders. However, this type of over-utilization of the rangeland resource has caused considerable deterioration of both soil and vegetation. Uprooting of perennial plants for fuel has resulted in the clearing of bush cover from large areas (it has been estimated, for instance, that an average Bedouin household annually uses shrubs for fuel from approximately 4.4 ha.).

Erosion pavements with limited infiltration and large active gullies are common and there has been a decrease in the numbers of the preferred forage plants such as *Salsola vermiculata, Poa sinaica, Plantago albans, Stipagrostis* and even *Artemisia herba alba* which are often replaced by plants of lower forage value such as *Perganum harmala, Noa mucronata* and *Anabasis syriaca* (Mirreh, 1999).

The project was started in February 1996, attempting to heal part of the severely degraded Al Badia steppe. It covers some 130,000 ha including 22,000 ha of wildlife reserve and 108,000 ha of land from three cooperatives with 426 households and 3,050 people owning 95,000 head of sheep.

Activities undertaken to date have included direct seeding and shrub planting on 5,000 ha, closing off areas to livestock grazing (3,000 ha), data

[2] Extracted from: Reynolds et al. 1999. Perspectives for Rangeland Development – the Food and Agriculture Organization of the United Nations (FAO). In: Proceedings, VIth International Rangelen Congress, vol. 1, 160-165.

collection and monitoring, the preparation of range management plans, assessment of alternative energy sources for cooking and heating and income generating activities. In addition, gazelle and oryx have been re-introduced to a newly created reserve and management plans developed. Major emphasis is also given to training and capacity building among the national staff, and to the full involvement of the local communities in range management planning and implementation.

2.2.2 Results

Gazelle and oryx have been re-introduced and released into a fenced 1,000 ha area within the 22,000 ha Talila Reserve, which is currently managed for both camels and wildlife. Benefiting from two years of better than average rains, the project has been able to demonstrate excellent regrowth in exclosure areas and also excellent germination and growth of plants in improved areas.

The improved areas served as demonstration plots with standing biomass of more than 400 kg/ha of dry matter from annuals alone. In December 1998 and January 1999, more than 7,000 tons of dry matter were cropped from the project area saving the Bedouins some 26 to 50 M Syrian pounds in terms of replacement feeds.

2.2.3 Lessons learned

- The use of native species (*Salsola and Atriplex leucoclada*) demonstrated higher potential than introduced species;
- Adoption of simple and cheap techniques of direct seeding rather than transplanting are preferable (the planting cost of seedlings was more than four times the cost of direct seeding) and machines used should be adapted to the local soil conditions;
- Demonstrations of successful rehabilitation of rangeland areas may convince local herders to cooperate in plans to better utilize not only improved areas but also to develop plans for utilization of all range areas. Direct benefits for the herders appeared to be the key incentive;
- The project has demonstrated that rehabilitation of rangelands is possible with two successive good rainfall years, but training and capacity building among the local Bedouins is a long term issue;
- Since there are certain to be further droughts in the future, Government should consider replicating the experience of the project and establish protected areas and prepare drought contingency plans and an early warning system for all rangelands in Syria.

The project has been able to demonstrate to Government that, with the cooperation of the local Bedouin population, rehabilitation of large areas of the Al-Badia Steppe is possible, but also to illustrate the necessity of urgently addressing the issues associated with free access of all to all areas of land, the effects of drought and the need for preparing drought contingency plans.

The total number of sheep, overstocking and issues associated with subsidized feed are other matters that will need to be addressed. The focus of future project activities will be on selected rangeland site management improvement and rehabilitation, with the selected areas managed by Bedouin organizations. A large IFAD funded project, which is currently being planned, will endeavour to develop larger areas of the Al-Badia rangelands.

2.3 Preliminary Assistance for Date Industry Development in Namibia

2.3.1 Background

Namibia is a vast country with a total land area of 824,295 km^2 and inhabited by only 1,700,00 people, of whom about 70% live in communal areas. The agricultural sector's contribution to the GDP (1980-88 average 10%) contrasts with its role as the dominant employer and occupation (70%). The country is predominantly arid with low agricultural potential, limited irrigation water and available irrigable land. Large areas of the country are classified as desert environment with climatic and soil conditions that are not conducive to crop production.

The focus of Government policy for the agricultural sector has been to increase and diversify crop production, with the goal of improving the country's food-sufficiency ratio. The diversification strategy for agricultural production involves the determination of which crops tolerate drought and which perform best under irrigation.

Date palms have grown in dry areas next to rivers for more than 90 years and it has been proven that the climate is suited to date production. The existing plants originated from seeds planted by early settlers. Vegetative growth is satisfactory as are flowering, pollination, fruit set and natural maturing of fruits. No major pests and diseases are found.

Local technical knowledge of modern date propagation, protection and production was until recently virtually non-existent. Dates were not produced for commercial reasons and no effort was made to acquire knowledge or conduct research. Some practices commonly implemented, like burning of leaves, were detrimental to fruit production and the survival of the palms. No organised plantations were found where proper cultural techniques were practised. The date palms were, in most cases, not even irrigated and their survival depended on the existence of groundwater within reach of their root system. The few tons of dates produced under natural conditions in rural areas were consumed locally and no fruits were channelled into any formal markets.

The Government of Namibia requested FAO to investigate the potential of date production in the country. A project proposal was compiled and efforts were made to secure foreign funding. These efforts were however unsuccessful and the Namibian Government decided to fund the project from its own resources.

The basic aims of the project are to transfer knowledge to local Namibian experts and farmers in date palm propagation, production and protection, as well as to support the establishment of commercial plantations. The Programme can thus be regarded as an "umbrella" programme that oversees and supports related date palm activities. Support is focused on the following components/aspects:

- Establishment of three local date production projects to investigate the viability of date palm cultivation and evaluation of introduced and local germplasm;
- Practical on-farm research on issues that will probably influence the viability of date palm production in Namibia such as: a) evaluation of selected and local varieties; b) water requirements; c) pollination and fruit thinning experiments; d) processing and packaging of fruits;
- Establishment of a Tissue Culture Laboratory to: a) ensure the establishment of an economically viable production unit and b) initiate selection of varieties for future propagation.
- Support to private farmers as: a) project planning guidelines and advice on project initiation; and b) training and technical advice on-farm during scheduled visits.
- Training for national team members and farmers through visits to producing countries, field demonstrations and advice on propagation, production and protection, as well as extension activities through the production of leaflets, specialised documents and illustrative photographs.

2.3.2 Results

The situation changed drastically with the implementation of the project. Date palm is now considered as a viable alternative and the prospects are supported by the Government of Namibia and private sector entrepreneurs. A tremendous increase in know-how has taken place and can be regarded at this stage to be on a level comparable to countries where modern date palm cultivation is practised. Interest in date production was expressed by other countries in the region and support to them is given by the Namibian team. The initial steps have been taken to initiate a regional project based on the success in Namibia.

The existing local date palm plantations have changed considerably with the introduction of new techniques. Monitoring and selection of locally promising varieties is being conducted and future propagation can be envisaged.

The marketing potential is very promising as Namibia is situated in the southern hemisphere whereas all the other major date producing countries are located in the northern hemisphere. Namibia will thus be able to produce and supply dates to all the major markets during the traditional off-season. The lack of major diseases and pests add a value to the Namibian dates since no chemicals are used and consequently fruits are natural/organic products.

The successful establishment of date plantations with modern technology and the creation of a functional date palm tissue culture laboratory which is producing plants for Namibian farmers are examples of the co-ordinated work of Government, private sector and an international organization (FAO).

2.3.3 Future Activities

One of the main objectives of the Date Production Support Programme is the settlement of landless people. Two settlement projects were written where date production is included as a production component. The project proposals were submitted to the Government for approval and implementation.

For the date industry to be sustainable, Namibia needs to maximise sales revenue by targeting the higher priced markets of Europe and the Middle East. In order to reach these markets, a modern processing and packing house facility needs to be established.

As proposed in the Project Document and also as highly recommended by different international consultants, the organisation of an International Date Palm Symposium will be of great benefit to the Namibian date industry. The event will be held in February, 2000 and will gather scientists, businessmen and date growers from all over the world. An International Date Information Network is planned.

Namibia has already embarked on the establishment of a date palm gene bank, which can be of benefit to all date growing countries, as Namibia is free of all major pests and diseases.

3. THE CONTRIBUTION OF ROOT AND TUBER CROPS

Cassava (*Manihot esculenta* Crantz) and sweetpotato (*Ipomea batatas*) are major staple food crops and the most important source of calories for approximately 500 million people throught the tropical and sub-tropical areas of Africa, Asia and Latin America. The two crops are particularly important in "hungry months" and provide flexible source of food for households. Small farmers who do no rely on purchased inputs produce both crops.

The majority of cassava and sweet potato are grown in marginal, low-fertility acid soils with annual rainfall that ranges from less than 600 mm in the semiarid tropics to more than 1,500 mm in the sub-humid and humid tropics (El-Sharkawy et al., 1993). In many of these areas rainfall is not uniform and there are prolonged (4-7 months) dry seasons.

Historically, cassava and sweet potato have been able to offer the potential for expanded production in stressful environments because of their ability to give reasonable yields under harsh environments, particularly where prolonged drought and poor soils are major constraints (Porto et al., 1991). Under dry conditions cassava partially closes its stomata while maintaining reasonable daily photosynthetic rates (Porto 1983) whilst sweet potato reduces its leaf canopy and top growth at the onset of stress, hence reducing water loss but continuing to accumulate dry matter in their storage roots. Moreover, the two crops are capable of slowly extracting deep water when it is available. These characteristics are highly advantageous under drought conditions or in conditions of sporadic rainfall encountered in cassava and sweet potato growing areas such as Sub-Saharan Africa and North-East Brazil.

It is believed that crop diversification using well adapted crops such as cassava and sweet potato can lead to sustainable drought mitigation and hence to food security and sustainable development. They are therefore strategically highly appropriate crops for risk aversion. This important role is being widely appreciated both at grassroots and government policy levels.

Strategic actions for the mitigation of desertification based on root and tuber crops should cover the following:

- *Initial rapid response to the prevailing adverse farming conditions in drylands*. This will include, on one hand, the use of local knowledge and farmers' practices for extending cropping systems, conserving their varieties and producing starter planting materials. On the other hand, it will consist of rapid multiplication of planting materials and distribution from farmer to farmer to facilitate the development of informal farmer-based system in the absence of formal multiplication and supply services.

- *Medium to longer term measures* to build capacity at small-scale farmers and processor level, to achieve integration of cassava and sweet potato into the farming and food systems of drought prone areas. Action would focus on expanding the multiplication and distribution system of planting materials at country and regional level with emphasis changing from short term mitigation to promotion of changes in farming, utilisation and consumption habits. Promotion of adapted germplasm, evaluation of new improved and/or introduced varieties, and promotion of improved production, protection and processing technologies with a high expectation of adoption would be central to the development orientation of this action area. The more long-term action will look ahead to future interventions relating to commercial development of cassava and sweet potato, encompassing food, feed and other agro-industrial sectors.

Sustainable development in drylands can be achieved by crop diversification though expansion of cassava and sweet potato in dry areas thus fixing farmers to their land and avoiding land degradation. The promotion of these crops in drylands will be done through the following interventions:

- Strengthening national systems of germplasm exchange, collection and utilisation of genetic resources;
- Supporting and strengthening multiplication and distribution schemes to ensure availability of planting material;
- Dissemination of information on all aspects of cassava and sweet potato production, processing and utilisation;
- Providing technical support to national programmes;
- Advising member countries in the implementation of training activities related to the improvement of both crops.

4. CACTUS PEAR: A VIABLE ALTERNATIVE TO DRYLANDS

A crop which should receive increased attention is cactus pear (*Opuntia spp.*). Of interest are the conclusions reached by experts from different countries gathered in Arequipa, Peru, in May 1998[3], summarised below:

[3]International Symposium, Nopal (*Opuntia* spp.): its use and application to combat desertification. May 26-28, Arequipa, Peru. 3p.

The species of the *Opuntia* spp. subgenus have developed phenological, physiological and structural adaptations favourable to their development in arid environments. Among these adaptations stand out its asynchronous reproduction, and its CAM metabolism, which combined with structural adaptations such as succulence allow this plant to continue the assimilation of carbon dioxide during long periods of drought and in this way reach acceptable productivity levels even in years of severe drought. Besides:

Biomass generation per unit of water is on average about 3 times higher than for C_4 plants and 5 times higher than for C_3 plants.

Opuntia spp. can develop in severely degraded soils, which are inadequate for other crops. *Opuntia* spp. have a great capacity for adaptation and are ideal for responding to global environmental changes. Their root characteristics avoid wind and rain erosion, encouraging their growth in degraded areas.

The establishment of sustainable systems of production based on *Opuntia* spp. may contribute to the food security of populations in agriculturally *Opuntia* spp. are some of the best plants for the reforestation of arid and semiarid areas because can resist scarce and erratic rainfall and high temperatures.

The use of Opuntia as forage, vegetable, fruit and industrial raw material is variable. As forage it can be produced in many areas where few other crops can grow and its cladodes present high levels of palatability and digestibility, associated with a high water content which reduces the need for water supply to animals. As a vegetable it is consumed fresh, mainly in Mexico. Fruits are consumed fresh and have high market potential both in America and Europe where it is considered an exotic product. Nopalitos are considered as having medicinal properties as experimental results have shown that it can be efficient in treatment against diabetis, gastritis and obesity.

In 1993 the FAO's International Network on Cactus Pear (CACTUSNET) was created in Guadalajara, Mexico, to increase cooperation among scientists of different countries and to facilitate the exchange of information, knowledge and technical cooperation (FAO, 1995). The Organization recognises the importance and potential of Opuntia for the development of dry areas where cactus pear can be economically exploited with benefitial consequences for food security and environmental conservation.

A Technical Cooperation Project (TCP) on "Technology Transfer on Cactus Pear (*Opuntia* spp*)* Production and Utilization" has been recently approved by FAO to be implemented in Eritrea. The objectives of the project are basically to provide technical assistance, training and planting materials as a basis for cactus pear domestication, improvement and full utilisation. It is hoped that the project will achieve its objectives and, as a result, promote the full utilisation of Opuntia as a means to promote income generation and its rational exploitation.

ACKNOWLEDGEMENTS

The author would like to thank the professional Officers of the Crop and Grassland Service, especially Dr. Stephen Reynolds, Dr. Ne Bambi Lutaladio and Mr. Enrique Arias, as well as the Secretariat of FAO's Inter-departmental Working Group for Combating Desertification for provision of information material used in the preparation of this paper.

REFERENCES

1. Cock, J. H. (1985). *Cassava: new potential for a neglected crop.* Westview, Boulder, CO. 109 p.
2. El-Sharkawy, M.A. (1993). *Drought-tolerant cassava for Africa, Asia and Latin America.* BioScience, Vol. 43, No 7:441-451.
3. FAO, (1995). *Agro-ecology, cultivation and uses of cactus pear.* FAO Plant Production and Protection Paper N° 132. Rome, Italy. 219p.
4. FAO (1997). Chapter 12: *Managing fragile ecosystems: combating desertification and drought.* Progress Report FAO. 6p.
5. FAO (1999a.) Drylands. Background paper for the FAO/Netherlands Conference on the Multifunctional Character of Agriculture and Land. Maastrich, the Netherlands, 12-18, September, 35p.
6. FAO (1999b). Report of the Food and Agriculture Organization of the United Nations (FAO) for the Third Conference of Parties of the Convention to Combat Desertification CCD). November, 14p.
7. FAO (1993). *Sustainable development of drylands and combating desertification.* 4p.
8. Mirreh, M. M., (1999). GCP/SYR/003/ITA. An illustrated booklet of project activities and progress. Project GCP/SYR/003/ITA. 33p.
9. Porto, M.C.M. (1983). *Physiological mechanisms of drought tolerance in cassava (Manihot esculenta* Crantz). University of Arizona, Tucson. 200p. Ph.D. Dissertation.
10. Porto, M.C.M.; Asiedu, R.; Dixon, A. and Hahn, S.K., (1991). *An agroecologically oriented introduction of cassava from Latin America into Africa.* In: Ofori, F. and Hahn, S.K. (Eds) (1991). Tropical Root Crops in Developing Economy. Proceedings of the Ninth Symposium of the International Society for Tropical Root Crops, 20-26 October, Accra, Ghana. 118-129.
11. Reynolds, S.; Batello, C. and Baas, S. (1999). *Perspectives on Rangeland Development - the Food and Agriculture Organisation of the United Nations (FAO).* In: Proceedings, VI International Rangeland Congress, vol.

COMBATING DESERTIFICATION WITH TREES

The Uganda Experience

Ruth Mubiru

UWTPM, PO Box 10351, Kampala, Uganda, Tel: 256-41-235602, Fax: 256-41-345597, email: uwtpm@infocom.co.ug

1. INTRODUCTION

Uganda is a small land locked country in Eastern Africa lying astride the equator. It occupies about 236,000 km² of the central African plateau. The country has a diverse climate, influenced by the country's latitudinal position between 1° South and 4° North and its altitude and topography. Seasonal movements of the Inter-Tropical Convergence Zone (ITCZ) determine the general pattern of rainfall. Although much of the country receives 1,000 – 1,500 mm of rain per annum, there are some dry parts in the northern and northeastern region which receives less than 500 mm and are prone to long periods of drought which adversely affect agricultural and other rain dependent activities in the country. A bigger portion of this region is either semiarid or arid due to the Sudan-Sahelian climatic influence. It is also known as the cattle corridor because together with some parts of the Southwestern they are heavily used by nomadic pastoralists who subject them to seasonal bush-burning to encourage the growth of palatable pasture during the rainy season. This together with the large stocking of livestock has contributed to land degradation, decimation of tree cover, soil erosion, drought and desert like conditions.

Uganda is still a predominately agricultural nation and 83% of its labor force is engaged in agriculture. Agricultural products account for nearly all the country's export earning and contribute 60 % to the Gross Domestic Product (World Bank 1986). In common with other Africa nations, Uganda's human population has risen sharply during this century with an average

Combating Desertification with Plants, edited by Dov Pasternak and Arnold Schlissel.
Kluwer Academic/Plenum Publishers, New York, 2001.

annual growth rate of 2.8%. The population is predominately rural with 8.7% of the people living in urban centers (Anon, 1982). The urban population growth rate is in the order of 6.4% per annum (compared to the rural growth rate of approximately 2.4%.

2. THE ROLE OF TREE FORESTS IN THE NATIONAL ECONOMY

Forests and trees growing on agricultural and pastoral land, play a crucial role in the national economy both in satisfying energy and industrial products needs and in providing essential environmental services that support Uganda's agriculture, sustain her water supplies and protect her soils. In terms of satisfying energy requirements it is estimated that 96% of Uganda's current energy consumption is provided by wood fuels, equivalent to 18.3 million m^3 of wood per annum (World Bank, 1986). This is by far the greatest pressure on the forest resources and the greatest challenge to those responsible for tree planting programmes in the country.

Wood fuels are used domestically in most rural and urban homes, and they provide the primary source of energy in the tea, tobacco and brick burning industries. In 1985 for example, firewood (which generate about 15% of the total wood fuel demands) generated over 3 million man-days of employment and equipment to about US $ 5 million in rural incomes (World Bank, 1986).

The social role of forests is frequently over looked or under valued, but they nevertheless, contribute sustainably to rural communities living in their vicinity. Construction poles are used for rural dwellings throughout the country and may account for as mush as 400,000 m^3 of timber annually (World Bank, 1986). Forests and trees also provide a number of other commodities and benefits whose values are difficult to quantify in cash terms, yet they represent a significant contribution to community welfare and the national economy.

Equally difficult to quantify, but of no less importance, is the role forests play in providing irreplaceable environmental services that support other sectors of the economy. By protecting water catchment areas, forests and trees ensure supplies of domestic water, maintain down stream fisheries and in some cases sustain hydroelectric power generation. In a country so heavily dependent on agriculture, the role of forests in ameliorating local climatic conditions is also likely to be vitally important, though as yet little understood. There are approximately 14,900 km^2 of gazetted forest reserves in the country. (Aluma, 1987). This constitutes about 10.7% of Ugandan's land surface. Much of the deforestation takes place on private land especially during land clearing for agriculture. Trees felled in such operations are converted into charcoal while a small fraction is sold as firewood and the rest burnt in-situ.

Studies carried out recently have revealed that land clearing for agriculture and livestock farming is occurring at such a vast scale that for every cubic meter of wood felled and used as firewood or to produce charcoal another two cubic meters are felled and burnt in-situ. At this rate of clearing, effects of drought and desertification are being compounded by the day. The National Biomass Study of the Forest Department has estimated that total woody biomass production for charcoal, commercial and collected wood fuel is in the region of 28.6 million m^3 solid.

Localised deficits occur in areas of high demands and limited regeneration (e.g. around major urban centers and around densely populated rural areas with high levels of economic activity). This situation is compounded by the fact that woody biomass demand (urban and rural combined) is increasing by over 3% per annum in the order of 400,000 tones. Thus even if the woody biomass equation is presently in balance at the national level, it cannot be in the long-term unless wood production, conservation and efficient end use are accelerated.

3. THE CCD PROCESS IN UGANDA

Uganda fully participated in the negotiations of the Convention to Combat Desertification (CCD). It was one of the few countries that were selected to conduct a case study to elaborate on the problem, which elaboration was used in the convention negotiations. Uganda signed the Convention on 21 November ,1994 thereby becoming a party to the Convention and later on 25th June1997 ratified it thereby becoming legally bound to implement it.

The CCD is considered by many to be one of the best international conventions, especially with regard to its elaborate provisions for implementation and its emphasis on bottom-up and participatory approaches as well as partnership arrangements. Community participation in the design and implementation of programmes to combat desertification and/or mitigate the effects of drought has right from the beginning been a pre-requisite in Uganda based on the understanding that problems of desertification cannot be effectively tackled unless the people most affected are fully mobilised and committed. It was necessary to decide on how the different stakeholders were to participate in the National Action Plan formulation processes and what role they had to play taking into consideration their capacities (strengths, opportunities and weaknesses of each).

4. CONSULTATION WITH STAKEHOLDERS

The National CCD Secretariat in the Ministry of Agriculture, Animal Industry and Fisheries (MAAIF) organised a series of consultations at district level early 1998. Focus was on the districts making up the drylands which stretch from north-east to south-west. A total of nine districts out of the forty five in the country (i.e. one fifth of the country) fall within this area. These consultations helped the areas requiring intervention to combat desertification to be identified and prioritized in their order of degradation given the limited funding.

Early in 1993, the government of Uganda through the Inter-governmental Committee to Combat Drought and Desertification (INCD) process carried out a case study in three semiarid and arid parts of the country to solicit the views and opinions of the local people. This exercise was followed by a two-day national workshop on drought and desertification in November, 1993. From both the case study and national workshop a number of interventions were identified and presented to the National Study in support of the Inter-Governmental Negotiating Committee on Drought and Desertification. Both short-term and long-term actions were estimated to require a medium term incremental investment of US $ 300 million from external sources. However, due to Uganda's extreme level of indebtedness it was recommended that this type of investments must largely be in terns of grants.

In line with the INCD process, there was need to translate the case study results and deliberations of the National workshop into an action plan. The National Plan of Action Combat to Drought and Desertification (NAP CCD) was the conclusion of this process. The objective of the NAP-CCD was to define a set of policy actions, related investments and institutional strengthening activities to make Uganda's developmental strategies more viable through sound management of the country's extensive drylands which are a home to up to half of the country's population.

5. NATIONAL ACTIONS TO COMBAT DESERTIFICATION

One of the outcomes of the district consultations was the identification of different stakeholders at the national and district level who are considered relevant to the NAP implementation. Four categories of stakeholders were identified. These were:

- Government Departments
- Non-Governmental Organisations (NGOs)

- Local councils
- Community Based Organisations

At the center of all these players are individuals whose various activities impact and/or are impacted upon by drought and desertification. Government departments at the national level are also stakeholders since their activities and mandates link up with those at the district level. Some NGOs operate at national level but they aim at having impact of their activities at the grassroots.

5.1 Government Interventions

In recognition of the important role fuelwood plays in providing energy in the country and taking into consideration the alarming rate of tree felling which was not being matched with new plantings, the government of Uganda in 1992 launched an ambitious National Tree Planting Programme (NTPP) which aimed at establishing 3950 tree nurseries at parish level in the 38 districts of the country at that time. The government also announced new environmental regulations where both individual land owners and communities were compelled to plant trees in specific locations such as on steep slopes, along river banks or as border fences and in specific quantities (set at a minimum of 10 % of any private land holding).

Although the implementation of the National Tree Planting Programme was expected to be the boldest step government had taken to address the problem of deforestation and land degradation which were contributing to drought and desertification, financial constraints derailed this otherwise well intentioned programme.

However, with the limited resource at its disposal, government has annually commemorated a National Tree Planting Day during which individuals have been provided with seedlings for planting on their land holdings and at designated venues at district level during the main rainy season.

In 1994 the government enacted a Local Government Act which devolved power from the center to Local Authorities at district and lower levels. Government's affirmative action has been translated under the decentralization process with a provision for women to constitute one third of the local councils from village to district level. This has enabled easy mobilisation of women to take part in environmental activities which aim at combating desertification and effects of drought. These have mainly concentrated on tree planting activities using multipurpose tree species.

Since women and children provide the biggest fraction of farm labor which includes the gathering of firewood, their participation in tree planting activities have gone a long way in availing this resource on farms which has alleviated the burden of women and children walking long distances in search of firewood. The time they were spending in such endeavors can be profitably utilised in the production of food and carrying out other income generating activities which are environmentally friendly.

Government has also enacted a new Land Act (1998) to rationalise land ownership in the country. Land ownership issues have been major constraints on tree planting programmes. Many individuals and organisations have been reluctant to invest in tree planting due to uncertainty over future ownership of the trees. The new Land Act (1998) recognises customary ownership and allows squatters to acquire certificates of occupancy which protects their investment on the land as long as they pay a nominal annual rent fee to the Land Lord. As a whole, the Land Act (1998) has created an enabling environment for tree planting by peasant farmers and other bigger organisations.

5.2 Non-Governmental Organisations Interventions

Activities of NGOs are normally complementary to those of government. Since they tend to be community based they have good coordination link with various stakeholders at community level. They are also good agents in community mobilisation and capacity building for implementation of grassroots activities.

During the stakeholder consultations, important roles and responsibilities for NGOs in the implementation of NAP were identified. It is within the framework of these roles and responsibilities that many environmental NGOs operate. Many of tem operate in districts within the drylands of the country to promote tree-planting activities by local communities. They carry out sensitisation and training of the local people, provide them with seeds and implements of raising tree seedlings. Their target has been women groups and institutions which use a lot of firewood as their major source of energy e.g. Schools, prisons, lime and brick makers and fish processors.

The Uganda Women Tree Planting Movement is one of the local NGOs engaged in helping women to grow their own firewood on their land holdings using agro-forestry technologies. This NGOs was founded in 1985 to address problems of access to firewood and other tree products being faced by Ugandan rural women. The main objective of Uganda Women Tree Planting Movement is to mobilise rural women in Uganda and offer them training in nursery and woodlot establishment and management in order to provide easily accessible sources of firewood and help in the fight against desertification by conserving the natural tree vegetation through the provision of alternative sources of firewood on the farms.

Currently UWTPM is operating in six pilot districts out of a total of forty-five in the country. This has been dictated by financial constraints and the need to have greater impact on the ground. So far a total of 600 women have been trained in tree husbandry. They have been provided with tree seed and nursery equipment and encouraged to raise a variety of multi-purpose trees for themselves and for sale at a modest cost to others interested in tree planting.

Of late, Uganda Women Tree Planting Movement has embarked on training its members in the pilot districts in other inter-related topics such as sustainable agriculture, family planning, reproductive health/rights, of rural women, HIV/AIDS awareness and formation of cooperatives. These topics have a bearing on rural women livelihoods which in the short term have an impact on the environment in several ways. Uganda Women Tree Planting Movement has the potential to expand in other districts if external funding can be obtained.

A number of other NGOs are operating in a number of other districts to help local people raise and plant trees. Most of these NGOs are foreign and are well funded from external sources. They operate at grassroots in collaboration with the local administration to mobilise and train grassroots organisations of women, the youth and others in tree husbandry. They have registered good success levels in their areas of operation.

Despite the poverty level in the rural areas, people are beginning to appreciate the value of growing trees. Those who grew trees at the start of the NGO operations are beginning to reap financial benefits from the sale of poles and firewood which are on high demand in the urban areas due to the building boom which has resulted from favourable policies of government.

An important issue which these NGOs should be addressing is the sustainability of their operations. One solution would be to build capacity in Community Based Organisations and other viable grassroots organisations to implement National Action Plan activities once these have been agreed upon by local communities. They should also embark on training government extension agents, Community Based Organisations, the private sector and individuals in appropriate technologies aimed at combating desertification e.g. multipurpose tree planting, agroforestry, soil and water conservation, fuel saving cook stoves and income generating activities.

Many of the NGOs have been working in partnership with government (both at the centre and local level) to design innovative projects for implementation by local communities in order to combat desertification and to popularise such innovations among the populace. Some have directly been involved with the communities in planning, implementation, monitoring and evaluation of programmes and activities to combat desertification. Such activities have included tree planting, water harvesting, sustainable farming practices and commercial and income generating activities which are environmentally friendly.

5.3 The Private Sector

The private sector in Uganda has a big role to play in efforts to combat desertification and mitigate effects of drought. Many trees are cut down by the private sector annually to produce sawn timber and for use as firewood in tea, tobacco, sugar, lime and brick manufacture. A number of these players in the private sector have invested massively in tree planting programmes after being convinced that their operations contribute to deforestation and may lead to desertification. Tobacco, sugar and tea industries have established plantations of mainly eucalyptus to provide their fuelwood requirements. They also facilitate out growers who have enough land to grow eucalyptus woodlots and sale firewood to these industries which in the past relied heavily on bush firewood.

Government has also put in place a new policy which allows private entrepreneurs interested in growing industrial saw logs to acquire long-term permits (of up to 50 years) in gazetted government forest reserves in savannah woodlands. This is within government policy of privatisation of many of the public operations. Government last funded the establishment of industrial plantations in 1978. With financial constraints being faced by Uganda the participation of the private sector in tree growing has been long over due and will contribute to the efforts to combat desertification.

REFERENCES

1. Bagoora F.D.K (1998) *Participatory Approach among stakeholders in the implementation of NAP to Combat Desertification in Uganda.* Paper presented at the first National Forum on NAPCCD
2. Energy for Sustainable Development (ESD) Ltd (1995): *A study of Woody Biomass Derived Energy supplies in Uganda.*
3. Environment Liaison Centre International (ELCI – 1994): *Réseau International D'ONG Sur la Désertification* RIOD.
4. Forest Department (1992): *National Biomass Study Phase 1* Technical Report
5. Howard P.C. (1991): *Nature Conservation in Uganda's Tropical Forest Reserves.*
6. Ministry of Agriculture, Animal Industry and Fisheries (MAAIF-1998): Proceedings of a Planning Workshop for District Team Headers on the elaboration of a NAP for Uganda.
7. MAAIF (1998): Proceedings of a Workshop on Gender Consideration in the formulation of NAP for Uganda.
8. National Secretariat for CCD (1998): The National Action Programme to Combat Desertification and Drought for Uganda.

PROMOTING INDIGENOUS TREE SPECIES PLANTATION IN THE SAHEL, WITH SPECIAL REFERENCE TO BURKINA FASO

Albert Nikiema
Centre National de Semences Forestieres, 01 BP 2682 Ouagadougou , Burkina Faso, Tel: +226 356111, Fax: +226 356110, email : a.nikiema@fasonet.bf

1. INTRODUCTION

Indigenous species in the Sahelian regions provide an important part of the basic needs of the local people such as fruits, vegetables, medicines, fire-wood, building material etc. Despite their socio-economical importance, local people do not traditionally plant indigenous species. They grow naturally and are merely protected within the agroforestry parklands by farmers.

Change in landscape management of the rural areas due to population pressure and the need for more agricultural land, is affecting the vegetation and reducing the species diversity specially in the intensively cultivated areas.

Promoting indigenous tree plantations in the rural areas is therefore becoming more and more useful in the Sahelian countries as a mean to contribute to a sustainable management of the forest resources. It makes available in the neighbourhood of the villages the useful plants commonly needed for their fruits, their leaves, their wood, or whatever part used locally.

The vegetation of Burkina Faso belongs to the Sudan-Zambezian region. The climate is Sudan Sahelian with two seasons, a long and continuous dry season of 6 to 9 months and a rainy season with 6 to 3 consecutive months.

Combating Desertification with Plants, edited by Dov Pasternak and Arnold Schlissel.
Kluwer Academic/Plenum Publishers, New York, 2001.

The rainfall ranges from 300 mm per year in the north corresponding to the Sahelian region to 1100 mm in the south corresponding to the south Sudanian region.

2. IMPORTANCE OF INDIGENOUS TREE PRODUCTS IN THE SAHEL

Tree products contribute significantly to the nutrition and the income generation of the population. Bernard et al. (in Hall et al. 1997) found that the total revenue (207 000 CFA Francs) of a village in northern Côte d'Ivoire, *Parkia biglobosa* contributes for 4% comparing with 2% from *Vitellaria paradoxa* seeds, 25% from cereals and 18% from cotton. In terms of revenue from woody species in Burkina Faso, *Parkia biglobosa* provides 73% of the total income, compared with 14% for *Vitellaria paradoxa*, and 4% for *Adansonia digitata* (Ki in Hall et al., 1997).

Table 1. Important tree species and their usage in the Sahel (adapted from FAO, 1999)

Species	Food	Medicine	Fodder	Timber & Fire wood	Agroforestry	Others
Adansonia digitata	3	1	2		3	1
Acacia senegal	3		2	1		3
Acacia nilotica		1		2	1	3
Anogeissus leiocarpus				3		1
Balanites aegyptiaca	3		2	1	2	1
Borassus aethiopum	3			3		3
Faidherbia albida		1	3	2	3	
Khaya senegalensis		1	2	3		
Parkia biglobosa	3	2	1	3	3	
Prosopis africana				3		
Pterocarpus erinaceus			3	3	1	
Tamarindus indica	3	3			3	3
Vitellaria paradoxa	3	3		3	3	3
Ziziphus mauritiana	3		2			

1 low usage
2 important usage
3 very important usage

Human diseases used to be treated with plants by the local communities in the Sahel. Nowadays, according to Guissou in Kere (1998) around 90% of the population in Burkina rely on traditional medicines for their health care.

Examples of highly valued indigenous tree products from the Sahel are:

- Tamarind juice
- Shea butter
- Gum from *Acacia senegal*
- The fermented nere seeds called *soumbala*

3. GENDER AND NTFP

Women usually do harvesting and transformation of NTFP. They consider the activities as a source of income and the cheapest way to make food provisions and to gather other basic needs for the household consumption. In Burkina Faso, shea nuts are collected and processed into shea butter by women, the same goes for the transformation of nere seeds into soumbala (fermented seeds). Women possess most of the traditional medical recipes used to cure babies' diseases.

Women are also involved in handicraft making. In most of the sahelian countries a lot of tools and useful items are made from wood or leaves of different species. Leaves of *Borassus aethiopium*, *Hiphaenea tebaica*, (dattier) are used to make nattes, beds, baskets (Nikiema et al., 1997).

4. THREATS TO INDIGENOUS SPECIES

4.1 The shortening and suppression of fallow periods

Population growth and the introduction of cash crops in the farmers' production habits in the Sahel have led to a scarcity of the agricultural land. The immediate consequences are that farmers clear more land for agriculture and furthermore, fallow periods are reduced or simply suppressed. These changes in the farming system have a negative impact on the vegetation since the regeneration niches of some important species are destroyed. Typical Sudanian agroforestry species populations such as *Parkia biglobosa*, *Vittelaria paradoxa*, *Adansonia digitata* and *Faidherbia albida*, are not renewed.

Some useful plants disappear or become rare in the agroforestry parklands. The fallow vegetation no longer constitutes an improvement of soil fertility.

4.2 Bush fire

The frequency of bush fires is difficult to control and seems to have increased in the savannah region of West Africa, which affects the regeneration of some woody species. Some species are non-fire-tolerant, especially at the young stages e.g. *Ceiba pentandra, Adansonia digitata, Parkia biglobosa.*

It was found in the central part of Burkina Faso that more than 80% of the indigenous species were lost from the permanently cultivated land compared to a protected area.

The species found on cultivated land were: *Acacia seyal, Acacia nilotica, Adansonia digitata, Anogeissus leiocarpus, Azadiracta indica, Balanites aegyptiaca, Bombax costatum, Cassia siamea, Eucalyptus camaldulensis, Ficus sp. Khaya senegalensis, Mitragyna inermis Sclerocarya birrea, Terminalia avicennoides, Vitex doniania,* (Nikiema,1993).

4.3 Grazing

Cattle grazing in some area may have a negative impact because of the fragile ecosystem or because of the inappropriate attitudes specially the selective felling or coppicing of trees to feed animals. As a consequence of these practices some species become endangered. The following species are cited in the Sahel:

- *Pterocarpus erinaceus*
- *Faidherbia albida*
- *Acacia senegal.*

4.4 Harvesting of Firewood

More than 80% of the inhabitants in the sahelian countries use wood as a source of energy (Saw, 1990). Firewood is harvested from the natural vegetation, the contribution of the plantations is still negligible.

The reduction of some specific tree species populations is considered to be the consequence of wood consumption of certain species by the village and town households (Kere in Wittig & Guinko, 1998).

4.5 Drought

The effect of droughts due to the cumulated deficiency of rainfall in the Sahel, combined with soil degradation, has led to a large mortality in tree species populations.

Diallo, 1990 wrote that tree mortality is perceptible in the northern Sudan and Sahel regions of Burkina Faso. He concluded that more than 10% of the standing trees were dead in the Sahelian zone of Burkina Faso. Most of the dead trees were *Pterocarpus lucens, Dalbergia melanoxylon, Balanites aegyptiaca, Adansonia digitata* and many Acacias.

5. THREATS TO INDIGENOUS SPECIES

5.1 Seed procurement

The availability of good seeds for tree growing is the starting point of the afforestation programmes in the Sahel. Since the early 1980s efforts have been developed by some of the Sahelian countries to create seed centres, (e.g. Burkina Faso, Senegal, Niger).

The progress made so far by these centres consists of having located the natural seed stands and the selection of superior mother trees for seed harvesting. Seeds harvested are cleaned and stored at room temperature and in a cool room when required.

Thus seed of around 130 species are harvested and made available at the seed centre of Burkina Faso (CNSF, 1998).

5.2 Nursery techniques

Nursery techniques are progressively developed for growing local species, thus the problems of low and heterogeneous germination of some species is solved by different seed treatments.

In most of the nurseries in the Sahel, the containers used to raise seedlings are plastic bags of about 25 to 30 cm long and 7 cm diameter. These containers are not adapted to all species because of the rapid development of root systems at the seedling stage. Some species require other sizes of bags.

Suitable transplantation size from nursery to plantation area is not yet well defined for all-important species, although some research is carried out in this regard by some national research institutes in the Sahelian countries. In Burkina Faso, a field guide for nursery techniques of 43 species have been published (Nikiema, 1993).

5.3 Transplantation and survival rate

Transplantation techniques applied in the Sahelian regions are still to be improved because of the cost involved for site preparation and the low survival rates frequently observed.

Survival rate is sometimes observed to be less than 50% in the plantations. Drought is the main cause, but the survival of the seedlings can be improved if appropriate plantation techniques are applied and protection from the grazing animals efficient.

5.4 Tree improvement programme

The production of various species in terms of fruits, legumes, medicines, firewood or timber can be improved by the means of domestication and the development of tree improvement programmes.

Table 2. Actual state of research on local species in Burkina Faso

Area of research	Concerned species
Survey on species distribution areas	*Faidherbia albida, Acacia nilotica, Acacia senegal, Acacia seyal, Acacia raddiana, Anogeissus leiocarpus, Balanites aegyptiaca, Bauhinia rufescens, Khaya senegalensis, Parkia biglobosa, Ziziphus mauritiana*
Germplasm collection & conservation	Seeds of 130 species are collected and dispatched or available for sale; Herbarium specimens are collected and kept in the national herbarium
Genetic diversity study	*Parkia biglobosa, Faidherbia albida, Acacia senegal, Sclerocaria birrea, Tamarindus indica, Anogeissus leiocarpus, Khaya senegalensis, Acacia macrostachya*
Plant propagation	Nursery techniques and vegetative propagation techniques are developed for more than 40 of the commonly planted species.
Provenance and/or progeny trials	*Acacia senegal, Khaya senegalensis, Parkia biglobosa, Sclerocaria. Birrea, Faidherbia albida, Vitellaria paradoxa, Tamarindus indica..*

The potential economic value of species such as *Tamarindus indica, Vitellaria paradoxa, Acacia senegal, Khaya senegalensis* can easily justify investment to improve their production. As a matter of fact these species are already recognised by many Sahelian countries to be economically important (FAO, 1999).

6. CONCLUSION

Indigenous species actually play an important role for the population in the arid zones of West Africa by providing them with food, fuel, medicines, shelter, wood and many other products for the inhabitants. Despite this social and economic value the resources are not sustainably used. Domestication by way of selection in available germplasm of some of the important species becomes therefore an important issue. The development of appropriate sylvicultural techniques adapted to the sahelian climatic conditions are needed. The success of the small scale plantations done by farmers in the rural areas could be improved by applying adapted technology.

Table 3. Priority research area for local species adopted in Burkina Faso (adapted to the actual development in tree plantation)

	ECOZONES		
ACTIVITIES	SAHEL	NORTH SUDAN	SOUTH SUDAN
Distribution area assessment of species	*Adansonia digitata, Maerua crassifolia, Sclerocaria birrea*	*Adansonia digitata Vitellaria paradoxa, Sclerocaria birrrea*	*Adansonia digitata Borrassus aethiopum Tamarindus indica Vitelaria paradoxa*
Collection and conservation of vegetal material	*Maerua crassifolia*	*Vitellaria paradoxa*	*Borassus aethiopum Vitellaria paradoxa*
Study on eco-physiology of the species	*Acacia senegal Adansonia digitata Balanites aegyptiaca Bauhinia rufescens Maerua crassifolia Ziziphus mauritiana*	*Anogeissus leiocarpus Detarium microcarpum Parkia biglobosa Tamarindus indica Vitellaria indica*	*Anogeissus leiocarpus Borassus aethiopum Khaya senegalensis Parkia biglobosa Prosopis africana Tamarindus indica Vittellaria paradoxa*
Study on reproductive biology	*Acacia senegal Adansonia digitata Balanites aegyptiaca Faidherbia albida Maerua crassifolia*	*Detarium microcarpum Khaya senegalensis Parkia biglobosa Sclerocaria birrea Tamarindus indica*	*Borassus aethiopum Khaya senegalensis Parkia biglobosa Prosopis africana Tamarindus indica Vitellaria paradoxa*
Ecology of the species	*Acacia raddiana Acacia senegal Adansonia digitata Balanites aegyptiaca Bauhinia rufescens Boscia senegalensis Pterocarpus lucens Faidherbia albida Maerua crassifolia*	*Adansonia digitata Balanites aegyptiaca Detarium microcarpum Faidherbia albida Khaya senegalensis Parkia biglobosa Tamarindus indica Vitellaria indica*	*Afzelia africana Borassus aethiopum Khaya senegalensis Parkia biglobosa Prosopis africana Tamarindus indica Vitellaria paradoxa*

Table 3. Priority research area for local species adopted in Burkina Faso (continued)

	ECOZONES		
ACTIVITIES	SAHEL	NORTH SUDAN	SOUTH SUDAN
Genetic diversity	*Acacia senegal* *Balanates aegyptiaca* *Maerua crassifolia* *Ziziphus mauritiana*	*Adansonia digitata* *Parkia biglobosa* *Prosopis africana* *Tamarindus indica* *Vitellaria paradoxa*	*Borrassus aethiopum* *Khaya senegalensis*
Propagation and Sylviculture	*Adansonia digitata* *Acacia senegal* *Balanates aegyptiaca* *Bauhinia rufescens* *Maerua crassifolia*	*Bauhinia rufescence* *Detarium microcarpum* *Prosopis africana* *Tamarindus indica* *Vitellaria paradoxa*	*Borassus aethiopum*
Root symbiosis study	*Ziziphus mauritiana*	*Prosopis africana* *Tamarindus indica* *Vitellaria paradoxa* *Ziziphus mauritiana*	*Prosopis africana* *Tamarindus indica* *Vitellaria paradoxa*
Pest and disease resistance	*Bauhinia rufescens*	*Bauhinia rufescens* *Vitellaria paradoxa* *Parkia biglobosa*	*Parkia biglobosa* *Vitellaria paradoxa* *Khaya senegalensis* *Borassus aethiopum*
Selection	*Ziziphus mauritiana*	*Parkia biglobosa* *Tamarindus indica* *Vitellaria paradoxa* *Ziziphus mauritiana*	*Parkia biglobosa* *Tamarindus indica* *Vittellaria paradoxa*

ACKNOWLEDGEMENT

I am very grateful to Pr. Jos Van der Maesen who helped to improve the paper with his comments and his corrections.

REFERENCES

1. C.N.S.F. 1998. Catalogue 1998-2000.

2. F.A.O. 1999. Plan d'action sous regional sur les ressources genetiques forestieres de la zone Sahelienne et Nord-Soudanienne. (draft).
3. Guinko, S. 1998. *La diversité des ressources genetiques forestières du Burkina Faso, quelques aspects de leur conservation et de leurs utilisations.* Atelier Régional de Formation sur la Conservation et l'Utilisation Durable des Ressources Genetiques forestières (en Afrique de l'Ouest, Central et Madagascar), Ouagadougou, 16-27 mars 1998. 20 pp.
4. Hall, J. B., Tomlinson, H. F., Oni, P. I., Buchy, M., Aechischer D. P. 1997. *Parkia biglobosa: a monograph.* School of Agricultural and Forest Sciences, Publication Number 9, University of Wales, Bangor. 107 pp.
5. Kere, U. 1998. *Végétation et utilisation des plantes spontanées dans la région de Tenkodogo (BURKINA FASO).* Etudes flor. Vég. Burkina Faso 4, 3-55.
6. Nikiema, A. 1993. *Regeneration of Parkia biglobosa* (Jacq. R. Br. Ex G. Don) in an agroforestry system, M.Sc. Thesis, Department of Forestry, Wageningen Agricultural University, Netherlands.
7. Nikiema, A., Ouedraogo, S., Boussim J. 1998. *Conservation et utilisation des ressources génétiques forestières.* Rapport national du Burkina faso. Atelier Regional de Formation sur la conservation et l'utilisation durable des ressources génétiques forestières en Afrique de l'Ouest, Central et Madagascar. Ouagadougou, 16 - 27 mars 1998. CNSF.
8. Nikiema, A., Sanon, M., Fraiture de, A., Tolkamp, W. 1993. Fiches de production de plants en pepiniere. Centre National de Semences Forestieres. Notes Techniques n°4.
9. Saw, H. 1990. Le bois-énergie au Sahel. Environnement et développement. ACCT, CTA, KARTHALA.
10. Wiersum, K. F., Slingerland, M. 1998. *Use and management of two multipurpose tree species (Parkia biglobosa and detarium microcarpum) in agrosilvopastoral land use systems in Burkina Faso.* Antenne Sahelienne de l'Université Agronomique de Wagenigen (Pays Bas) et de l'Université de Ouagadougou (Burkina Faso). Document de projet 41.

INCORPORATING TREE AND SHRUB SPECIES INTO SMALL SCALE FARMING SYSTEMS IN THE DRY REGION OF INDONESIA

Extension Strategy and the Challenges for Institutional Development

Anthony P.Y. Djogo
LPPSP, Jalan Maumere No. 18, Nefonaek, Kupang, NTT 85227-Indonesia.
Tel/fax:62-380-832292, email:djogo@asia-mail.com or djogo@yahoo.com

1. INTRODUCTION

Indonesia is generally classified as a humid tropical region with large areas of rain forest[1]. There are, however, some regions which are dry, receiving an average annual rainfall of 1000 mm to 1200 mm during a 3-4 month rainy season, with an ETP of 1800-2200 mm per year. These areas are important resources for around ten million people and were once among the main areas of livestock raising in Eastern Indonesia. It is a poor area, with a fragile environment and a complex social and economic situation, with the rural communities populated primarily by peasant farmers.

This paper focuses on the East Nusa Tenggara Province (NTT). Various projects have been tried to combat the increasing degradation of this dry region, mainly in the areas of soil and water conservation, reforestation and re-greening[2]. Agroforestry (*wanatani*) and community forestry (*hutankema-syarakatan*)[3] methods are included in these designs, as they offer the added benefit of products such as fruit, fuel, fodder, and building materials.

The government has a number of projects designed to improve the physical, environment and socio-economic aspects of the region. Non-governmental organisations on the other hand seem to have more influential impact on conservation and development efforts at the community level. The success of the programmes was not determined by the

Combating Desertification with Plants, edited by Dov Pasternak and Arnold Schlissel.
Kluwer Academic/Plenum Publishers, New York, 2001.

sophistication in technology, type of species introduced or the budget available, but primarily by their community approach and extension strategy, and by local level institutional development and empowerment.

This paper presents the experiences and results of projects based on farmer-to-farmer extension in agroforestry and community forestry development promoted by two NGOs in Eastern Indonesia. While farmer based extension strategy is important, it is not sufficient to sustain the development endeavour and needs to be supported by sound institutional arrangements. The community approach and extension strategy as part of empowerment and institutional development are crucial for the success of incorporating trees and shrubs species at the field level. At the higher level of planning, decision making and policy leverage, the institutional dimension is critical as well[4].

2. TREE AND SHRUB SPECIES IN RURAL FARMING SYSTEMS

NTT is dominated by small-scale dryland farming systems, which are changing from subsistence to market oriented, a process which influences the use of tree and shrub species. In the traditional system, when natural resources were still vastly available, more species were involved in farming activities and livelihood, but that number decreased significantly during the transition. A large number of local species used in the past has been identified as being of potential use for agroforestry and community forestry[5]. However, they are not fast growing, and even though ecologically well adapted, farmers do not actively incorporate them on their farms. Most of the "improved" agroforestry models employ fast growing exotic species. For the subsistence farmers the goal is to reduce fallow and generate tree products for consumption, while for the market farmer it is fodder banks, plantation or timber species.

3. AGROFORESTRY AND COMMUNITY FORESTRY

3.1 Agroforestry

3.1.1 Indigenous agroforestry

Although the formal science and practice of agroforestry was introduced to the province towards the end of the 1980s, indigenous models of agroforestry constitute a centuries-old tradition.

The number of species involved in indigenous agroforestry was high compared to introduced agroforestry. The system tended to be stable, productive and ecologically sustainable at the subsistence level, but proved unable to withstand pressures of population growth and to respond to the demand of the market.

3.1.2 Introduced agroforestry

Interesting agroforestry models were introduced in the 1930s, 1960s and 1980s. In NTT the most successful models have been the Amarasi, the Sikka, and the Diversified Models. Several other approaches are also being promoted, but farmers are still resistant to change and hesitate to risk possible crop reductions should the new systems not meet expectations.

3.1.2.1 Amarasi

The Dutch introduced this model in the Amarasi area of West Timor in the 1930s. This model has contributed significantly to the improvement of the ecosystem and to the local economy. The system is based on shrub type leucaena (*Leucaena leucocephala*, previously called *Leucaena glauca*) as the main multipurpose species, which provides support for stabilising the local agroecosystem (Metzner, 1983).

The leucaena was introduced to combat the land degradation caused by shifting cultivation and free grazing and to contain the spread of *Lantana camara*. Leucaena based agroforestry system assisted the transition from subsistence farming with shifting cultivation techniques to permanent agriculture with livestock as the main source of income.

The success of this model was determined not only by the strength of the fast growing *L. leucocephala* but by the transformation of the institutional setting as well. The local ruler has the authority to permit local people to cultivate the land, and farmers were not allowed to open new land before planting leucaena on their previous farms.

With the introduction of leucaena as fodder shrubs, cattle production changed from free grazing to the cut and carry system. This process promoted a more equitable distribution of cattle ownership and reduced destruction of the ecosystem caused by free grazing cattle.

3.1.2.2 The Sikka Model

The Sikka model was developed in the 1960s by a Catholic Church based-NGO, the local government and local communities in response to the increasing population pressures in the district (Flores Island) where land capability for supporting agriculture was very limited. Both poverty and food insecurity reached alarming proportions (Metzner, 1976).

Collaborating with the local community and associating agroforestry with productive agricultural crops such as coconut, cocoa, and cashew allowed the degraded areas to be converted into productive agricultural lands. *Leucaena leucocephala* was used as the single multipurpose tree species planted in the hedgerow form.

The success of this project was not necessarily due to the *L. leucocephala* as a miracle species. The charismatic role of the church and its leaders, the guiding role of the local government and the acceptance by the local people were all important contributory factors.

3.1.2.3 Diversified Model (Yayasan Tana Nua / Yayasan Geo Meno Approach)[6]

With the devastation of leucaena in 1985/1986 due to the attack by psyllid (*Heteropsylla cubana*), several NGOs in NTT working with soil and water conservation helped local people with fodder, fuel wood, timber and living fences using diversified species. While the Amarasi and the Sikka Models used *L. leucocephala*, the diversified model incorporated other species such as *Gliricidia sepium*, *Cassia siamea*, *Sesbania gradiflora*, *Calliandra callothyrsus*, *Zappoteca tetragona*, *Acacia villosa*, *Flamegia congesta*, *Acacia angustissima* etc.

The projects were initially developed in Sumba island by the Yayasan Tana Nua (YTN) and in Flores Island by the Yayasan Geo Meno (YGM). Although they started with a small-scale project, they have significantly changed the farming systems from shifting cultivation or fallow-based farming system into permanent agriculture with better productivity and minimized dependence upon external input.

The strength of these projects was rooted in the community based extension strategy. Farmer to farmer extension strategy, the key component of the development, was open to modification depending on the situation at the local level. However, the projects and local communities were subject to the rules and regulations imposed by the local government, as well as to social and economic conditions imposed by the market.

3.2 Community Forestry

In the early 1990s some NGOs and government projects developed various community forestry models based on planting of potential timber trees either on farmers' properties, common properties or state properties. YTN and YGM are two major NGOs working with this approach, both addressing private property. In this *family forestry* model, each farmer interested in joining the community forestry project receives technical training in nursery development and seedlings. Farmers are free to select the trees, generally *Swietenia mahogany*, *Tectona grandis*, *Gmelina arborea*,

Acacia auriculiformis, and *Senna siamea* (*Cassia siamea*). Some fruit trees can also be incorporated to the family forest.

The other project developed by the State Forest Cooperation (in West Timor) incorporated fast growing timber species on government forest land, and involved farmers only in the first 2 to 3 years when they were allowed to grow annual food crops in rows between the timber trees. The project has been a success in terms of silvicultural techniques and forest management, but has resulted in some serious conflicts with the local communities, primarily related to land tenure issues. If institutional adjustments are not forthcoming, the poverty of the local people will almost certainly drive them to illegal logging and encroachment.

4. COMMUNITY APPROACH AND EXTENSION STRATEGY

The integration of tree and shrub species into small-scale agricultural production system is most successful when a community approach and extension strategy are used, as demonstrated by a number of NGO projects.

Among them are projects developed by the YTN and theYGM. They share an approach to technology development, testing and transfer to the rural community through farmer-to-farmer or community based extension. Their projects begin with small-scale activities, involving a small number of farmers at the outset (possibly fewer ten farmers) from diagnostic to farm planning, and small scale farm trials, each stage supported by cross visits, quarterly meeting, farmers' workshops and local level training.

Since the dry region face particularly critical challenges owing to its environmental characteristics, sufficient production of seedlings in terms of quantity, continuity and quality is crucial. The appropriate time for seedlings to be available should be well planned. Due to the limited availability of water for nurseries and the scattered location of farmers' fields, the project encouraged the establishment of nurseries by groups and individuals. Compared to the government projects, which tend to provide seedlings, these projects provide seeds so that every individual or group of farmers can establish his/its own nursery. This approach fosters responsibility and commitment on the part of each participant.

The increasing number of farmers interested in these projects is due not so much to the extension process as to the transfer of information from farmer to farmer and to the sharing of seeds between villages or hamlets. Experience also shows that farmer based extension is also very effective in encouraging self-sustained technology development, minimizing external support, and use of local organisations at the decision making level, generating potential solutions to problems related to property rights issues and collective action (Djogo, Asa, and Nurak, 1992 and Djogo 1990). Many projects developed with conventional agricultural extension systems have failed to achieve their objectives (Fisher, 1992).

However, compared to government projects, which usually target a large number of farmers (100 to 200 families) and wide areas (100 to 250 ha per village, and sometimes 1000 ha), implementation is on a much smaller scale[7].

5. INSTITUTIONAL DEVELOPMENT: THE CHALLENGES FOR THE FUTURE

The development of community based agroforestry (or community forestry) is generally challenged by latent uncertainties. Besides the environmental factor, some of the risks come from human opportunistic behavior, conflicts about property rights issues, social injustice, market and population pressures, as well as political changes.

The second challenge is how to develop market-based agroforestry and community forestry. In the villages where we[8] are working, there are usually no economic institutions to support the newly opened access to the market.

Many subsistence farmers are peasants slowly moving to market oriented activities. But they are not strong enough to get involved in markets and contracts. They lack information and bargaining power, business skill and knowledge, capital for investment, savings and management capabilities. The middlemen and traders often control them. They tend to be weak in negotiations and transactions. A case study in Flores Island showed that some farmers successfully grew timber trees but made a very bad deal out of it. Although still far from harvesting time (more than 20 years for *Swietenia mahogany* and more than ten years for *Gmelina arborea*), traders approached them to purchase the trees at an early stage, so as to have the right to harvest the trees when mature. Of course prices and profits would be much higher then. Since the farmers do not have enough cash, the large sums offered by the traders are a big temptation (Djogo, 1999).

The other main problem, as widely discussed, is the land tenure issue. Any development project relates to the conservation and development of natural resources management, which involves state property or common property. Most of the time, the government objectives do not match the farmer's goals. The land entitlement for individual farmers is often weak and tends to discourage the adoption of long term species through agroforestry or community forestry.

Control of free grazing, fire and theft, illegal logging, encroachment etc need firm enforcement based on the agreed rules and regulations, policies and laws. Many projects have failed due to an inability to overcome these drawbacks. How can the government with its formal rules and regulations support local rules and regulations?

Local level management calls for collective action. There are traditional farmers' working groups in some districts but none in many others, and there is a lack of social cohesion. In order to control the inappropriate behaviour of government personnel, fight abuse of power and enforce local rules and regulations, there have to be changes in social relationships, family relationship and joint property, as well as a move to contractural arrangements, markets and individual property and territory (Agrawal and Gibson 1999). Serious transformations in informal institutions involving constructive support from the formal institutions will have to take place.

Government contribution in the form of extension services, law enforcement and institutional changes in these locations is still very weak. While non-governmental organisations and local institutions have shown their potential role in capacity building and management coalition, unfortunately they lack the power of regulation.

Some NGOs have tried to promote the conflict resolution approach, which goes beyond the short term perspectives and takes into account not only local perspectives but also the link with national level policy making. Such an approach not only improves the performance of local organisations, it also affects empowerment, management coalition, extension strategy and information delivery, etc.

How could local rules and regulations be enforced and if possible legitimised? The development approach in Indonesia (particularly the rural development and poverty alleviation programme) has been criticized for focusing excessively on physical infrastructure development and neglecting social infrastructure/social capital. In fact the current situation has shown that institutional infrastructure is a crucial dimension.

6. IMPLICATIONS FOR COMBATING DESERTIFICATION

From the ecological standpoint, the experiences from agroforestry and community forestry development in the semidry region of Indonesia would seem to be irrelevant to arid regions, although desertification can occur in semiarid or sub humid lands as well (MASHAV, 1995). However, the socio-economic aspects of the efforts to develop small scale farming systems involving peasant farmers with a poor resource base and complex social-political institutional issues might be useful for comparative studies. Desertification (in Indonesia we often use the term environmental deterioration) is not purely caused by climatic, physical/geological or biological process, but also by human factors. Social and economic aspects can be the cause of land degradation as well as offering solutions to the problem. Such solutions include management practices, land tenure rules, settlement patterns and indigenous knowledge systems, gender relationship, etc (Stiles, 1995). The causes of desertification include human activities such as land use management, generally overgrazing, over cultivation and deforestation (Thomas and Middleton, 1994, Mainguet, 1994).

There are many new approaches to facilitate the development of rural community and environmental conservation. They include community based development, farmer to farmer extension strategy, participatory development, and a combination of conservation and development. The farmers-first paradigm has been widely adapted to challenge the preconceptions dominating rural development (Chambers, 1983 and 1993). Unfortunately, the farmers-first premise has been more a matter of academic discourse than application in the field. Very few NGOs have incorporated this approach into their extension strategy. We have tried for almost ten years to work with a farmer to farmer development process promoted by World Neighbours. We achieved a certain level of progress, which from our point of view is a good foundation for sustainable development, with a small-scale development project. Our farmers encounter complex political, social, economic and environmental problems, one of the most important issues being the institutional environment and arrangements[9]. Government rules and regulations often do not harmonize with local level decision-making, rules and regulations. If there had been good collaboration between government and the NGO for facilitating the development of local institutions, we would achieved a better progress.

From the environmental perspective, the selection of trees and shrubs species needs to take into account the combination of indigenous species and exotic species. Local species with drought resistance and multipurpose characteristics would be suitable for the reversal of resource degradation. However, local farmers and project implementing agencies tend to prefer fast growing exotic or commercial species.

Different types of project would require species with different characteristics. The key question is, what species are locally superior from the environmental point of view and could be used as pioneer trees or shrubs to challenge the local harsh climatic and edaphic conditions? Can farmers accept incorporation of these species t into their farming systems, and what advantages (direct or indirect) might farmers gain? Then, what species are economically promising or could provide basic needs for local people, while possessing the potential to be combined into improved farming systems with non-economic oriented species in the form of agroforestry or community forestry? What forms of institutional arrangement and mechanism could be employed in combining conservation (incorporating non economic locally adapted drought resistant species) and development (involving cash crops and commercial timber trees)?

7. CONCLUSION

Community based projects which attempt to combat resource degradation through small-scale agroforestry and community forestry seem to be promising. Our experience has shown that incorporating tree and shrub species is an important entry point and that as small-scale projects in farmers' fields they can then be expanded to common property or scaled up to include more farmers or villages in the project.

Farmer-to-farmer extension has shown its potential in the process of transfer of technology. However, risks and uncertainty may be encountered, particularly from social, economic and political perspectives.

There are possibilities for reorienting the approach in working with peasant farmers living in the dry regions of Indonesia. Institutional issues in terms of property rights, rules and regulations and the development of economic oriented institutions will need particular attention.

Farmer based extension strategy is appropriate for short and mid term perspectives, but for the long-term orientation, institutional development will have to be taken into consideration.

REFERENCES

1. Agrawal, A. and Gibson, C.C. (1999). *Enchantment and disenchantment: The role of community in natural resource conservation.* World Development Vol. 27(4): 629-650
2. Bromley, D.W. (1989). *Economic Interests and Institutions: The conceptual foundations of public policy.* Basil Blackwell. New York.
3. Bunch R. (1985). *Two Ears of Corns: a guide to people-centred agricultural improvement.* World Neighbours. Oklahoma.
4. Chambers, R. (1983). *Rural Development: Putting the Last First.* Longman. London.
5. Chambers, R. (1993). *Challenging the Professions. Frontiers for Rural Development.* IT Publications. London.
6. Djogo, T. (1990). *Yayasan Geo Meno and Its Work in the Village Gerodhere, Flores, Indonesia.* Social Forestry Network No. 10. Summer. 1990. ODI. London U.K.
7. Djogo, A.P.Y. (1992). *The Possibilities of Using Local Drought Resistant Multipurpose Tree Species as Alternatives to Lamtoro (Leucaena leucocephala) for Agroforestry and Social Forestry in West Timor.* Working Paper No. 32. EAPI East West Center. Honolulu, Hawaii, USA.
8. Djogo, A.P.Y., Asa, Y. and Nurak, V. (1992). *Promoting Agroforestry Techniques: Experiences with participatory development process and farmer-based extension strategy of an NGO from NTT.* Paper presented to APAN Regional Expert Consultation on Farmer to Farmer Agroforestry Extension. Cebu, The Philippines. 1992.
9. Djogo, A.P.Y. (1993) *Diagnosa Kebutuhan Pengembangan Agroforestry (Diagnose The Needs for Agroforestry Development).* Paper presented to the First National Agroforestry Workshop. 24-26 August, 1993. In Gintings et.al. (1994). Prosiding

Lokakarya Nasional Agroforestry. P3HKA dan APAN. Bogor. Indonesia. pp. 103-117

10. Djogo, A.P.Y. (1997). *Domestication of Indigenous and Naturalised Species for Reforestation and Agroforestry.* Paper presented to Regional Workshop on Domestication of Agroforestry Trees. Yogyakarta, Indonesia. 4 - 7 November 1997. Jointly organised by ICRAF, WINROCK, UGM, and ACIAR.
11. Djogo, A.P.Y. (1999). *Forestry Sector Policy in Indonesia: Implications for Community Forestry and Rural Economy Development.* MSc Dissertation, Wye College, University of London.
12. Fisher, L.A. (1992). *Agricultural Development in the Uplands: How conventional agricultural extension systems have failed resource-poor farmers, and what to do about it?* Paper presented to the Regional Meeting on Basic Education and the Environment, Berastagi, North Sumatera, Indonesia. 5 - 9 October 1992.
13. Fox, J.J. (1977). *Harvest of the Palm.* Harvard University Press.
14. Fox, J.J. (1988). *The Historical Consequences of Changing Patterns of Livelihood on Timor.* In: Wade-Marshall, D. And Loveday, P. (Eds.). Contemporary Issues in Development. North Australian Research Unit, Canberra. Pp 259-279.
15. Mainguet, M. (1994). *Desertification. Natural background and human mismanagement.* 2nd Ed. Springer-Verlag. Berlin.
16. MASHAV - Centre for International Cooperation. (1995). State of Israel, Ministry of Foreign Affairs. *Combating Desertification: The Israeli Experience.* Romema, Jerusalem 91950. Israel. http://www.israel-mfa.gov.il/mfa/.
17. Metzner, J.K. (1976). *Agriculture and Population Pressure in Sikka, Isle of Flores.* Development Studies Center. Monograph No. 28. The Australian National University
18. Metzner, J.K. (1983). *Innovations in Agriculture Incorporating Traditional Production Methods. The Case of Amarasi (Timor).* Bulletin of Indonesian Economic Studies. Vol. XIX No. 3. December 1983.
19. Sinaga, N. and Wodicka, S. (1996). *Farmers-based extension in the marginal uplands of Sumba, Indonesia: a case study of Tananua experience.* Agricultural Research and Extension Network. Network Paper No. 59a. January 1996. ODI
20. Stiles. D. (Ed.). (1995). *Social Aspects of Sustainable Dryland Management.* John Wiley and Sons. Chichester and New York. Publication of UNEP.
21. Thomas, D.S.G. and Middleton, N.J. (1994). *Desertification: Exploding the Myth.* John Wiley and Sons. Chichester and New York.
22. UNEP. (1991). *Status of Desertification and Implementation of the United Nations Plan of Action to Combat Desertification.* UNEP. Nairobi. (United Nations Environment Programme).
23. Nabli, M. and Nugent, J. (1989). *The new institutional economics and its applicability to development.* World Development Vol. 17. No. 9:1333-1347.
24. North, D.C. (1990). *Institutions, Institutional Change and Economic Performance.* Cambridge University Press, Cambridge.
25. Ostrom, E. (1985). *Formulating the elements of institutional analysis.* Paper presented to a Conference on Institutional Analysis and Development Washington DC May 21- 22. 1985.
26. Uphoff, N. (1986). *Local Institutional Development.* West Hartford. CT. Kumarian Press.
27. Williamson, O.E. (1985.) *Economic Institutions of Capitalism: Firms, markets, relational contracting.* The Free Press. Macmillan, New York, London

FOOTNOTES

1. In Indonesia an area classified as semiarid could turn out to be sub-humid or dry sub-humid from the standpoint of precipitation. The following two tables show the dry region classification based on precipitation and the Thornthwaithe moisture index.

Classification of dry region based on the annual precipitation

No.	Dry Ecosystem	Rainfall (mm/year)
1	Arid	<50 - 150
2	Semiarid	150 - 600
3	Sub-humid (with marked dry season)	600 - 1200

Source: Mainguet, M. (1994). Desertification. Natural background and human mismanagement. 2nd Ed. Springer-Verlag, Berlin.

Climatic region based on the Thornthwaite moisture index of the ratio of precipitation to potential evapo-transpiration.

No.	Climatic region	P/ETP
1	Hyper-arid	<0.05
2	Arid	0.05-0.20
3	Semiarid	0.21-0.50
4	Dry sub-humid	0.51-0.65
5	Moist sub-humid	>0.65

Source : UNEP. 1991. Status of Desertification and Implementation of the United Nations Plan of Action to Combat Desertification. UNEP, Nairobi.

2. Reforestation (*reboisasi*) and re-greening (*penghijauan*) are government terminology. Reforestation takes place on government land with forest species, while re-greening would be on non-forest lands (private or communal lands) involving non-timber species. The two NGOs (YGM and YTN) tend to use the term soil and water conservation (*konservasi tanah dan air*) and family forestry (*hutan keluarga*)

3. In Indonesia, community forestry (*hutan kemasyarakatan*) is a form of social of (*perhutanan sosial*). Social forestry could be community based forest management in any type of land whether state owned land, common properties or private properties. Community forestry is usually implemented in government forestland. In other countries these difference does not matter. Both could be community forestry or both could be social forestry. In this paper I used the term community forestry for both schemes.

4. By institutions' in this paper is meant: organisations, rules and regulations, codes of conduct, social constraints, markets, contracts etc. Included are: transaction costs, property rights, opportunistic behaviour, asymmetry in information, collective action. See: North (1990 and 1995), Williamson, Nabli and Nugent (1989), Uphoff (1986), Bromley (1989), Ostrom (1985 and 1996). Management coalition or co-management schemes include the participatory and community based development process, or local level management. Institutional analysis could be viewed from in the light of New Institutional Economics (NIE) or Institutional Analysis for Development (IAD).

5. Since 1985 we have identified a large number of drought resistant local species for agroforestry and social forestry or community forestry and established their potential for domestication (Djogo,1992 and 1997).

6. YTN (Yayasan Tana Nua) or Tana Nua Foundation and YGM (Yayasan Geo Meno) or Geo Meno Foundation are two NGOs that are adopting farmers for the farmers' extension strategy for soil and water conservation and the development of community forestry (locally called family forestry). YGM has phased out the project in 1997 while YTN continues to expand their activities on three islands in the NTT (Sumba, Flores and Timor). These two organisations are working with support from World Neighbours (WN), Oklahoma, USA. See http://www.wn.org

7. This is to be differentiated from the quantitative measure where the number of farmers and the areas developed should be large to fulfil the target the GO's project planning. Quality could be assessed through the consistency of the programmes farmers have adopted, farm planning, design and its application, productivity, stability, and autonomy, as well as solidarity amongst community members.

8. I worked with YGM in 1989-1997 and have also had a good collaboration with YTN for consultation, sharing ideas and experiences, materials (such as seeds, information, etc), training.

9. Institution Environment is one way of analysis of rules and regulations in New Institutional Economics proposed by North (1990). Institutional arrangement according to Williamson (1985) is an arrangement between economic units that govern the way in which these units can co-operate and/or compete. Ostrom (1985), from the perspective of Institutional Analysis for Development classifies institutional arrangements as operational rules for governing resource use and collective choice rules. Institutional infrastructure consists of information, property right and governance. Institutional mechanism may be effective rules and restraints, competitive pressures from the market, etc.

OBSTACLES TO THE ADOPTION OF COVER CROPS

The Mucuna in Benin

Emile N. Houngbo
GRAAP-DESR (FSA/UNB), 05 BP 774, Cotonou, Rep. du Benin, Tel: 229-330183, email: enomh@hotmail.com

1. INTRODUCTION: AGRICULTURE VERSUS THE BATTLE AGAINST DESERTIFICATION

In 1969 Thorbecke established that the discussions regarding development strategies had had a profound effect on the analyses of agricultural growth. He indicated that more and more economists, who had been almost totally ignorant of the strategic importance of agriculture in economic growth, were shifting the blame for the failure of Third World development strategies on to the theories which had advocated accelerated industrialization. This once again raises the problem of the place to be accorded to agriculture in development, in particular in the Third World where it is still the predominant sector of production and employment. The role of agriculture is therefore vital in any development system which is meant to be durable insofar as it represents the very prototype of an economic production entity where the trilogy of the production factors - Land /Capital/Labor - has not been sufficiently substantiated. Land and capital are closely intertwined, so much so that some identify land as capital at the level of the land resource. Land as capital is comparable to the value in land. Indeed it is particularly in agriculture that land cannot automatically be considered as a finite natural resource. In this economic entity, humankind uses input and cultivation methods to improve the land so that it becomes a capital which can be amassed. The use in agro-forestry of cover crops such as the *Mucuna pruriens* enhances the agricultural production capacity of land-capital. It

Combating Desertification with Plants, edited by Dov Pasternak and Arnold Schlissel.
Kluwer Academic/Plenum Publishers, New York, 2001.

follows therefore that the finite nature of land is not obvious in agriculture as is the case in the industrial sector where mining resources (oil, diamonds, phosphates) are bound to come to an end sooner or later. This quality of agriculture where almost all production resources can be improved by humankind is so fundamental in connecting other sectors of the economy that agro-centrists such as Ojala, Kuznets and Nicholls believe that "the good health of an economy, whatever its level of development, always depends on the good health of agriculture."

However, agriculture, which itself also exploits natural resources (soil, water, plants, air), leads to the depletion of the land's agronomic value if the cultivation systems developed fail to improve or at least respect the land-capital which is available from the beginning. Such systems inevitably lead to chemical and physical degradation of the soil, a vector of desertification in the long-term.

Unfortunately cultivation systems which show little respect for land as capital (and therefore little respect for the environment) are widespread in Africa. The Republic of Benin is a typical example in this regard.

After all is said and done, agriculture can be a precursor of desertification or a remedy if properly implemented. It is in this sense of agriculture as a barrier to desertification that the search for cover crops has been made in Benin and elsewhere. But finding cover crops is not sufficient; methods must also be found to ensure that their use is adapted. This is not always a simple task as in this case study from the sub-prefecture of Sakété in Benin.

2. DESCRIPTION OF LOWER BENIN AND THE SAKÉTÉ SUB-PREFECTURE

Situated in West Africa on the coast of the Gulf of Guinea, the Republic of Benin lies between a latitude of 6 - 12 degrees north and a longitude of 1 and 3 degrees east. It is a small country (112,622 km^2) with some 6 million inhabitants (1998), with an annual per capita income of US$ 370 making it one of the Less Developed Countries as defined by the World Bank (1997).

In terms of administrative and territorial division, Lower Benin covers an area of 17,920 km^2 or 16% of the country with population density estimated at 220 inhabitants per km^2 (as compared with 45 inhabitants per km^2 for all Benin) and fewer than 20 inhabitants per km^2 in the four departments of the north (Borgou, Alibori, Donga and Atacora).

Lower Benin contains the entire rain forest zone and part of the zone of forest in transition. The fields derived from this vegetation have already been invaded by undesirable species such as the *Imperata cylindrica* and the *Commelina benghalensis*. The zone of forest in transition comprises two sub-zones in the south and north, of which the one in the south is situated on the terminal continental plateau where the soil is more impoverished owing to over-exploitation (high population density).

The Sakété sub-prefecture located in the continental plateau covers an area of 350 km^2. Densely populated with a primarily farming community (approximately 75%), it enjoys a tropical humid Sudanese-Guinean climate with average annual rainfall of between 800 mm and 1400 mm. Ethnic groups are, in order of importance, the Nagot, Goun, Torri and Holli. In 1992 population density was 197 inhabitants per km^2.

3. AGRONOMIC VALUE OF THE *MUCUNA*

The *Mucuna pruriens* (known in English as the velvet bean) is an annual rampant leguminous plant which probably originates from Asia (Purseglove, 1968). The *utilis* variety has been selected as the cover crop to restore the land in a short period of time after one or several cultivation cycles instead of the traditional -often long - fallow.

Aside from the production of a large quantity of organic matter (from 7 to 9 tons of dry matter and 100 kilograms of nitrogen per hectare), the cover of the *Mucuna* improves certain of the soil's physical characteristics such as porosity (better aeration), reduced density (better root development), and water retention capacity (less sensitivity to drought) (Hulugalle et al., 1986). Caveness (1988) believes that the parasite nematode population is also reduced. This last characteristic of the *Mucuna* is frequently a weakness of many other cover crops which enlarge the nematode population, thereby increasing the likelihood of crop sickness (Koudokpon & Versteeg, 1996). The results of studies conducted in Benin and elsewhere indicate that a one-year fallow for the *Mucuna* almost doubles the corn yields in the control plots.

Apart from the fact that *Mucuna* improves soil fertility and structure, it also controls undesirable species such as couch grass *(Imperata cylindrica)* and *Cynodon dactylon*. Results of studies conducted at the station have demonstrated that the *Mucuna* is more effective than the herbicide "Glyphosate" (trade name - "Round-up"), or mechanical methods such as periodic cutting and ridging after plowing (Koudokpon & Versteeg, 1996).

All in all, cultivation of the *Mucuna* enhances impoverished soil colonized by the *Imperata*:

- by introducing organic matter which improves the soil's fertility and physical structure;
- by introducing the nitrogen necessary for subsistence crops, particularly corn;

- by choking the population of the *Imperata* until it is practically harmless or, at best, causing its elimination.

4. OBSTACLES TO THE ADOPTION OF THE *MUCUNA* IN THE SUB-PREFECTURE OF SAKÉTÉ

Despite all the above-mentioned benefits, *Mucuna* has not been widely adopted by Benin's many farmers. This is true in Sakété where cultivation systems have already led to advanced soil degradation. The production systems predominating in Benin derive initially from itinerant cultivation practices which comprise a cultivation phase (one to three years) and a fallow (four to twenty years). These traditional systems, although proven to be stable and effective, require much cultivable land.

However, their ongoing implementation, together with human pressure on land resources, has led to the reduction and even the elimination of the fallows, the degradation of most of the land, and falling yields. During the past twenty years this problem has therefore caused farmers to develop a certain "intensification" of cultivation systems by modifying cultivation techniques consisting for example of progressively replacing corn with cassava and cassava with niebe and groundnuts, moving from clearing the vegetation by machete to the intensive ridging of soil.

However it should be noted that these techniques have not been developed to improve or maintain soil fertility levels and yields. Rather, they constitute a means of extracting the last mineral reserves from the soil. This is a type of agriculture which resembles mining in that it mines the soil without compensation just like the mining of ore. The consequence is the widespread soil degradation in the sub-prefecture of Sakété and Lower Benin with an accompanying invasion of the soil by couch grass (*Imperata cylindrica*).

Since the farmers confronting this problem are deeply integrated into a barter economy, we expected that in order to increase their income from progressively less farming, they would adopt agro-forestry technologies proposed by research and development organizations: *Mucuna pruriens, Senna siamea, Gliricidia sepium, Cajanus cajan*. Paradoxically however the farmers reacted quite differently. Few technologies have been adopted, although the farmers are well aware of the degradation of their lands. This is proving to be a serious headache for all the developers and researchers who take an interest in the agricultural and environmental situation of Benin.

Still concerned about this problem, in 1995 we became aware, during a study of "The problems of allocating resources linked to the durable production of cassava in Benin" which was part of the sub-regional project ESCaPP (Ecologically Sustainable Cassava Plant Production), initiated by the IITA (International Institute of Tropical Agriculture), that the farmers of Sakété had begun trading in couch grass. This is still true today.

In fact the couch grass straw had been customarily used in the Sakété sub-prefecture. However the high cost of construction materials was a problem, especially after the 1994 devaluation of the CFA franc, and so the pressure for couch grass straw which is more and more in demand for the construction of roofing increased.

Demand intensified, landlords whose degraded lands are colonized by couch grass began to charge for its harvest. It is harvested by cutting it very short. This trading phenomenon is reinforced by the inheritance-dominated method of land tenure which leads to progressive division of lands over the years and even to the appearance of "farmers without land", and the fact that not all fields possess exploitable quantities of couch grass.

So, aside from the fact that owners of couch grass fields - which they maintain - can use the grass for roofing, in particular for dwellings and animal enclosures, they can also sell it. Measurements have been specially devised for this purpose: the "egbi" (a small bundle of couch grass straw) and the "eru" (a large bundle equivalent to 12 small bundles of couch grass straw) are the units currently in use. For each collection the customer is obliged to give a liter of Sodabi (a local alcoholic beverage distilled from palm wine). In addition, this beverage must be accompanied by between 2,000 and 10,000 CFA francs when the collected quantity exceeds 25 erus of straw. It should be noted that to cover a dwelling of 9 meters by 4 meters, 60 to 80 erus of straw are required.

This method of using couch grass appears to be a veritable source of trading income since the grass grows back in the same six-month period during which it was cut. In fact, the *Imperata cylindrica* is a perennial and enduring poaceous species with extremely resistant rhizomes which develop in clumps of erect ribbon-like leaves pointed at the tip with a rigid and cylindrical stubble.

Bearing in mind the fact that not everyone is adept at the collection (which is difficult) of this straw, it should be noted that the couch grass provides work and therefore generates extra revenue for certain specialist farmers.

Mining nothing, couch grass is becoming a source of wealth in the sub-prefecture of Sakété. One therefore wonders if this state of affairs is not one of the reasons for the low adoption rate of the *Mucuna* fallow which was introduced in Benin in 1987.

In any event, before launching an in-depth study, one may believe this for the moment and dwell on this new order for at least four reasons:

- The assessment is surprising in its way and demonstrates that reasons which had not previously been thought of are strengthening the current position of the farmers.
- The *Mucuna* is endowed with agronomic advantages exceeding those of many other introduced plants such as the *Senna siamea* which for example does not form nodules on its roots and therefore does not fix nitrogen in the air.
- The *Mucuna,* which must do battle with the *Imperata,* is still regarded by farmers as adventitious to the extent that it is neither edible nor marketable and so cannot constitute an economic rival to couch grass.

The increased knowledge behind this assessment makes it possible to discover the range of obstacles which could explain the low adoption rate of the *Mucuna* in the entire Lower Benin and also the intrinsic qualities which must now constitute an objective of agro-forestry innovation.

5. CONCLUSION: MORE IN-DEPTH STUDIES REQUIRED

Lower Benin in general and the sub-prefecture of Sakété in particular are confronting serious problems of advanced soil degradation. On the one hand, this is one of the causes of the growing pauperization of farmers. On the other hand it is the consequence of the irresistible increase in land pressure coupled with the continuation of unsuitable agricultural practices.

The fortuitous marketing of couch grass practiced by the farmers of the sub-prefecture of Sakété is without doubt an interesting situation, which must constitute a point of departure for more in-depth studies. These would enable a global improvement in the adoption rate of numerous agroforestry technologies where mining agriculture presently reigns.

A study of the socio-economic determinants underlying the paradoxical continuation of a mining agriculture in land-starved Benin would be rather useful, as would efforts to improve the organoleptic qualities and/or the processing of the *Mucuna* seeds, in order to allow for their consumption and marketing.

REFERENCES

1. Adegbidi, A & G, Biaou (1994): *L'agriculture durable au Bénin: Rôle de l'Etat.* Rapport de séminaire sur l'agriculture durable au Bénin.Amsterdam.

2. Caveness, E. (1988): *Observations on soil nematode population development under a Mucuna utilis fallow.* Internal Report 10 - 1987/1988 Root and Tuber Improvement Program IITA, 6 p.
3. Encyclopedie Economique, Préface de Michel Albert. Economica, Paris. 1984. 1223 p
4. Floquet, A & R, Mongbo (1998): *Des paysans en mal d'alternatives: Dégradation des terres, restructuration de l'espace agraire et urbanisation au Bas Bénin.* Université de Hohenheim. Margraf Verlag, Weikersheim.
5. Houngbo, N. E. (1996): *Les problèmes d'allocation de ressources liés à la production durable du manioc au Bénin: Etude de cas de la Sous-préfecture de Sakété Ouémé.* Thèse d'Ingénieur agronome. FSA/UNB. Abomey-Calavi, Bénin.
6. Houngbo, N. E. (1997): *Les obstacles à l'adoption du Mucuna.* In Coraf Action N°5: 9, 10. Lettre d'information pour la recherche et le développement agricoles en Afrique de l'Ouest et du Centre de la CORAF (conférence des responsables de recherche agricole en Afrique de l'Ouest et du Centre). Dakar-Yoff, Sénégal.
7. Houngbo, N. E. (1999): *Repenser l'agriculture durable.* In Nouvelles du REDAD N°7: 8, 9. Bulletin de liaison du Réseau de développement d'agriculture durable (REDAD). Cotonou, Bénin.
8. Hulugalle, N. R. R; Lal & C. H, Ter Kuile (1986): *Amelioration of soil physical properties by Mucuna after mechanised land clearing of a tropical rain forest.* Soil Sci. 141: 219-224
9. INRAB-Institut National des Recherches Agricoles du Bénin (1994): *Enquête diagnostic extensive de la saison sèche.* Rapport préliminaire. Projet ESCaPP-Bénin, IITA. Ibadan, Nigéria. 36p
10. Koudokpon, V & M, Versteeg (1996): *Mucuna: Solution pour les terres pauvres et les champs envahis par les chiendents.* Ministère du développement rural et de l'action coopérative (MDRAC). Cotonou, Bénin. 12p
11. MDR - Ministère du développement rural (1999): *Rapport général de l'atelier national de lancement de l'initiative sur la fertilité des sols* (IFS). Cotonou, Bénin. Pp 39-55
12. Mounier, A (1992): *Les théories économiques de la croissance agricole.* Inra-Economica, Paris.
13. Pursegolve, J. W. (1968): *Tropical crops.* Dicotyledons I: Longmans, Green & Co. LTD, London.
14. Thorbecke, E. (1969) *The role of Agriculture in Economic Development.* National Bureau of Economic Research. Columbia University Press, New York.

INDEX

C

D

E

F

G

H

I

J

K

L

R

S

T

U

V

W

Z

MIX
Papier aus verantwortungsvollen Quellen
Paper from responsible sources
FSC® C105338

If you have any concerns about our products,
you can contact us on
ProductSafety@springernature.com

In case Publisher is established outside the EU,
the EU authorized representative is:
Springer Nature Customer Service Center GmbH
Europaplatz 3, 69115 Heidelberg, Germany

Printed by Libri Plureos GmbH
in Hamburg, Germany